Instrumentation Electronics

Instrumentation Electronics

P. P. L. Regtien

Prentice Hall

NEW YORK · LONDON · TORONTO · SYDNEY · TOKYO · SINGAPORE

First published 1992 by
Prentice Hall International (UK) Ltd
Campus 400, Maylands Avenue
Hemel Hempstead
Hertfordshire, HP2 7EZ
A division of
Simon & Schuster International Group

© 1987 by Vereniging Voor Studie-en Studentenbelangen te Delft

All rights reserved. No part of this publication may be
reproduced, stored in a retrieval system, or transmitted,
in any form, or by any means, electronic, mechanical,
photocopying, recording or otherwise, without prior
permission, in writing, from the publisher.
For permission within the United States of America
contact Prentice Hall Inc., Englewood Cliffs, NJ 07632

Typeset in 10/12pt Century Old Style
by Keyset Composition, Colchester, Essex

Printed and bound in Great Britain by
Dotesios Printers Limited, Trowbridge, Wiltshire

Library of Congress Cataloging-in-Publication Data

Regtien, P. P. L.
   Instrumentation electronics / P. P. L. Regtien.
     p. cm.
   Includes index.
   ISBN 0-13-473562-5 :
   1. Electronic instruments. 2. Measuring instruments. I. Title.
TK7878.4.R43   1992
621.381′54—dc20   92-11963
                                                                                    CIP

British Library Cataloguing in Publication Data

A catalogue record for this book is available from
the British Library

ISBN 0-13-473562-5 (pbk)

1 2 3 4 5 96 95 94 93 92

Contents

Preface *ix*

1 Measurement systems *1*
1.1 System functions *1*
1.2 System specifications *6*
Summary *13*
Exercises *14*

2 Signals *15*
2.1 Periodic signals *15*
2.2 Aperiodic signals *22*
Summary *34*
Exercises *35*

3 Networks *38*
3.1 Electric networks *38*
3.2 Generalized network elements *44*
Summary *47*
Exercises *48*

4 Mathematical tools *51*
4.1 Complex variables *51*
4.2 Laplace variables *55*
Summary *61*
Exercises *62*

5 Models *64*
5.1 System models *64*
5.2 Signal models *73*
Summary *78*
Exercises *79*

6 Frequency diagrams 82
6.1 Bode plots 82
6.2 Polar plots 86
Summary 91
Exercises 92

7 Passive electronic components 94
7.1 Passive circuit components 94
7.2 Sensor components 100
Summary 112
Exercises 113

8 Passive filters 115
8.1 First- and second-order RC filters 116
8.2 Filters of higher order 124
Summary 128
Exercises 129

9 PN diodes 131
9.1 Properties of pn diodes 131
9.2 Circuits with pn diodes 137
Summary 146
Exercises 147

10 Bipolar transistors 150
10.1 Circuits of bipolar transistors 150
10.2 Circuits with bipolar transistors 154
Summary 166
Exercises 167

11 Field-effect transistors 170
11.1 Properties of field-effect transistors 170
11.2 Circuits with field-effect transistors 177
Summary 181
Exercises 182

12 Operational amplifiers 185
12.1 Amplifier circuits with ideal operational amplifiers 185
12.2 Non-ideal operational amplifiers 192
Summary 198
Exercises 199

Contents   vii

13 Frequency-selective transfer functions with operational amplifiers *202*
13.1 Circuits for time-domain operations *202*
13.2 Circuits with high frequency selectivity *209*
Summary *216*
Exercises *217*

14 Non-linear signal processing with operational amplifiers *220*
14.1 Non-linear transfer functions *220*
14.2 Non-linear arithmetic operations *227*
Summary *236*
Exercises *238*

15 Electronic switching circuits *241*
15.1 Electronic switches *241*
15.2 Circuits with electronic switches *249*
Summary *258*
Exercises *258*

16 Signal generation *262*
16.1 Sine wave oscillators *262*
16.2 Voltage generators *268*
Summary *275*
Exercises *276*

17 Modulation and demodulation *279*
17.1 Amplitude modulation and demodulation *281*
17.2 Systems based on synchronous detection *289*
Summary *295*
Exercises *296*

18 Digital-to-analog and analog-to-digital conversion *299*
18.1 Parallel converters *299*
18.2 Special converters *310*
Summary *316*
Exercises *317*

19 Digital electronics *319*
19.1 Digital components *319*
19.2 Logic circuits *331*
Summary *344*
Exercises *346*

20 Microprocessor systems *348*
  20.1 Semiconductor memories *348*
  20.2 Structure of microprocessors *356*
  Summary *364*
  Exercises *365*

21 Measurement instruments *367*
  21.1 Electronic measurement instruments *367*
  21.2 Computer-based measurement instruments *380*
  Summary *385*
  Exercises *386*

22 Measurement errors *389*
  22.1 Types of measurement errors *389*
  22.2 Measurement interference *393*
  Summary *401*
  Exercises *402*

Appendix

  A.1 Notation *405*

  A.2 Examples of manufacturers' specifications *407*

  Answers to exercises *420*
  Index *482*

Preface

Electronics has penetrated deeply in daily life. When looking around in houses, offices or plants, electronic systems can be observed almost everywhere. There are hardly any appliances, tools or instruments that do not contain electronic parts. Designers of technical systems must be fully aware of the possibilities and limitations of modern electronics, in order to be competitive with or even ahead of competing companies. Users of electronic systems also need at least some basic knowledge of electronic principles to exploit fully the possibilities of an instrument, to be aware of its limitations, to interpret correctly measurement results or to decide soundly on the purchase, repair, expansion or replacement of electronic equipment.

This book offers the basic knowledge to obtain these skills. The scope covers basic properties of both analog and digital components and circuits, provides insight into the features of electronic measurement systems and tries to impart a critical attitude when using such instruments.

This book is based on courses in electronics and electronic instrumentation for students of the Departments of Mechanical Engineering, Aeronautical Engineering and Mining Engineering at the Delft University of Technology, The Netherlands. The organization of the book makes it suitable for a much wider user group. To meet the various demands, a modular approach is followed. Each chapter discusses a particular subject and is divided into two parts; the first provides the basics, the second more specific information. Each chapter ends with a summary and some exercises; worked answers are given at the end of the book. This approach supports self-tuition and allows the composition of tailor-made course programmes. The diagram on page xi may help the reader to find proper sequences. Starting from the top left (Section 1.1), you can follow any track downwards or to the right. Dotted tracks are optional. Conversely, for an arbitrary chapter, understanding of the subject matter in chapters positioned to the left and upwards is required. For instance, successful study and application of Section 7.2 requires at least the knowledge found in Sections 1.1, 1.2, 2.1, 3.1, 4.1, 5.1, 6.1 and 7.1. Section 21.1 requires Sections 19.1, 19.2 and 18.1 (and upwards).

The required background knowledge is basic mathematics and physics on an average first-year academic level. No background knowledge of electronics is needed to read the book. For further study on particular subjects the reader is referred to the widely available course books on electronics, measurement techniques and instrumentation.

I am indebted to all people who have contributed to the realization of this book. In particular, I thank Johan van Dijk, who carefully refereed the original Dutch text. I further acknowledge Reinier Bosman for working out all the exercises, G. van Berkel for making over 600 illustrations and Jacques Schievink of the VSSD for the processing of the Dutch editions of this book and for his mediation with the publisher.

P. P. L. Regtien
Delft, October 1991

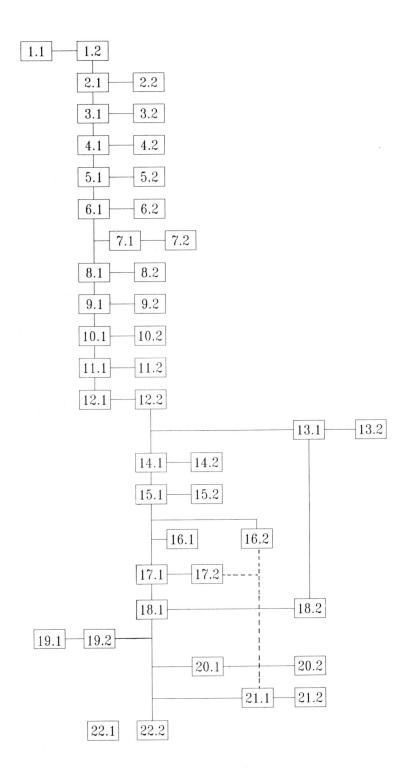

1 Measurement systems

A measurement system is an instrument that is designed to obtain information about a physical process and to present that information in a suitable way to an observer or another technical system. In an electronic measuring system, the various instrument functions are realized with electronic components.

In the first part of this chapter some basic system functions are introduced. The degree to which an instrument meets the specified requirements is characterized by the system specifications, which are discussed in the second part of the chapter.

1.1 System functions

A measuring system can be considered as being a transport channel for the exchange of information between a measurement object and a target object. In such a system three main functions can be distinguished: data acquisition, data processing and data distribution (Figure 1.1).

- *Data acquisition:* information about the measurement object is acquired and converted into electrical measurement data. The multiple input indicates the possibility for simultaneous measurements. The single output of the data acquisition block indicates that all data are transferred via a single connection to the next block.
- *Data processing:* this block includes processing, selecting or otherwise manipulating measurement data according to the prescribed program. Often a processor or a computer is used to perform this function.
- *Data distribution:* the supply of measurement data to the target object. The multiple output indicates the possible presence of several target instruments, such as a series of control valves in a process installation.

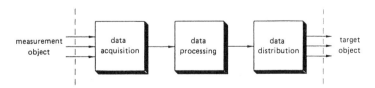

Figure 1.1 The three main functions of any measuring system.

2 Measurement systems

This subdivision is not always possible; sometimes part of the system may be classified under both data acquisition and data processing. Some authors call the entire system in Figure 1.1, a data acquisition system, claiming that the data have not been obtained until reaching the target object.

The data acquisition and data distribution parts are subdivided into smaller functional units. Since most physical measurement quantities are non-electric, they should first be converted into an electrical form in order to allow electronic processing. This conversion is called transduction and is performed by a transducer or sensor (Figure 1.2). In this book, a sensor is considered as the smallest technical unit in which data conversion from the non-electric to the electric domain takes place. A transducer is a unit containing the sensor, wiring, connections and part of the signal processing if so desired, all put together in a casing. In general, the transducer is separate from the main instrument and can be connected to it by a special cable.

The sensor or input transducer connects the measuring system to the measurement object; it is the input port of the system through which the information enters the instrument.

Many sensors or transducers produce an analog signal, that is a signal whose value, at any moment, is a measure of the quantity to be measured: the signal follows continuously the course of the input quantity. However, a great deal of processing equipment can only handle digital signals, which are binary-coded signals. A digital signal contains a finite number of distinguishable codes, usually a power of 2 (for instance $2^{10} = 1024$).

The analog signal must be converted into a digital one. This process is called analog-to-digital conversion or AD conversion. AD conversion comprises three processes. The first one is sampling: at discrete time intervals, samples are taken from the analog signal. Each sampled value is maintained for a certain time interval, during which the following processes can take place. The second step is quantization. This is the rounding off of the sampled value to the nearest of the limited number of digital values. Finally, this quantized value is converted into a binary code.

Both sampling and quantization may introduce loss of information. Under certain conditions, however, this loss can be limited to an acceptable minimum.

The output signal from a transducer is not necessarily suitable for conversion by an AD converter; the input of the converter should satisfy certain conditions. The required signal processing to fulfil such conditions is called signal conditioning. The

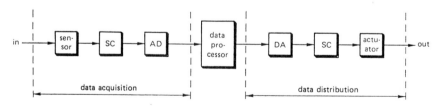

Figure 1.2 A single-channel measuring system. SC = signal conditioning; AD = analog-to-digital conversion; DA = digital-to-analog conversion.

various processing steps necessary to achieve the proper signal conditions are explained in separate chapters. The major steps will be explained briefly below.

- Amplification: to increase the signal's magnitude or its power content.
- Filtering: the removal of non-relevant signal components.
- Modulation: modification of the signal shape in order to enable signal transport over a long distance or to reduce the sensitivity to interference during transport.
- Demodulation: the reverse operation of modulation.
- Non-linear and arithmetical operations: such as logarithmic conversion and multiplication of two or more signals.

Evidently, none of the above operations should affect the information content of the signal.

After being processed by the (digital) processor, the data are submitted to a reverse operation (Figure 1.2). The digital signal is converted into an analog signal by a digital-to-analog converter (DA converter). Then it is supplied to an actuator (other names are effector, excitator and output transducer), which transforms the electrical signal into the desired non-electric quantity. If the actuator cannot be connected directly to the DA converter, the signal is conditioned first. This conditioning usually consists of signal amplification.

The actuator or output transducer connects the measurement system to the target object; it is the output port of the instrument, through which the information leaves the system.

Depending on the goal of the measurement, the actuator performs various functions: indication, for instance a digital display; registration (storage), such as a printer, a plotter or a magnetic disk; and process control, e.g. by a valve, a heating element or an electric drive.

The diagram of Figure 1.2 refers to only one input variable and one output variable. For processing more than one variable, one could take a set of single channel systems. Obviously, this is not efficient and unnecessary. In particular, the processor in Figure 1.2 is able to handle a large number of signals, thanks to a high data-processing speed. Figure 1.3 shows the layout of a multichannel measuring system, able to handle multiple inputs and outputs, with only one central processor. Central processing of the various digital signals is possible by means of multiplexing. The digital multiplexer in Figure 1.3 alternately connects the output of each AD converter to the processor. The multiplexer can be considered as being an electronically controlled multistage switch, controlled by the processor. This type of multiplexing is called time multiplexing, because the channels are scanned and their respective signals are transferred successively in time to the processor. Another type of multiplexing, frequency multiplexing, is discussed later.

At first sight, the concept of time multiplexing has the disadvantage that only the data from the selected channel is processed while the information from the non-selected channels is blocked. It can be shown that, when the time between two successive selections for a particular channel is made sufficiently short, the

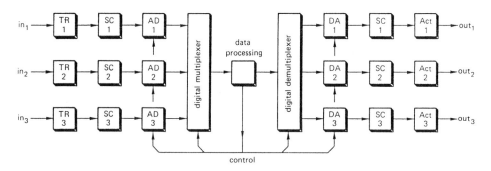

Figure 1.3 A three-channel measuring system with one central processor. TR = input transduction; SC = signal conditioning; AD = analog-to-digital conversion; DA = digital-to-analog conversion; Act = actuator or output transduction.

information loss will be negligible. In Section 2.2 it is explained what is meant by sufficiently short.

Figure 1.3 clearly shows that a system with many sensors or actuators will also contain large numbers of signal processing units, resulting in a high price. In such a case, the principle of multiplexing can be applied also to the AD and DA converters. Figure 1.4 shows the layout of such a measurement system, in which all conditioned signals are supplied to an analog multiplexer. It is even possible to put a central signal conditioner after the multiplexer, to reduce further the number of system components. The process of centralizing instrument functions can be extended towards the output of the system. An analog multiplexer distributes the converted analog signals over the proper output channels. Multiplexing of the output signal conditioner is not common practice because (de)multiplexers are usually not designed for large power signals.

Although analog and digital multiplexers have similar functions, their design is completely different. Digital multiplexers handle only digital signals which have a better noise and interference immunity than have analog signals. Therefore, digital multiplexers are far less critical (and less expensive) than analog multiplexers. The same holds for the AD converters. In Figure 1.3, each AD converter has a full cycle period of the multiplexer to perform a conversion. In the system of Figure 1.4, the conversion should be finished within the very short time during which a channel is connected to the processor. So this system configuration requires a high speed (and higher-priced) converter. The centralized system contains a reduced number of more expensive components. The choice for either a centralized or a distributed system depends strongly on the number of channels.

In some situations, the measurement signals and control signals have to be transported over a long distance. This part of the instrumentation is called telemetry. A telemetry channel consists of an electric conductor (for instance a telephone cable), an optical link (like a glass fiber cable) or a radio link (e.g. via a communication

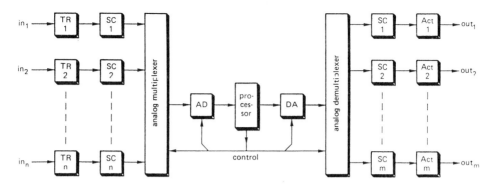

Figure 1.4 A multichannel measuring system with a centralized processor and AD and DA converters. For the explanation of the abbreviations see Figure 1.3.

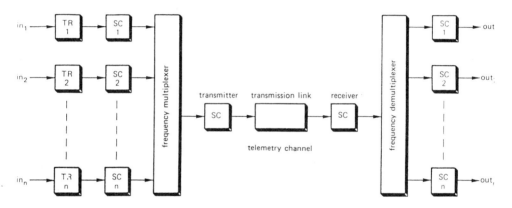

Figure 1.5 A multichannel measuring system with frequency multiplexing. TR = transduction; SC = signal conditioning.

satellite). To reduce the number of (often expensive) lines, the concept of multiplexing is used (Figure 1.5). Instead of time multiplexing, telemetry systems use frequency multiplexing. Each measurement signal is converted to a frequency band assigned to that particular signal. If the bands do not overlap, the converted signals can be transported simultaneously over a single transmission line. Upon arriving at the destination, the signals are demultiplexed and distributed to the proper actuators. More details on this type of multiplexing are given in Section 17.1.

Signals can be transmitted in analog or digital form. Digital transport is preferred if a high noise immunity is required, for instance at very long transport channels or links through a noisy environment.

1.2 System specifications

A measurement system is designed to perform measurements in agreement with its specifications. These specifications tell the user of the instrument to what degree the output is in accordance with its input. The specifications reflect the quality of the system.

The system functions correctly if it meets the specifications given by the manufacturer. It fails when that is not the case, even if the system is still functioning in the technical sense. Any measuring instrument and any subsystem accessible to the user have to be fully specified. Unfortunately, many specifications show lack of clarity and completeness.

The input signal of the single channel system of Figure 1.6 is denoted by x, its output signal by y. The relationship between x and y is called the transfer of the system.

By observing the output, the user draws conclusions about the input. Therefore, the user has to know precisely the transfer of the system. Deviations of the transfer cause uncertainties about the input, resulting in measurement errors. Such deviations are allowed, but within certain limits, the tolerances of the system. These tolerances are also part of the specifications. In the forthcoming text the main specifications of a measurement system will be discussed.

First, the user should know the operating range of the system. The operating range is given by the measurement range, the required supply voltage, the environmental conditions and possibly other parameters.

Example 1.1
A manufacturer of a digital temperature measuring instrument gives the following description of the operating range:

- measuring range: −50 to 200°C;
- allowed operational temperature: −10 to 40°C;
- storage room temperature: −20 to 85°C;
- mains voltage: 220 V ±15%, 50 . . . 60 Hz; switchable to 115 V, 127 V, 240 V ±15%, 50 . . . 60 Hz;
- analog outputs: 0–10 V (load > 2 kΩ) and 0–20 mA or 4–20 mA (load < 600 Ω).

All other specifications only apply under the condition that the system is not or has not been outside its operating range.

The resolution indicates the smallest detectable change of the input quantity. Many system parts show a limited resolution. A few examples: a wire-wound potentiometer for the measurement of angles has a resolution set by the windings of the helix: the resistance between the slider and the helix changes leapwise with rotation; a display presenting a measurement value in numerals has a resolution equal to the least significant digit.

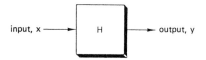

Figure 1.6 Characterization of a system with input x, output y and transfer H.

The resolution is expressed as the smallest detectable change in the input variable: Δx_{min}. Sometimes this parameter is related to the maximum value x_{max} that can be processed, the so-called full-scale value or FS of the instrument, resulting in the resolution expressed as $\Delta x_{min}/x_{max}$ or $x_{max}/\Delta x_{min}$. This mixed use of definitions seems very confusing. However, it is easy to see from the units or the value itself which definition is used.

Example 1.2
The resolution of a four-digit decimal display with a fixed decimal point on the third position from the left is 0.1 unit. The maximum indication apparently equals 999.9 units, which is about 1000. Hence, the resolution of this display is 0.1 unit or 10^{-4} or 10^4.

The inaccuracy is a measure for the total uncertainty in the measurement result, caused by all kinds of system errors. It comprises calibration errors, long- and short-term instability, component tolerances and other uncertainties that are not separately specified. Two definitions are distinguished: absolute inaccuracy and relative inaccuracy. The absolute inaccuracy is expressed in units of the measuring quantity concerned, or as a fraction of the full-scale value. The relative inaccuracy relates the error to the actual measuring value.

Example 1.3
The data sheet of a voltmeter with a four-digit indicator and a full-scale value of 1.999 V specifies the instrument inaccuracy as $\pm 0.05\%$ FS $\pm 0.1\%$ of the indication $\pm \frac{1}{2}$ digit. The absolute inaccuracy of a voltage of 1.036 V measured with this instrument equals: ± 0.05 of 2 V (approximate value of FS) plus $\pm 0.1\%$ of 1 V (approximate value of the indication) plus ± 0.5 of 1 mV (the weight of the last digit), that is ± 2.5 mV in total. The relative inaccuracy is the absolute inaccuracy divided by the indication, hence $\pm 0.25\%$.

Inaccuracy is often confused with accuracy, the latter being the complement of the inaccuracy. When a specification list gives an accuracy of 1%, it should mean an inaccuracy of 1% or an accuracy of 99%.

The sensitivity of a measuring system is defined as the ratio between a change in the output value and the change in the input value that causes that output change. The sensitivity of a current-to-voltage converter is expressed in V/A, that of a linear position sensor in, for instance, mV/μm and that of an oscilloscope in, for instance, cm/V.

A measuring system is usually also sensitive to changes in other quantities than the intended input quantity, such as the ambient temperature or the supply voltage. These unwelcome sensitivities should be specified as well when necessary for a proper interpretation of the measurement result. A better insight in the effect of such a false sensitivity is that it is related to the sensitivity of the measurement quantity itself.

Example 1.4
A displacement sensor with voltage output has a sensitivity of 10 mV/mm. Its temperature sensitivity is −0.1 mV/K. Since −0.1 mV corresponds with a displacement of −10 μm, the temperature sensitivity can also be expressed as −10 μm/K. A temperature rise of 5°C results in an apparent displacement of −50 μm.

Example 1.5
The sensitivity of a temperature sensor including the signal conditioning unit is 100 mV/K. The signal conditioning part itself is also sensitive to ambient temperature and appears to create an extra output voltage of 0.5 mV for each °C rise in ambient temperature (not necessarily the sensor temperature). So, the unwanted temperature sensitivity is 0.5 mV/K or 0.5/100 = 5 mK/K. A change in ambient temperature of ±10°C gives an apparent change in sensor temperature equal to ±50 mK.

Mathematically, the sensitivity is expressed as $S = dy/dx$. If the output y is a linear function of the input x, the sensitivity does not depend on x. In the case of a non-linear transfer function $y = f(x)$, S depends on the input or output value (Figure 1.7). Users of measuring instruments prefer a linear response, because then the sensitivity can be expressed in a single parameter. The transfer of a system with a slight non-linearity may be approximated by a straight line, to specify its sensitivity by just one number.

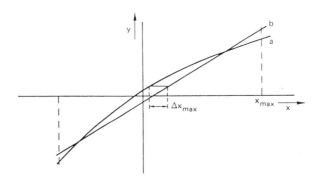

Figure 1.7 Example of a non-linear transfer characteristic: (a) real transfer; (b) linear approximation.

The user should still know the deviation from the actual transfer, which is specified by the non-linearity.

The non-linearity of a system is the maximum deviation of the actual transfer characteristic from a predescribed straight line. Manufacturers specify non-linearity in various ways, for instance as the deviation in input or output units: Δx_{max} or Δy_{max}, or as a fraction of FS: $\Delta x_{max}/x_{max}$. They may use different settings for the straight line: through the end points of the characteristic, the tangent through the point $x = 0$, the best fit (least-squares) line, to mention a few.

Figure 1.8 depicts some particular types of non-linearity occurring in measuring systems: saturation, clipping and dead zone (sometimes called cross-over distortion). These are examples of static non-linearity, appearing even at slowly changing inputs. Figure 1.9 shows another type of non-linearity, called slew rate limitation, occurring only at relatively fast-changing input values. The output is unable to follow the quickly

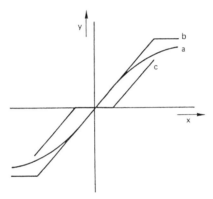

Figure 1.8 Some types of static non-linearity: (a) saturation; (b) clipping; (c) dead zone.

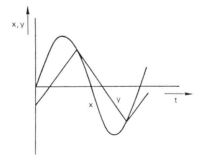

Figure 1.9 The effect of slew rate limitation on the output signal y at a sinusoidal input x.

changing input, resulting in a distortion at the output. The slew rate is specified as the maximum rate of change in the output of the system.

Most measurement systems are designed so that the output is zero at zero input. If the transfer characteristic does not intersect the origin ($x = 0$, $y = 0$), the system is said to have offset. The offset is expressed in terms of the input or the output quantity; specifying the input offset is preferred, to facilitate a comparison with the real input quantity. A non-zero offset arises mainly from component tolerances. Most electronic systems offer the possibility to compensate for the offset, either by manual adjustment, or by means of manually or automatically controlled zero-setting facilities. Once adjusted to zero, the offset may nevertheless change, due to temperature variations, changes in the supply voltage or ageing effects. This relatively slow change in the offset is called zero drift. In particular, the temperature-induced drift (the temperature coefficient or t.c. of the offset) is an important item in the specification list.

Example 1.6
A data book on instrumentation amplifiers contains the following specifications for a particular type of amplifier:

input offset voltage : maximum ± 0.4 mV, adjustable to 0
t.c. of the input offset : maximum ± 6 μV/K
supply voltage coefficient : 40 μV/V
long-term stability : 3 μV/month

There are two ways to determine the offset of a system. The first method is based on setting the output at zero by an adjustable input. The input value for which the output is zero is the negative value of the input offset. The second method consists of measuring the output at zero input. When the output is still within the allowable range, the input offset is simply the measured output divided by the sensitivity.

Sometimes a system is deliberately designed with offset. Many industrial transducers have a current output ranging from 4 to 20 mA (see Example 1.1). This facilitates the detection of cable fractures or a short-circuit, producing a zero output clearly distinguishable from a zero input.

The sensitivity of an electronic system may be increased to an almost unlimited level. There is, however, a limit to the usefulness of doing this. On increasing the sensitivity of the system its output offset will grow as well, up to the limits of the output range. Even at zero input voltage, an ever-increasing sensitivity is of no use, due to system noise. Electrical noise is a collection of spontaneous fluctuations in currents and voltages. They are present in any electronic system and arise from thermal motion of the electrons and from the quantized nature of electric charge. Electrical noise is also specified in terms of the input quantity, to show its effect relative to that of the actual input signal.

The sensitivity of a system depends on the frequency of the signal to be processed. A measure for the useful frequency range is the frequency band. The upper and lower limits of the frequency band are defined as those frequencies for which the power transfer has dropped to half the nominal value. For voltage or current transfer the criterion is ½√2 of the nominal voltage and current transfer, respectively (Figure 1.10). The lower limit of the frequency band may be zero; the upper limit has always a finite value. The extent of the frequency band is called the bandwidth of the system, expressed in hertz (Hz).

A frequent problem in instrumentation is the determination of the difference between two almost equal measurement values. Such a situation occurs, for instance, when large noise or interference signals are superimposed upon the relatively weak measurement signals. For this kind of measurement problem, the differential amplifier has been developed (Figure 1.11). Such an amplifier, usually a voltage amplifier, has two inputs with respect to ground and one output. Ideally, the amplifier is not sensitive to equal signals on both inputs (common mode signal), only to a difference between the two input signals (differential mode signals). In practice, any differential amplifier exhibits a non-zero transfer for common mode signals. A quality measure with respect to this property is the common mode rejection ratio (CMRR), defined as the ratio between the transfer for differential mode signals, v_o/v_d, and common mode

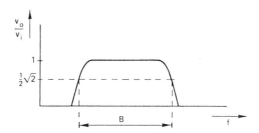

Figure 1.10 A voltage transfer characteristic, showing the boundaries of the frequency band; the nominal transfer is 1, its bandwidth is B.

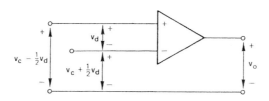

Figure 1.11 An ideal differential amplifier is insensitive to common mode signals (v_c) and amplifies only the differential signal v_d.

signals v_o/v_c. In other words, the CMRR is the ratio of a common mode input signal and a differential mode input signal, both giving equal outputs. An ideal differential amplifier has a CMRR which is infinite.

Example 1.7
A system whose CMRR is 10^5 is used to determine the difference between two voltages, both about 10 V. The difference appears to be 5 mV. The inaccuracy of this result due to the finite CMRR is ±2%, because the common mode voltage produces an output voltage that is equivalent to that of a differential input voltage of $10/10^5 = 0.1$ mV.

The last system property to be discussed in this chapter is related to the reliability of the system. There is always a chance that a system will fail after a certain period of time. Such a property should be described by probability parameters. One of these parameters is the reliability $R(t)$ of the system. It is defined as the probability that the system will function correctly (in accordance with its specifications) up to the time t (provided the system has operated within the allowed ranges). It should be clear that R diminishes as time elapses; the system becomes increasingly less reliable.

The system parameter R has the disadvantage that it changes with time. Better parameters are the mean time-to-failure (MTTF) and the failure rate $\lambda(t)$. The MTTF is the mean time that passes until the system fails; it is its mean lifetime.

Example 1.8
An incandescent lamp is specified for 1000 burning hours. This means that lamps from the series to which this lamp belongs burn 1000 hours on the average. Some lamps fail (much) earlier, others may burn longer.

The failure rate $\lambda(t)$ is defined as the fraction of failing systems per unit of time relative to the total number of systems that are functioning properly at time t. The failure rate appears to be constant during a large part of the system's lifetime. If the failure rate is constant with time, it is equal to the inverse of the MTTF.

Example 1.9
Suppose an electronic component has an MTTF equal to 10^5 hours. Its failure rate is the inverse, 10^{-5} per hour or 0.024% per day or 0.7% per month. So, from a certain collection of components still functioning correctly, 0.024% will fail daily.

The failure rate of electronic components is extremely low, if operated under normal conditions. For example, the failure rate of metal film resistors with respect to an open connection is approximately $5 \cdot 10^{-9}$ per hour. The reliability of many electronic components is well known. However, it is very difficult to determine the reliability of a complete electronic measurement system from the failure rates of the individual components. This is a reason why the reliability of complex systems is seldom specified.

SUMMARY

System functions

- The three main functions of an electronic measurement system are data acquisition, data processing and data distribution.
- The conversion of information about a physical quantity into an electrical signal is called transduction. It is performed by an input transducer or sensor. The inverse process is performed by an output transducer or actuator.
- The major operations on analog measurement signals are amplification, filtering, modulation, demodulation and analog-to-digital conversion.
- AD conversion comprises three elements: sampling, quantization and coding.
- Multiplexing is a technique that allows the transport of various signals at the same time over a single channel. There are two methods: time multiplexing and frequency multiplexing. The inverse process is demultiplexing.

System specifications

- The major specifications of a measurement system are operating range (including measuring range), resolution, (in)accuracy, sensitivity, non-linearity, offset, drift and reliability.
- Some types of non-linearity are saturation, clipping, dead zone, hysteresis and slew rate limitation.
- The bandwidth of a system is the frequency span between those frequencies at which the power transfer has dropped to half the nominal value, or at which the voltage or current transfer has dropped to $\frac{1}{2}\sqrt{2}$ of the nominal value.
- The common mode rejection ratio is the ratio between the transfer of differential mode signals and that of common mode signals, or the ratio between a common mode input and a differential mode input both producing equal outputs.
- Noise is the phenomenon of spontaneous voltage or current fluctuations occurring in any electronic system. It sets a fundamental limitation to the inaccuracy of a measurement system.
- The reliability of a system can be specified in terms of the reliability $R(t)$, the failure rate $\lambda(t)$ and the mean time-to-failure (MTTF). For systems with constant failure rate, $\lambda = 1/\text{MTTF}$.

Exercises

System functions

1.1 What is meant by multiplexing? Describe the process of time multiplexing.
1.2 Discuss the difference between the requirements for a multiplexer for digital signals and one for analog signals.
1.3 Compare an AD converter in a centralized system versus that in a distributed system, with respect to the conversion time.

System specifications

1.4 What would be the reason for putting a factor $\frac{1}{2}\sqrt{2}$ in the definition of the bandwidth for voltage transfer, instead of a factor $\frac{1}{2}$?
1.5 What is a differential voltage amplifier? What is meant by the CMRR of such an amplifier?
1.6 The CMRR of a differential voltage amplifier is specified as CMRR $> 10^3$; its voltage gain if $G = 50$. The two input voltages have values $V_1 = 10.3$ V, $V_2 = 10.1$ V. What is the possible range of the input voltage?
1.7 The slew rate of a voltage amplifier is 10 V/μs and its gain is 100. The input is a sinusoidal voltage with amplitude A and frequency f.
   (a) Suppose $A = 100$ mV; what is the upper limit of the frequency for which the output shows no distortion?
   (b) Suppose $f = 1$ MHz; up to what amplitude can the input signal be amplified without distortion?
1.8 A voltage amplifier is specified as follows: input offset voltage at 20°C: <0.5 mV; temperature coefficient of the offset: <5 μV/K. Calculate the maximum input offset that might occur within the temperature range 0–80°C.
1.9 The relationship between the input quantity x and the output quantity y of a system is given as: $y = \alpha x + \beta x^2$, with $\alpha = 10$ and $\beta = 0.2$. Find the non-linearity relative to the line $y = \alpha x$, for the input range $-10 < x < 10$.

2 Signals

Each physical quantity containing a detectable message is called a signal. The information carrier of an electrical signal is a voltage, a current, a charge or any other electric parameter.

The message contained in such a signal can be the result of a measurement, but also a code for an instruction or a location (for instance the address of a memory location). The nature of the message cannot be seen from its appearance. The processing techniques for electronic signals are not related to the contents or nature of the message.

The first part of this chapter starts with a characterization of signals and explains the various values of a signal in the time domain. Signals can also be characterized in the frequency domain. For periodic signals, the frequency spectrum is derived using the Fourier expansion.

The second part of this chapter deals with aperiodic signals, in particular noise, stochastic and sampled signals.

2.1 Periodic signals

2.1.1 Classification of signals

There are many ways to classify signals. One of these classifications is based on the dynamic properties of the signal.

- Static or DC signals (DC = direct current, a term that is also applied to voltages): the value of the signal is constant during the measuring time interval.
- Quasi-static signals: the signal value varies very little, according to some physical quantity. Drift is an example of a quasi-static signal.
- Dynamic signals: the signal value varies significantly during the observation time. Such signals are also referred to as AC signals (AC = alternating current, but also alternating voltages).

Another classification of signals is based on the distinction between deterministic and stochastic signals. A stochastic signal is characterized by the fact that it is

16 Signals

impossible to predict its exact value. Most measurement signals and interference signals, like noise, belong to this group. Examples of deterministic signals are:

- Periodic signals, characterized by $x(t) = x(t + nT)$, where T is the time of one signal period and n an integer.
- Transients, like the response of a system to a pulse-shaped input; the signal can be repeated (hence predicted) by repeating the experiment under the same conditions.

A third possibility is the division into continuous and discrete signals. The continuity may refer to both the time scale and the amplitude scale (the signal value). Figure 2.1 shows the four possible combinations. Figure 2.1b represents a sampled signal and Figure 2.1c a quantized signal, as introduced in Chapter 1. A quantized signal with only two levels is called a binary signal.

Finally, we should mention the distinction between analog and digital signals. As with many technical and in particular electronic terms, these words have a rather fuzzy meaning. In ordinary terms digital signals are sampled, time-discrete, binary-coded signals, as used in digital processors. Analog signals refer to time-continuous signals with a continuous or a quantized amplitude.

2.1.2 Signal values

The most complete signal description is given by an amplitude–time diagram like in Figure 2.1, representing the signal value for each moment within the observation interval. In many cases, it is not necessary to give that much information about the signal, and only an indication of a particular signal property would be sufficient. Some of such simple characteristic signal parameters are listed below. The parameters are valid for an observation interval $0 < t < \tau$.

peak value: $\quad x_p = \max\{|x(t)|\}$

peak-to-peak value: $\quad x_{pp} = \max\{x(t)\} - \min\{x(t)\}$

mean value: $\quad x_m = \dfrac{1}{\tau} \displaystyle\int_0^\tau x(t)\,dt$

mean absolute value: $\quad |x|_m = \dfrac{1}{\tau} \displaystyle\int_0^\tau |x(t)|\,dt$

root-mean-square value: $\quad x_{rms} = \sqrt{\dfrac{1}{\tau} \displaystyle\int_0^\tau x^2(t)\,dt}$

mean signal power: $\quad P_m = \dfrac{1}{\tau} \displaystyle\int_0^\tau x^2(t)\,dt$

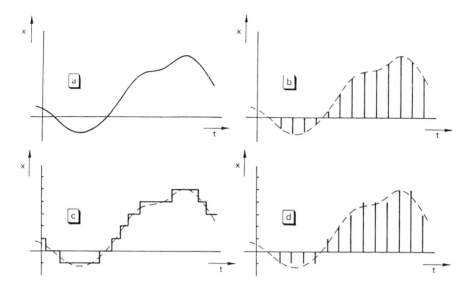

Figure 2.1 Continuous and discrete signals: (a) continuous in time and in amplitude; (b) time discrete, continuous amplitude (sampled signal); (c) discrete amplitude, continuous in time (quantized signal); (d) discrete both in time and amplitude.

The peak and peak-to-peak values are important in relation to the limits of the signal range of an instrument. The mean value is used when only the DC or quasi-DC component of a signal is of importance. The rms value is a parameter related to the signal power content. An AC current of arbitrary shape, whose rms value is I (A), flowing through a resistor, produces exactly as much heat in that resistor as does a DC current whose (DC) value is I (A). Note that the rms value is the square root of the mean power.

Example 2.1
The mathematical description of a sinusoidal signal is:

$$x(t) = A \sin \omega t = A \sin \frac{2\pi}{T} t$$

where A is the amplitude, $f = \omega/2\pi$ the frequency and $T = 1/f$ the period time. Figure 2.2 shows one period of this signal, and indicates the characteristic parameters

18   Signals

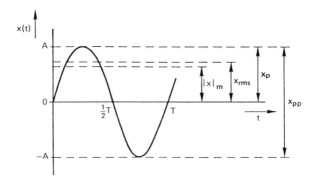

Figure 2.2 Signal values for a sine wave signal.

defined previously. Applying these definitions to the sine wave results in the following values:

$$x_p = A \qquad |x|_m = \frac{2A}{\pi}$$

$$x_{pp} = 2A \qquad x_{rms} = \frac{1}{2}A\sqrt{2}$$

These values apply also for a full periodical sine wave, as all periods have equal shapes.

Many rms voltmeters do not really measure the rms value of the input signal. Instead, they measure the mean of the absolute value, $|x|_m$, which value can be realized using a very simple electronic circuit. However, both these values are not the same. To obtain an indication in terms of rms, such instruments are calibrated in rms. As both signal parameters depend on the signal shape, the calibration is only valid for the particular signal used during calibration. Generally, rms meters are calibrated for sinusoidal inputs. Example 2.1 shows that the mean absolute value should be multiplied (internally) by $\frac{1}{4}\pi\sqrt{2}$, that is about 1.11, to obtain the rms value. Such instruments indicate only the proper rms value for sine-shaped signals.

Some voltmeters indicate the 'true rms' value of the input voltage. A true rms meter functions differently from those discussed earlier. Some of them use a thermal converter to obtain directly the rms value of the input signal. The indication is true for (almost) all types of input voltages.

2.1.3 Signal spectra

Any periodic signal can be split up into a series of sinusoidal subsignals. If the time of one period is T, the frequencies of all the subsignals are multiples of $1/T$. There are no components with other frequencies. The lowest frequency is just equal to $1/T$ and is called the fundamental frequency of the signal.

The subdivision of a periodic signal into its sinusoidal components is called the Fourier expansion of the signal; the resulting series of sinusoids is the Fourier series. The Fourier expansion is described mathematically as follows:

$$\begin{aligned} x(t) &= a_0 + a_1 \cos \omega_0 t + a_2 \cos 2\omega_0 t + a_3 \cos 3\omega_0 t + \ldots \\ &\quad + b_1 \sin \omega_0 t + b_2 \sin 2\omega_0 t + b_3 \sin 3\omega_0 t + \ldots \\ &= a_0 + \sum_{n=1}^{\infty} (a_n \cos n\omega_0 t + b_n \sin n\omega_0 t) \\ &= a_0 + \sum_{n=1}^{\infty} c_n \cos(n\omega_0 t + \varphi_n) \end{aligned} \quad (2.1)$$

These three representations are identical; the second is a short form of the first, and in the third expression, the corresponding sine and cosine terms are combined into a single cosine having a new amplitude c and phase angle φ, satisfying the relations:

$$c_n = \sqrt{(a_n^2 + b_n^2)}; \qquad \varphi_n = \tan^{-1}(b_n/a_n)$$

The coefficients a_n, b_n and c_n are the Fourier coefficients of the signal. Each periodic signal can be considered as an assembly of sinusoidal signals with amplitudes given by the Fourier coefficients and frequencies that are multiples of the fundamental.

The term a_0 in Equation (2.1) is nothing else than the mean value of the signal $x(t)$; the mean value must be equal to that of the complete series, and the mean of each sine signal is zero. All sine and cosine terms of the Fourier series have a frequency that is a multiple of the fundamental, f_0; they are called the harmonic components or the harmonics (making the signal audible by a loudspeaker a perfect 'harmonic' sound would be heard). The component with frequency $2f_0$ is the second harmonic, that with $3f_0$ the third harmonic, etc.

The shape of a periodic signal is reflected in its Fourier coefficients. We can draw a diagram of the Fourier coefficients as a function of the corresponding frequency. This picture is called the frequency spectrum of the signal (Figure 2.3). Usually, the amplitude of the combined sine and cosine terms (so the coefficients c_n in Equation (2.1)) is plotted versus frequency.

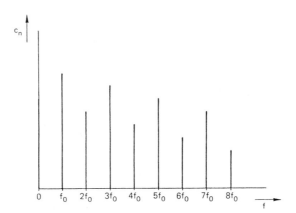

Figure 2.3 An example of a frequency spectrum of a periodic signal.

The Fourier coefficients are related to the signal shape. They can be calculated using the transformation formulas given in Equation (2.2):

$$a_0 = \frac{1}{T} \int_{t_0}^{t_0+T} x(t)\,dt$$

$$a_n = \frac{2}{T} \int_{t_0}^{t_0+T} x(t) \cos n\omega_0 t\,dt \qquad (2.2)$$

$$b_n = \frac{2}{T} \int_{t_0}^{t_0+T} x(t) \sin n\omega_0 t\,dt$$

These equations present the discrete Fourier transform for real coefficients. In general, the Fourier series has an infinite length. The full description of a signal by its spectrum requires an infinite number of parameters. Fortunately, the coefficients tend to diminish at increasing frequency. A remarkable property of the coefficients is that the first N elements of the series form the best approximation of the signal by N parameters.

Example 2.2

The Fourier coefficients of the square-shaped signal from Figure 2.4a, calculated with Equation (2.2), are:

$$a_0 = 0$$

$$a_n = 0$$

$$b_n = \frac{2A}{n\pi}(1 - \cos n\pi)$$

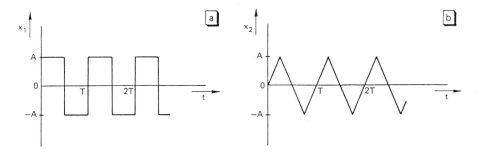

Figure 2.4 Examples of two periodic signals: (a) a square wave signal; (b) a triangular signal.

Apparently, its Fourier series is described by:

$$x_1(t) = \frac{4A}{\pi}\left(\sin \omega_0 t + \frac{1}{3}\sin 3\omega_0 t + \frac{1}{5}\sin 5\omega_0 t + \ldots\right)$$

The signal seems to be composed of only sinusoids with frequencies that are odd multiples of the fundamental.

Example 2.3
Using the same transformation formulas, the frequency spectrum of the triangular signal from Figure 2.4b is calculated as:

$$x_2(t) = \frac{8A}{\pi^2}\left(\sin \omega_0 t - \frac{1}{9}\sin 3\omega_0 t + \frac{1}{25}\sin 5\omega_0 t + \ldots\right)$$

and consists of components with identical frequencies, but different amplitudes.

According to the theory of Fourier, any periodic signal can be split up into sinusoidal components with discrete frequencies. The signal has a discrete frequency spectrum or a line spectrum. Obviously, one can also create an arbitrary periodic signal, by summing the required sinusoidal signals with the proper frequencies and amplitudes. This composition of periodic signals is used in synthesizers.

The Fourier transform is also applicable to aperiodic signals. It appears that such signals have a continuous frequency spectrum: the sinusoidal components may have any arbitrary frequency. However, in a continuous spectrum there are no longer individual components. Therefore, the strength of the signal is expressed in amplitude density rather than amplitude. A more usual presentation of a signal is its power spectrum, that is the spectral power (W/Hz) as a function of frequency.

Figure 2.5 shows the power spectra of two different signals. One signal varies slowly with time, the other much faster. One can imagine the first signal being composed of sinusoidal signals with relatively low frequencies; signal (b) contains components whose frequencies are higher. This is clearly seen in the corresponding frequency spectra of the signals: the spectrum of signal (a) covers a small range of frequency; its bandwidth is low. Signal (b) has a much wider bandwidth.

This relationship between the signal shape (time domain) and its spectrum (frequency domain) is also illustrated in Figure 2.6. It shows the spectrum of two periodic signals, one with very sharp edges (the rectangular signal) and another that varies more slowly (a rectified sine wave). Obviously, the high frequency components of the rectangular wave are much larger than those of the clipped sine wave.

The bandwidth of a signal is defined similar to that for systems. It is that part of the spectrum for which the spectral power density exceeds half of the nominal or maximal value. For an amplitude spectrum these boundaries are set at $\frac{1}{2}\sqrt{2}$ of the nominal amplitude density.

A measurement system can only handle signals adequately with a bandwidth up to that of the system itself. Signals with high frequency components require a wide band processing system; the bandwidth of the measuring instrument should match that of the signals to be processed.

Randomly varying signals or noise also have a continuous frequency spectrum. Some types of noise (in particular thermal-induced electrical noise) have a constant spectral power P_n (W/Hz); its power spectrum is flat, up to a certain maximum frequency. Such a signal is called white noise, analogous to white light, that contains equal wavelength components (colours) within the visible range. Noise is also specified in terms of spectral voltage or spectral current, expressed in $V/\sqrt{Hz}$ and $A/\sqrt{Hz}$, respectively.

Example 2.4
The spectral power, spectral voltage and spectral current density of white noise are P_n W/Hz, V_n V/$\sqrt{Hz}$ and I_n A/$\sqrt{Hz}$, respectively. The noise power, noise voltage and noise current of this signal, measured within a frequency band from 200 to 300 Hz, are, respectively: $100 \cdot P_n$ W, $10 \cdot V_n$ V and $10 \cdot I_n$ A.

2.2 Aperiodic signals

In this section we extend the Fourier expansion to aperiodic signals and use the result to derive the spectrum of sampled signals. Stochastic signals (whether continuous or discrete) can be described in three ways: in the time domain (e.g. time average, rms value), in the frequency domain (amplitude spectrum, power spectrum) and in the amplitude domain (expressing the signal value with probability parameters).

Aperiodic signals 23

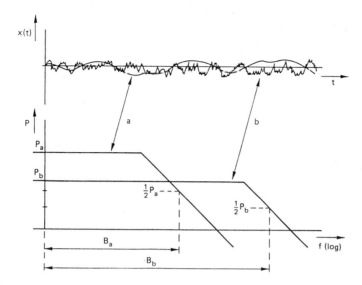

Figure 2.5 The amplitude–time diagram of two signals a and b, and the corresponding power spectra. Signal a varies slowly, and has a narrow bandwidth. Signal b moves quickly; it has a larger bandwidth.

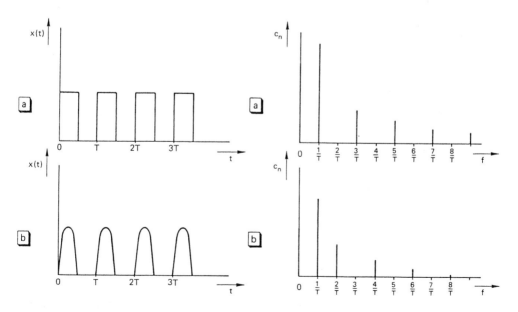

Figure 2.6 The amplitude–time diagram and the frequency spectrum of (a) a rectangular signal; (b) the positive half of a sine wave.

2.2.1 Complex Fourier series

In the first part of this chapter we introduced the Fourier expansion of a periodic signal into a series of real sine and cosine functions. The complex Fourier expansion is found using Euler's formula:

$$e^{\pm jz} = \cos z \pm j \sin z$$

Solving $\sin z$ and $\cos z$, and replacing the real goniometric functions in Equation (2.1) by their complex counterparts, we have:

$$x(t) = a_0 + \sum_{n=1}^{\infty} \left(\frac{a_n}{2}(e^{jn\omega t} + e^{-jn\omega t}) + \frac{b_n}{2j}(e^{jn\omega t} - e^{-jn\omega t}) \right)$$

By the substitutions $C_0 = a_0$, $C_n = \frac{1}{2}(a_n - jb_n)$ and $C_{-n} = \frac{1}{2}(a_n + jb_n)$ this can be simplified to:

$$x(t) = C_0 + \sum_{n=1}^{\infty} (C_n e^{jn\omega t} + C_{-n} e^{-jn\omega t}) = \sum_{n=-\infty}^{\infty} C_n e^{jn\omega t} \qquad (2.3)$$

This result is the complex Fourier series. Similarly, the complex form of Equation (2.2) becomes:

$$C_n = \frac{1}{T} \int_{t_0}^{t_0+T} x(t) e^{-jn\omega t} dt, \qquad n = 0, 1, 2, \ldots \qquad (2.4)$$

the discrete complex Fourier transform. The complex Fourier coefficients C_n can easily be derived from the real coefficients using the relations:

$$|C_n| = \frac{1}{2}\sqrt{a_n^2 + b_n^2}, \qquad n \neq 0$$

$$\arg C_n = \tan^{-1} \frac{-b_n}{a_n} \qquad (2.5)$$

As C_n is complex, the complex signal spectrum consists of two parts: the amplitude spectrum, a plot of $|C_n|$ versus frequency, and the phase spectrum, a plot of $\arg C_n$ versus frequency.

Example 2.5

The complex Fourier series of the rectangular signal in Figure 2.4a is calculated as follows: $C_0 = 0$, so $|C_0| = 0$ and $\arg C_0 = 0$. As $C_n = \tfrac{1}{2}(a_n - jb_n)$, its modulus and argument are:

$$|C_n| = \frac{1}{2}\sqrt{a_n^2 + b_n^2} = \frac{A}{n\pi}(1 - \cos n\pi), \quad n = 1, 2, \ldots$$

and

$$\arg C_n = \tan^{-1}\frac{-b_n}{a_n} = -\frac{\pi}{2}, \quad n = 1, 2, \ldots$$

The amplitude and phase spectra are depicted in Figure 2.7.

2.2.2 Fourier integral and Fourier transform

To derive the Fourier expansion of a non-periodic signal, we start with the discrete complex Fourier series for periodic signals, given by Equations (2.3) and (2.4). Consider one period of this signal. Replace t_0 by $-\tfrac{1}{2}T$ and let T approach infinity. Then:

$$x(t) = \lim_{T\to\infty} \sum_{n=-\infty}^{\infty} C_n e^{jn\omega t}$$

$$= \lim_{T\to\infty} \sum_{n=-\infty}^{\infty} e^{jn\omega t}\left(\frac{1}{T}\int_{-\frac{1}{2}T}^{\frac{1}{2}T} x(t)e^{-jn\omega t}\,dt\right)$$

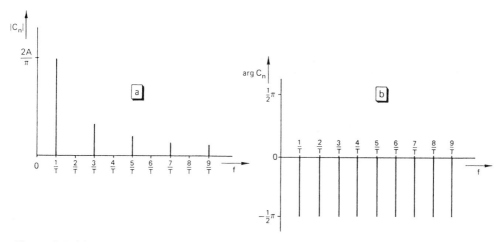

Figure 2.7 (a) Amplitude spectrum and (b) phase spectrum of the rectangular signal from Figure 2.4a.

Signals

When taking the limit for $T \to \infty$, the summation becomes an integration, $n\omega$ changes to ω and $T = \omega/2\pi$ becomes $d\omega/2\pi$:

$$x(t) = \int_{-\infty}^{\infty} e^{j\omega t} \left(\int_{-\infty}^{\infty} x(t) e^{-j\omega t} dt \right) \frac{d\omega}{2\pi}$$

With

$$X(\omega) = \int_{-\infty}^{\infty} x(t) e^{-j\omega t} dt \tag{2.6a}$$

this results in:

$$x(t) = \frac{1}{2\pi} \int_{-\infty}^{\infty} X(\omega) e^{j\omega t} d\omega \tag{2.6b}$$

$X(\omega)$ is the complex Fourier transform of $x(t)$. Both $x(t)$ and $X(\omega)$ give a full description of the signal, the first in the time domain, the other in the frequency domain. Equations (2.6a and b) transform the signal from the time domain into the frequency domain and vice versa. The modulus and the argument of $X(\omega)$ describe the frequency spectrum of $x(t)$. In general, this is a continuous spectrum, extending from $-\infty$ to $+\infty$, and also contains (in the mathematical sense) negative frequencies.

To find the Fourier transform of the product of two signals $x_1(t)$ and $x_2(t)$, we first define a particular function, the convolution integral:

$$g(\tau) = \int_{-\infty}^{\infty} x_1(t) x_2(\tau - t) dt$$

This is the product of $x_1(t)$ and the shifted and back-folded function $x_2(t)$ (Figure 2.8), integrated over an infinite time interval. The convolution function $g(\tau)$ is also denoted as:

$$g(\tau) = x_1(t) * x_2(t)$$

The Fourier transform of $g(\tau)$ is:

$$F\{g(\tau)\} = \int_{-\infty}^{\infty} g(\tau) e^{-j\omega \tau} d\tau$$

$$= \int_{-\infty}^{\infty} \left(\int_{-\infty}^{\infty} x_1(t) x_2(\tau - t) dt \right) e^{-j\omega \tau} d\tau$$

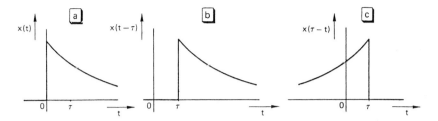

Figure 2.8 (a) Original function; (b) shifted over $t = \tau$; (c) shifted and back-folded.

Splitting the term $e^{-j\omega\tau}$ into $e^{-j\omega t} \cdot e^{-j\omega(\tau-t)}$ and changing the order of integration results in:

$$F\{g(\tau)\} = \int_{-\infty}^{\infty} x_1(t) e^{-j\omega t} \left(\int_{-\infty}^{\infty} x_2(\tau - t) e^{-j\omega(\tau-t)} d\tau \right) dt$$

$$= X_1(\omega) \cdot X_2(\omega)$$

So, the Fourier transform of two convoluted functions $x_1(t)$ and $x_2(t)$ equals the product of the individual Fourier transforms. Similarly, the Fourier transform of the convolution $X_1(\omega) * X_2(\omega)$ equals $x_1(t) \cdot x_2(t)$.

The Fourier transform is used to calculate the frequency spectrum of both deterministic and stochastic signals. The Fourier transform is only applicable to functions that satisfy the following inequality:

$$\int_{-\infty}^{\infty} |x(t)| dt < \infty \tag{2.7}$$

To calculate the frequency characteristics of functions that do not satisfy Equation (2.7), another kind of transformation should be used.

2.2.3 Description of sampled signals

In this section we will calculate the spectrum of a sampled signal. We consider sampling over equidistant time intervals. Sampling a signal $x(t)$ can be considered as the multiplication of $x(t)$ with a periodic, pulse-shaped signal $s(t)$, indicated in the left part of Figure 2.9. The sampling width is assumed to be zero.

As $y(t)$ is the product of $x(t)$ and $s(t)$, the spectrum of $y(t)$ is described by the convolution of their Fourier transforms $X(f)$ and $S(f)$, respectively. $S(f)$ is a line spectrum because $s(t)$ is periodical. The heights of the spectral lines are all equal when the pulse width of $s(t)$ approaches zero (their heights decrease with frequency at finite pulse width). $X(f)$ is limited in bandwidth, its highest frequency being B.

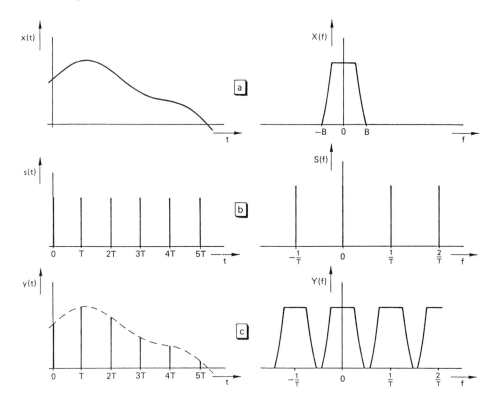

Figure 2.9 The amplitude–time diagram and the frequency spectrum of (a) an analog signal x(t), (b) a pulse-shaped signal s(t) and (c) the product y(t) = x(t)·s(t).

The first step in finding $Y(f) = X(f) * S(f)$ is back-folding $S(f)$ along a line $f = \xi$, to find the function $S(\xi - f)$, with ξ a new frequency variable. As $S(f)$ is a symmetric function, $S(\xi - f)$ is found simply by shifting $S(f)$ over a distance ξ along the f axis. Then, for each ξ, the product of the shifted version of $S(f)$ and $X(f)$ is integrated over the full frequency range. This product consists of a single line at $f = \xi$, as long as one pulse component of $S(\xi - f)$ falls within the band $X(f)$. Only that line contributes to the integral, because for all other values the product is zero. The convolution process results in periodically repeated frequency bands, the so-called alias of the original band (Figure 2.9c, right).

In the previous paragraph, it is assumed that the bandwidth B of $x(t)$ is less than half the sampling frequency. In that case the multiple bands do not overlap and the original signal can be reconstructed completely without information loss. At a larger bandwidth of $x(t)$ or at a sampling frequency below $2/B$, the multiple bands in the spectrum of the sampled signal overlap, preventing signal reconstruction without loss of information or signal distortion. The error due to this overlap is called aliasing error, occurring when the sample frequency is too low. The criterion for there to be no such aliasing errors is a sampling frequency of at least twice the highest frequency component of the analog signal. This result is known as the sampling theorem of Shannon, and gives the theoretical lower limit of the sampling rate. In practice, one would always choose a much higher sampling frequency, to facilitate the reconstruction of the original signal.

2.2.4 Description of stochastic signals

In this section we will describe stochastic signals in the amplitude domain, in terms of statistical parameters. At the end, some particular parameters are related to the signal parameters in the time domain.

We make a distinction between continuous and discrete stochastic signals or variables, denoted by x and $\underline{x}$, respectively. A discrete stochastic signal may be the result of converting a continuous stochastic signal into a digital signal by an AD converter. Again, a full description in the time domain of a stochastic signal requires a lot of information. For most applications, a rough description in terms of statistical parameters will suffice.

Consider a signal source with known statistical properties, generating a continuous, stochastic signal $x(t)$. Although it is impossible precisely to predict the signal, we can estimate its value, according to the nature of the source or the process that generates the signal. For instance we know the probability P that the signal value $x(t)$ at an arbitrary moment will not exceed a certain value x. This probability, which depends on the value x, is called the distribution function of $x(t)$, denoted as $F(x) = P\{x(t) < x\}$. Figure 2.10a gives an example of such a distribution function. From the definition, it follows immediately that $F(x)$ is a monotonically non-decreasing function of x, that $F(x \to \infty) = 1$ and that $F(x \to -\infty) = 0$.

Another important statistical parameter is the derivative of $F(x)$, the probability density (function): $p(x) = \mathrm{d}F(x)/\mathrm{d}x$. This function describes the probability of $x(t)$

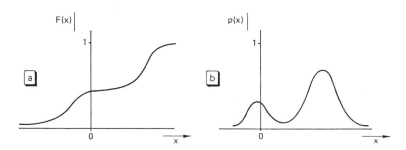

Figure 2.10 An example of (a) a distribution function and (b) the corresponding probability function.

having a value between x and $x + dx$ (Figure 2.10b). As the sum of the probability of all possible values is exactly 1, $p(x)$ satisfies:

$$\int_{-\infty}^{\infty} p(x)\,dx = 1$$

If $p(x)$ is known, $F(x)$ can be found through:

$$\int_{-\infty}^{x} p(x)\,dx = F(x) - F(-\infty) = F(x)$$

Many physical processes are governed by a normal or Gaussian distribution function. The probability density of the produced signals is written as:

$$p(x) = \frac{1}{\sigma\sqrt{2\pi}} e^{-(x-\mu)^2/2\sigma^2}$$

The meaning of the parameters μ and σ will be explained later in this section. The corresponding distribution function is:

$$F(x) = \int_{-\infty}^{x} p(x)\,dx = \frac{1}{\sigma\sqrt{2\pi}} \int_{-\infty}^{x} e^{-(x-\mu)^2/2\sigma^2}\,dx$$

For the normalized form, that is $\mu = 0$ and $\sigma = 1$, numerical values of this integral function can be found in mathematical tables.

The distribution function $F(x)$ of a discrete stochastic variable $\underline{x}$ is the probability that $\underline{x}$ does not exceed a value x, so $F(x) = P\{\underline{x} < x\}$.

Example 2.6
Suppose an electric voltage can have only two values: 0 V and 2 V, with a probability of ⅔ for the value 0 V. Figure 2.11 gives the distribution function and the corresponding probability density function of this binary signal.

Other statistical parameters to characterize a continuous or discrete stochastic variable are the mean or average value, the expectancy or first-order moment, the second-order moment and the standard deviation. We will first discuss these parameters for discrete variables.

The mean value of a discrete stochastic variable $\underline{x}$ is defined as:

$$\underline{x}_m = \frac{1}{N} \sum_{i=1}^{N} x_i$$

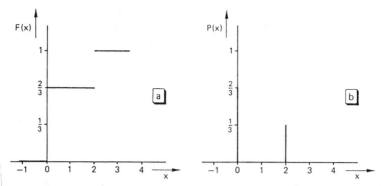

Figure 2.11 (a) The distribution function and (b) the probability density function of a binary signal of which $P(\underline{x} = 0) = 2/3$.

with N the number of all values x_i considered in a certain experiment. The expectancy or the first moment of $\underline{x}$ is defined as:

$$E(\underline{x}) = \sum_{i=1}^{N} p(x_i) x_i$$

which can be considered as a weighted average of all the values x_i. Only for an infinite number of values ($N \to \infty$) does the (algebraic) mean x_m approach the expected value $E(\underline{x})$. This can be seen as follows. The probability $p(x_k)$ of a certain output x_k is equal to N_k/N, where N_k is the number of outputs x_k. Suppose there are m different outputs. Then:

$$\lim_{N \to \infty} x_m = \lim_{N \to \infty} \frac{1}{N} \sum_{i=1}^{N} x_i$$

$$= \lim_{N \to \infty} \frac{1}{N} \sum_{k=1}^{m} N_k x_k$$

$$= \sum_{k=1}^{m} p(x_k) x_k = \sum_{i=1}^{N} p(x_i) x_i = E(\underline{x})$$

Example 2.7

The mean of the number of dots per throw after 100 throws with a die is simply the sum of all dots thrown divided by 100. The result is, for instance, $358/100 = 3.58$. As the probability for any result $i = 1$ to 6 is just $1/6$, the expectancy is:

$$E(\underline{x}) = \sum_{i=1}^{6} \frac{1}{6} i = \frac{21}{6} = 3.5$$

It is important to know not only the mean or expected value of a stochastic variable but also the expected deviation from the mean value. As this deviation may be positive or negative, the square of the deviation is usually taken, so $\{\underline{x} - E(\underline{x})\}^2$. The expectancy of this parameter is called the variance or the second moment:

$$\begin{aligned}\text{var}(\underline{x}) = E[\{\underline{x} - E(\underline{x})\}^2] &= E[\underline{x}^2 - 2\underline{x}E(\underline{x}) + E^2(\underline{x})]\\ &= E(\underline{x}^2) - 2E(\underline{x})E(\underline{x}) + E^2(\underline{x})\\ &= E(\underline{x}^2) - E^2(\underline{x})\end{aligned}$$

where E is assumed to be a linear operator. The square root of the variance is the standard deviation. This parameter has the same dimension as $\underline{x}$ itself, but it is not a stochastic variable.

Now we turn back to continuous variables. The first and second moment of a continuous stochastic variable are defined as:

$$E(x) = \int_{-\infty}^{\infty} x p(x) \, dx$$

$$\text{var}(x) = \int_{-\infty}^{\infty} (x - E(x))^2 p(x) \, dx$$

which are similar expressions to those for discrete variables. In particular, for the normal distribution function $x(t)$, the expected value is:

$$E(x) = \int_{-\infty}^{\infty} x \frac{1}{\sigma\sqrt{2\pi}} e^{-(x-\mu)^2/2\sigma^2} \, dx$$

$$= \frac{1}{\sigma\sqrt{2\pi}} \int_{-\infty}^{\infty} (y + \mu) e^{-y^2/2\sigma^2} \, dy = \mu$$

The parameter μ in the Gaussian distribution function is exactly the expected value and corresponds with the top of the probability density function (Figure 2.12). The variance is:

$$\text{var}(x) = E(x^2) - E^2(x)$$

$$= \frac{1}{\sigma\sqrt{2\pi}} \int_{-\infty}^{\infty} x^2 e^{-(x-\mu)^2/2\sigma^2} \, dx - \mu^2$$

$$= \frac{1}{\sigma\sqrt{2\pi}} \int_{-\infty}^{\infty} (y + \mu)^2 e^{-y^2/2\sigma^2} \, dy - \mu^2$$

$$= (\sigma^2 + \mu^2) - \mu^2 = \sigma^2$$

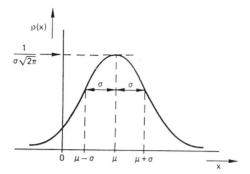

Figure 2.12 The probability density function of a normal or Gaussian distribution.

The parameter σ in the normal distribution function appears to be exactly the standard deviation, corresponding with the point of inflection in the probability density function (Figure 2.12).

We will now relate the parameters μ and σ with some of the signal parameters introduced in Section 2.1.2, where the mean or time average of a continuous signal was expressed as:

$$x_m = \frac{1}{\tau} \int_0^\tau x(t) \, dt$$

The time average of a stochastic continuous signal (such as a thermal noise voltage) is the same as its statistical mean value $E(x)$, if the statistical properties remain constant over the time interval considered. For signals having a normal distribution, this value equals μ, so $x_m = E(x) = \mu$. If during sampling or quantizing of the signal, the statistical parameters do not change, the same holds for discrete stochastic signals.

The power time average of a continuous signal $x(t)$ equals $P_m = 1/\tau \int_0^\tau x^2(t)\,dt$ (see Section 2.1.2). If the time average of the signal is zero, the mean power equals the square of the rms value. Consider the time signal as a continuous stochastic variable. The variance appeared to be $\int_{-\infty}^{\infty}(x - E(x))^2 p(x)\,dx$. If $E(x) = 0$, $\text{var}(x) = \int_{-\infty}^{\infty} x^2 p(x)\,dx$ which is σ for a normally distributed variable. For such signals, $P_m = \text{var}(x) = \sigma^2$, if the statistical properties do not change during the time of observation. Hence, the rms value is identical to the standard deviation σ (if the mean is zero).

Finally, we make some remarks with respect to the relation with the description in the frequency domain. The Fourier transform $F(\omega)$ of a time function $x(t)$ gives a complete description of a particular signal during its time of observation. Another signal would result in another Fourier function, although it might have identical statistical properties. So, the Fourier transform of $x(t)$ does not account for the statistical properties of the signal. The power density spectrum describes how the signal power is distributed over the frequency range. It can be proven that the power spectrum $S(f)$ is independent of a particular signal shape and is, therefore, a measure for the statistical properties of the signal.

SUMMARY

Periodic signals

- Possible partitionings of signals are:
  - static, quasi-static and dynamic signals;
  - deterministic and stochastic signals;
  - continuous and discrete signals;
  - analog and digital signals.
- Important signal characteristics are peak value, peak-to-peak value, mean or time average value, root-mean-square (rms) value and mean power.
- Any periodic signal with period time T can be split (expanded) into a series of sinusoids the frequencies of which are multiples of the fundamental frequency $f_0 = 1/T$ (Fourier series).
- The amplitudes of the sinusoidal components (Fourier coefficients) can be found from the signal time function using the Fourier transformation formulas in Equation (2.2).
- The plot of the Fourier coefficients versus frequency is the frequency spectrum of the signal.
- A periodic signal has a discrete spectrum or line spectrum; a non-periodic signal has a continuous spectrum.
- White noise is a noise signal whose frequency spectrum is flat over a wide frequency range.

Aperiodic signals

- The complex discrete Fourier series is given as $x(t) = \Sigma_{n=-\infty}^{\infty} C_n e^{jn\omega t}$; a plot of C_n (the complex Fourier coefficients) versus frequency is the (complex) frequency spectrum.
- The Fourier transform $X(\omega)$ of $x(t)$ is $X(\omega) = \int_{-\infty}^{\infty} x(t) e^{-j\omega t} dt$; the inverse transformation is $x(t) = (1/2\pi) \int_{-\infty}^{\infty} X(\omega) e^{j\omega t} d\omega$.
- The convolution of two signals $x_1(t)$ and $x_2(t)$ is defined as $g(\tau) = \int_{-\infty}^{\infty} x_1(t) x_2(\tau - t) dt = x_1(t) * x_2(t)$.
- The Fourier transform of $x_1(t) * x_2(t)$ is $X_1(\omega) \cdot X_2(\omega)$; the Fourier transform of $x_1(t) \cdot x_2(t)$ is $X_1(\omega) * X_2(\omega)$.
- The frequency spectrum of a sampled signal with bandwidth B consists of multiple frequency bands positioned around multiples of the sampling frequency f_s. Each band is identical to the spectrum of $x(t)$, and is called an alias. If $f_s > 2/B$, the bands do not overlap.

- Shannon's sampling theorem: a signal with highest frequency B can be fully reconstructed after sampling if $f_s > 2/B$.
- The statistical properties of a stochastic signal are described by its distribution function $F(x) = P\{x(t) \leq x\}$ and its probability density $p(x) = dF(x)/dx$.
- The expected value or first moment of a discrete stochastic variable is $E(\underline{x}) = \sum_{i=1}^{N} p(x_i) x_i$; that of a continuous stochastic variable $E(\underline{x}) = \int_{-\infty}^{\infty} p(x) x \, dx$.
- The variance or second moment of a discrete stochastic variable is $\text{var}(\underline{x}) = E[(\underline{x} - E(\underline{x}))^2] = E(\underline{x}^2) - E^2(\underline{x})$; that of a continuous variable $\text{var}(\underline{x}) = \int_{-\infty}^{\infty} (x - E(x))^2 p(x) \, dx$. The standard deviation is the square root of the variance.
- In the expression for the probability density of a normal or Gaussian distribution function $p(x) = (1/\sigma\sqrt{2\pi}) e^{-(x-\mu)^2/2\sigma^2}$, μ is the mean value of σ the standard deviation.
- The mean power of a signal with Gaussian amplitude distribution equals $P_m = \text{var}(\underline{x}) = \sigma^2$; the rms value equals the standard deviation σ.

Exercises

Periodic signals

2.1 The following figure shows one period of three different periodic signals with period time $T = 6$ s. Find the peak-to-peak value, the time average and the rms value of all of these signals.

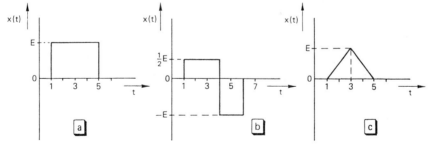

2.2 The crest factor of a signal is defined as the ratio between its peak value and rms value. Calculate the crest factor of the signal below.

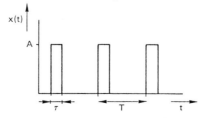

36 Signals

2.3 The signal from Exercise 2.2 is connected to an rms voltmeter that can handle signals with a crest factor up to 10. What is the minimum value of the signal parameter τ?

2.4 A voltmeter measures $|v|_m$ but is calibrated for the rms value of sinusoidal voltages. Predict the indication of this meter for each of the next signals:
(a) DC voltage of -1.5 V;
(b) sine voltage with amplitude 1.5 V;
(c) rectangular voltage (Figure 2.4a), with $A = 1.5$ V;
(d) triangular voltage (Figure 2.4b) with $A = 1.5$ V.

2.5 An rms meter is connected to a signal source producing a signal and noise; the indication appears to be 6.51 V; the signal is turned off, only the noise is left. Now the indication of the same meter is 0.75 V. Calculate the rms value of the measurement signal without noise.

2.6 Find the Fourier coefficients of the next periodic signals, using Equation (2.2).

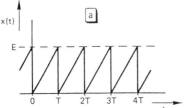
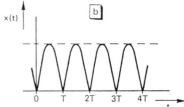

2.7 An electric resistor produces thermal noise with spectral power density equal to $4kT$ (Johnson noise); k is Boltzmann's constant ($1.38 \cdot 10^{-23}$ J/K), T is the absolute temperature (K). Calculate the rms value of the noise voltage across the terminals of the resistor, at room temperature (290 K), over a frequency range from 0 to 10 kHz.

Aperiodic signals

2.8 Derive the frequency spectrum (amplitude and phase diagrams) of the next signal (single-sided rectified sine) shown in the figure below.

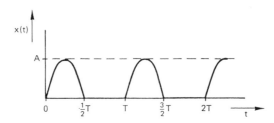

2.9 A signal $x(t)$ is characterized by:
$$x(t) = e^{-\alpha t} \text{ for } t > 0$$
$$x(t) = 0 \text{ for } t \leq 0$$

(a) Prove that the Fourier transform of $x(t)$ exists.
(b) Determine the Fourier transform.
(c) Draw the amplitude and phase spectrum.

2.10 Draw the distribution function $F(\underline{v})$ and the probability density function $p(\underline{v})$ of a stochastic signal with the following properties:

$\underline{v} = -5$ V, probability 0.2;
$\underline{v} = 0$ V, probability 0.5;
$\underline{v} = +5$ V, probability 0.3.

2.11 A signal $x(t)$ with Gaussian amplitude distribution has zero mean value. Derive an expression for the expected value $E(y)$ of a signal $y(t)$ that satisfies:

$y(t) = x(t)$ for $x > 0$;
$y(t) = 0$ for $x \leq 0$
(single-sided rectified signal).

3 Networks

This chapter gives a brief introduction to the theory of networks. The first part focuses on electric networks, composed of electric network elements. The theory can also be applied to networks consisting of non-electrical components. This will be demonstrated in the second part of the chapter.

3.1 Electric networks

The major signal quantities in an electronic measurement system are voltages and currents. These signals are processed by electronic components that are arranged and connected so that the system performs the desired processing.

We must distinguish between passive and active components. Active components offer the possibility of signal power amplification. The required energy is withdrawn from an auxiliary energy source, such as a battery or the mains. No power gain is possible with only passive components. These components may store signal energy but can never supply more than is stored in them. Examples of passive components are resistors, capacitors, inductors, transformers; an example of an active component is a transistor. Semiconductor technology allows the integration of many transistors and other components, resulting in very compact electronic building blocks, namely integrated circuits (ICs). Such an IC may be considered as a single but sometimes very complex electronic component, like an operational amplifier or a microprocessor.

Electronic systems, circuits and components are modelled by networks consisting of network elements. Figure 3.1 summarizes all existing electronic elements, and gives the corresponding relations between the currents and voltages.

We have explicitly distinguished between components and elements. Network elements are models for particular properties of physical components. A capacitor, for instance, has a capacitance C, but also dielectric losses, modelled by a parallel resistance R. Similarly, an inductor not only has a self-inductance L, but also a resistance of the wires and capacitance between the wires. The properties of transistors can be described only by current sources or voltage sources; an adequate characterization, however, requires an extended model with resistances and capacitances.

In modelling an electronic system of several components, the corresponding network elements are connected to each other to form a complete network model of

Electric networks 39

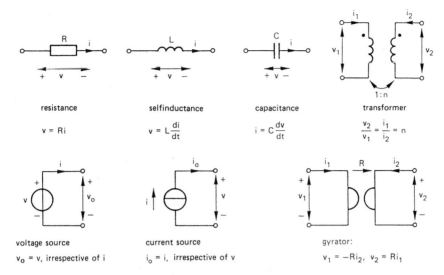

Figure 3.1 All electric network elements.

the system. A network contains nodes, branches and loops (Figure 3.2). A branch contains the network element; a node is the end of a branch. A loop is a closed path along arbitrary branches.

In a network, Kirchhoff's rules apply:

- rule for currents: the sum of all currents flowing towards a node is zero: $\Sigma_k i_k = 0$.
- rule for voltages: the sum of all voltages along a loop is zero: $\Sigma_m v_m = 0$.

All voltages and currents in a network can be calculated, using the voltage–current relations of the individual network elements and Kirchhoff's rules.

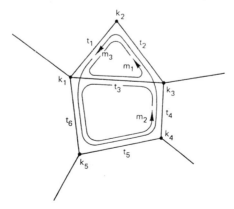

Figure 3.2 Nodes (k_i), loops (m_i) and branches (t_i) in a network.

Example 3.1

The node voltage v_k in the network of Figure 3.3 can be expressed as a function of the voltages v_1, v_2 and v_3 at the end points of the elements R, L and C (all voltages are with respect to a common reference voltage). To that end, we define three currents i_1, i_2 and i_3 (positive in the direction of the arrows). According to Kirchhoff's rule for currents $i_1 + i_2 + i_3 = 0$. Further, applying the voltage–current relations of the three elements:

$$i_1 = \frac{1}{R}(v_1 - v_k)$$

$$i_2 = C \frac{d(v_2 - v_k)}{dt}$$

$$i_3 = \frac{1}{L}\int (v_3 - v_k)\,dt$$

Eliminating the three currents from these four equations results in:

$$v_k + \frac{L}{R}\frac{dv_k}{dt} + LC\frac{d^2 v_k}{dt^2} = v_3 + \frac{L}{R}\frac{dv_1}{dt} + LC\frac{d^2 v_2}{dt^2}$$

The current direction can be chosen arbitrarily. If, for instance, i_2 in Figure 3.3 flows positive in the opposite direction, the result remains the same: the two equations with i_2 change to $i_1 - i_2 + i_3 = 0$ and $i_2 = C \cdot d(v_k - v_2)/dt$.

The relations between voltages and currents in a network are expressed as differential equations, which have the following properties:

- The differential equation is linear: the highest power of the signal quantities and their time derivatives is 1. It is the result of the linear voltage–current relation of the network elements. Linearity implies the use of superposition to facilitate the calculations; in a network with several sources, a current or voltage is found by calculating the contribution due to each of these sources separately.

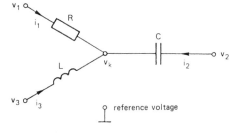

Figure 3.3 An example of the application of Kirchhoff's rule for currents.

- The differential equation has constant coefficients (they do not change with time). The reason is that the network elements are assumed to have constant parameters (like the resistance value). The consequence of linearity and constant coefficients is the preservation of frequency: sinusoidal signals keep their shape and frequency.
- It is an ordinary differential equation (not a partial one); time is the only independent variable.

These properties allow special calculation methods that facilitate solving the differential equations. Two of these methods are described in Chapter 4; they are based on complex variables and the Laplace transform, respectively.

The order of the differential equation determines the order of the system. The differential equation of the preceding example is of the second order: the network models a second-order system.

The user of an electronic system is probably not interested in knowing all the voltages and currents in the system. Only the signals on the terminals (the accessible points) are of interest, for example the voltage between the output terminals of a transducer, or the input and output currents of a current amplifier. We consider the system as a closed box, with a number of terminals. As a consequence, its model (an electric network) should be considered as a box as well; the only important nodes are those through which information exchange with the system's environment takes place. According to the number of external nodes, such a model is called a two-terminal (three-terminal, etc.) network (Figure 3.4). The terminals can be grouped two by two, each pair forming a port. So a two-terminal network is also called a one-port network; a four-terminal network is a two-port, etc. Many electronic instruments and circuits have two ports with one common terminal that is usually grounded (zero potential) (Figure 3.5). The port to which the signal source is connected is the input port (or the input) and that from which the signal is taken is the output port (the output for short). The corresponding voltages and currents are the input and output voltages (v_i and v_o) and the input and output currents (i_i and i_o), respectively. Referring to Figure 3.5, we can distinguish the following relations between the input and the output: the voltage transfer (or voltage gain) v_o/v_i, current transfer $-i_o/i_i$, the voltage-to-current transfer i_o/v_i, the current-to-voltage transfer (or transimpedance) v_o/i_i and the power transfer $p_o/p_i = -v_o i_o/v_i i_i$.

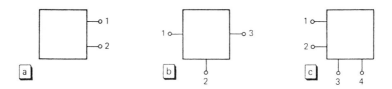

Figure 3.4 (a) A two-terminal network or one-port; (b) a three-terminal network; (c) a four-terminal network or two-port.

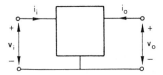

Figure 3.5 A three-terminal network connected as a two-port.

A network with n terminals can be fully characterized by a set of equations, relating all external currents and voltages. Such an n-terminal network may be built up in many different ways but nevertheless be characterized by the same set of equations. This property is used to simplify electric networks down to easily calculatable structures.

Example 3.2
Consider the network of parallel resistances in Figure 3.6a. For each branch k, $v = i_k R_k$; according to Kirchhoff's rule: $i = i_1 + i_2 + i_3 + \ldots + i_n$. Hence:

$$\frac{i}{v} = \frac{1}{v} \sum_k i_k = \sum_k \left(\frac{1}{R_k}\right)$$

So, a network consisting of parallel resistances is equivalent to a single resistance whose reciprocal value is $1/R_p = \Sigma_k (1/R_k)$. This value is always less than the smallest of the resistances.

For the network of Figure 3.6b, the equations $i = i_1 + i_2 + i_3 + \ldots + i_n$ and $i_k = C_k \cdot dv/dt$ apply, thus $i = \Sigma_k C_k \, dv/dt$. The network with parallel capacitances is equivalent to a single capacitance of value $C_p = \Sigma_k C_k$. This value is always larger than the largest capacitance.

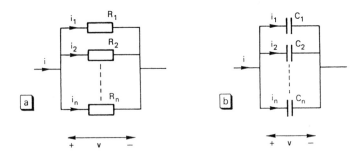

Figure 3.6 (a) Network consisting of parallel resistances; (b) network composed of parallel capacitances.

Example 3.3

Figure 3.7 shows two versions of a three-terminal network, both consisting of three resistances. For particular values of the resistances, the networks are equivalent. To find these conditions, we first calculate the resistance that can be measured between the terminals 1 and 2, leaving terminal 3 free (floating). Doing this for both networks, the results must be equal, so:

$$R_1 + R_2 = \frac{R_{12}(R_{13} + R_{23})}{R_{12} + R_{13} + R_{23}}$$

Similarly, two other relations can be found for the resistances between terminals 2 and 3 and between terminals 1 and 3. From these three equations, the conditions for equivalence can be found. The result is:

$$R_i = \frac{R_{ij} R_{ik}}{(R_{ij} + R_{ik} + R_{jk})}, i, j, k = 1, 2, 3, \text{cyclic}$$

The formula for the inverse network transformation is:

$$R_{ij} = \frac{R_i R_j + R_i R_k + R_j R_k}{R_k}, i, j, k = 1, 2, 3, \text{cyclic}$$

Example 3.4

Figure 3.8 shows a network with an input port and an output port.

To find the output voltage as a function of the input voltage, we assume a current i flowing through the loop. Elimination of i from the equations $v_i = iR_1 + iR_2$ and $v_o = iR_2$ results in:

$$v_o = v_i \cdot R_2/(R_1 + R_2)$$

Apparently, the output voltage is a fraction of the input voltage. This network is called a voltage divider.

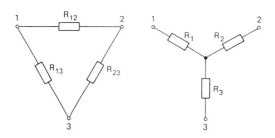

Figure 3.7 A triangular network can be converted into a star-shaped network and vice versa.

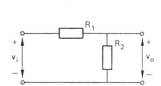

Figure 3.8 A voltage divider network.

3.2 Generalized network elements

In the first part of this chapter we showed that the relations between electrical quantities in a network are expressed as differential equations. This is also true for many other physical systems. Looking to the system equations in various disciplines, a remarkable similarity is noticed. As an example, compare Ohm's Law, $v = R \cdot i$, with the equation that relates the heat current q through a thermal conductor and the resulting temperature difference ΔT across that conductor, $\Delta T = R_{th} \cdot q$, or the equation relating the force F on a mechanical damper and the speed v of the damper, $v = (1/b)F$. Similarly, analog forms of the equations $i = C(dv/dt)$ and $v = L(di/dt)$ can be found in describing the properties of mechanical, hydraulic, pneumatic and thermodynamic systems.

The quantities that describe technical systems appear to belong to either of two classes: through variables and across variables. To explain this classification, we first introduce the term lumped element, which symbolizes a particular property of a physical component; that property is assumed to be concentrated in that element between its two end points or nodes. Exchange of energy or information only occurs through these terminals.

A through variable is a physical quantity which is the same for both terminals of the lumped element. An across variable describes the difference with respect to a physical quantity between the terminals. In an electronic system, the current is a through variable, the voltage (or potential difference) an across variable. For simplicity, we call a through variable an I variable and an across variable is a V variable. To indicate that a V variable always varies with respect to two points (nodes), we write subscripts a and b, so V_{ab}.

A lumped element is described by a relation between one I variable and one V variable. There are three basic relations:

$$I = C \frac{dV_{ab}}{dt} \tag{3.1a}$$

$$V_{ab} = L \frac{dI}{dt} \tag{3.1b}$$

$$V_{ab} = RI \tag{3.1c}$$

In these equations, the parameters C, L and R are the generalized capacitance, self-inductance and resistance, respectively. We will discuss each of these generalized network parameters separately.

■ *Generalized capacitance*
The relation between the I variable and the V variable for a generalized capacitance is given by Equation (3.1a). In the electrical domain, the relation between the current and the voltage of a capacitance is described by the equation:

$$i = C \frac{dv_{ab}}{dt}$$

Generalized network elements

In the thermal domain, the heat flow q towards a body and its temperature are related as:

$$q = C_{th}\frac{dT_{ab}}{dt}$$

Often, the reference temperature is 0 K, hence $q = C_{th}(dT/dt)$. C_{th} is the heat capacitance of the body, and indicates the rate of temperature change at a particular heat flow.

Newton's Law of Inertia, $F = ma$, can be rewritten as:

$$F = C_{mech}\frac{dv_{ab}}{dt}$$

It is apparent that mass can be considered as a 'mechanical capacitance'.

■ *Generalized self-inductance*

An ideal, generalized self-inductance is described by Equation (3.1b). In the electric domain:

$$v_{ab} = L\frac{di}{dt}$$

A mechanical spring is described as an analog:

$$v_{ab} = \frac{1}{k}\frac{dF}{dt}$$

with F the force on the spring, v_{ab} the difference in speed between the two end points, and k the stiffness. Likewise, the equation for a torsional spring is:

$$\Omega_{ab} = \frac{1}{K}\frac{dT}{dt}$$

with Ω_{ab} the angular velocity, T the moment of torsion and K the stiffness of rotation.

The thermal domain misses an element that behaves in an analogous way to a self-inductance.

■ *Generalized resistance*

The relation between the I and V variable of a generalized resistance is given by Equation (3.1c). In the electrical domain this is equivalent to Ohm's Law:

$$v_{ab} = R \cdot i$$

A mechanical damper is described analogously:

$$v_{ab} = \frac{1}{b} F$$

as mentioned before. The thermal resistance R_{th} is defined as the ratio between temperature difference and heat flow; the hydraulic resistance by the ratio between pressure difference and mass flow, etc.

There are even more similarities within the various groups of network elements. These refer to the stored energy and the dissipated energy. In a generalized capacitance, the V variable is responsible for the energy storage:

$$E = \int P\,dt = \int VI\,dt = \frac{1}{2} V_{ab}^2 C$$

In replacing C by, for instance, the mass m of a moving body, the result is the equation $E = \frac{1}{2}mv^2$, the kinetic energy. There is, however, an exception: the thermal capacitance. In the thermal domain, the I variable is a power quantity: q (W, J/s). The thermal energy is:

$$\int_{t_1}^{t_2} q\,dt = C_{th} T$$

In a generalized self-inductance, the I variable accounts for the energy storage:

$$E = \int P\,dt = \int VI\,dt = \frac{1}{2} LI^2$$

From this the energy stored in a torsional spring follows immediately: $\frac{1}{2}(1/K)T^2$.

The energy stored in pure C and L elements can be retrieved completely. This is not the case with R elements; they convert the electric energy into thermal energy (heat). That is why they are called dissipating elements. The energy uptake amounts to:

$$P = V \cdot I = \frac{V^2}{R} = I^2 R$$

The energy uptake of other R elements follows in the same way. Again, the thermal resistance is an exception; the energy uptake is the I variable q itself: $P = q = T/R_{th}$.

Any network that models a physical system in a particular domain can be transformed into a network for another domain, using the analogies previously discussed. The equations are the same, so the calculation methods, which will be discussed in Chapter 4, are applicable for electrical signals and in other domains.

Example 3.5

The mercury reservoir of a thermometer has a (concentrated) heat capacitance C_k; the heat resistance of the glass wall is R_g. Further, the temperature of the measurement object is T_a (relative to 0 K) and the temperature of the mercury is T_k. The (electric) model of this thermometer, measuring the temperature of a gas, is depicted in Figure 3.9a.

The model can be extended to account for the heat capacity of the glass reservoir and the heat transfer coefficient between the glass and the measurement object (the surrounding gas): k (W/m^{2}K), see Figure 3.9b. The thermal resistance between the glass and the gas equals $k \cdot A$, where A is the contact area between the glass and the gas.

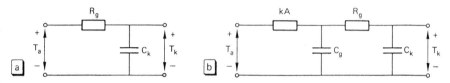

Figure 3.9 (a) A simplified electric analog model of a mercury-in-glass thermometer; (b) the model of the same thermometer, extended with the thermal capacity of the glass.

SUMMARY

Electric networks

- The most important network elements are resistance, capacitance, self-inductance, current source and voltage source. The voltage-to-current relations for a resistance, capacitance and self-inductance are, respectively, $v = Ri$, $i = C\,dv/dt$ and $v = L\,di/dt$.
- An electric network contains nodes, branches and loops. According to Kirchhoff's rules, the sum of all currents towards a node is zero, and the sum of all voltages around a loop is zero.
- Currents and voltages in an electric network are related by ordinary, linear differential equations; the order of the system corresponds to the order of the differential equation.
- A number of resistances connected in series is equivalent to a single resistance whose value is the sum of the individual resistance values. This summing rule also applies for self-inductances in series and for capacitances in parallel, as well as for voltage sources in series and current sources in parallel.
- A number of resistances connected in parallel is equivalent to a single resistance whose reciprocal value is the sum of the reciprocal values of the individual resistances. This reciprocal summing rule also applies for self-inductances in parallel and capacitances in series.

Networks

Generalized network elements

- Variables are partitioned into through variables or I variables and across variables or V variables.
- A lumped element is a model of a physical property, assumed to be concentrated between two terminals.
- A lumped element is characterized by a relation between an I variable and a V variable. There are three basic equations, corresponding to the three basic generalized elements capacitance, self-inductance and resistance:

$$I = C\frac{dV}{dt}; \quad V = L\frac{dI}{dt}; \quad V = IR$$

- The energy stored in a generalized capacitance is $\frac{1}{2}CV^2$, in a generalized self-inductance $\frac{1}{2}LI^2$. The energy dissipated in a resistive or dissipative element is $V \cdot I$.
- There is no thermal self-inductance.

Exercises

Electric networks

3.1 Replace each of the two-terminal networks a–f shown in the figure below by a single element and calculate their equivalent values.

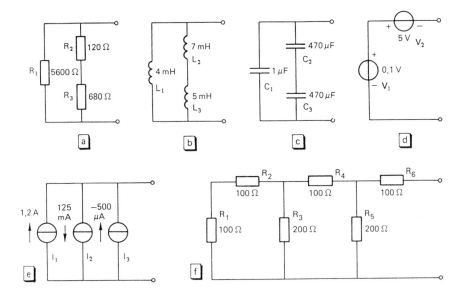

Exercises

3.2 The resistance of one edge of a cube made up of 12 wires (forming the edges) is 1 Ω. Calculate the resistance between the two end points of the cube's diagonal.

3.3 Calculate the transfer of the networks a–c shown in the figure below.

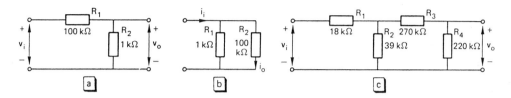

3.4 Calculate the transfer of the network shown in the figure below, using the transformation formulas for triangular- and star-shaped networks.

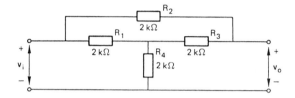

3.5 Derive the differential equations that describe the voltage transfer of the next networks a–f shown in the figure below.

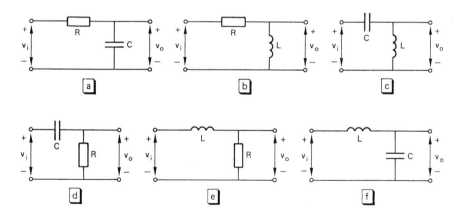

3.6 An ideal capacitor with capacitance C is charged by a constant current of 1 μA, starting from 0 V. After 100 s the voltage is 20 V. Find the capacitance C.

Generalized network elements

3.7 Which of the following elements or system properties is described by a generalized capacitance, which by a self-inductance and which by a resistance: thermal capacitance, mass, mechanical damping, moment of inertia, stiffness, thermal resistance?

3.8 In addition to V and I variables there are also time-integrated variables, for instance $x = \int v \cdot dt$. Assign one of the variables V variable, I variable, integrated V variable and integrated I variable to each of the next quantities: force, angular velocity, electric charge, heat flow, angular displacement, mass flow, heat, temperature.

3.9 Derive the electrical analog model of the mass–spring system of the figure below. What is the relation between F and x?

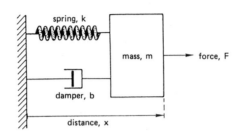

3.10 Derive the mechanical analog of the electric network in the figure below.

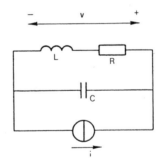

4 Mathematical tools

The computation of voltages and currents in an electric network means solving a set of differential equations. Even for relatively simple networks, solving the differential equations is rather time-consuming. In this chapter, two methods for facilitating the computations are discussed. In the first part, we introduce complex variables as a mathematical tool to calculate currents and voltages in a network, without the necessity to derive and solve the differential equations. The method is simple, but only valid for sinusoidal signals. In the second part, the Laplace transform is introduced as a mathematical tool for the computation of arbitrary signals.

4.1 Complex variables

4.1.1 *Properties of complex variables*

We start this chapter with a brief overview of the main properties of complex variables. A complex variable is defined as the sum of a real variable and an imaginary variable. The latter is the product of a real variable and the imaginary unit $i = \sqrt{-1}$. To avoid confusion with the symbol i for electric current, electrical engineers adopted the symbol j, thus $j = \sqrt{-1}$. A complex variable is written as $z = a + jb$, where a and b are real. The variables a and b are called the real and imaginary components of the complex variable z, and are also denoted as $\operatorname{Re} z$ and $\operatorname{Im} z$, respectively. So we can write a complex variable as $z = \operatorname{Re} z + j \operatorname{Im} z$.

Real variables are represented as points on a straight line. Complex variables are represented as points in the complex plane, their coordinates being positioned on the real and the imaginary axis (Figure 4.1).

This figure shows another representation of z, using a length and an angle as the two coordinates (polar coordinates). The distance from z to the origin (0, 0) is the modulus or absolute value of z, denoted as $|z|$; the angle between the 'vector' and the positive real axis is the argument of z, denoted as $\arg z$, or simply with a symbol for an angle, like φ. The relation between these two representations follows directly from Figure 4.1:

$$|z| = ((\operatorname{Re} z)^2 + (\operatorname{Im} z)^2)^{1/2}$$

$$\varphi = \arg z = \tan^{-1} \frac{\operatorname{Im} z}{\operatorname{Re} z}$$

52 Mathematical tools

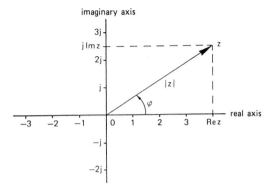

Figure 4.1 A complex variable is represented as a point in the complex plane.

Using the equations:

$$\operatorname{Re} z = |z| \cos \varphi$$
$$\operatorname{Im} z = |z| \sin \varphi$$

the complex variable can also be written as:

$$z = |z|(\cos \varphi + j \sin \varphi)$$

or, as $\operatorname{Re} z = |z| \cos \varphi$ and $\operatorname{Im} z = |z| \sin \varphi$, a third way to represent a complex variable is:

$$z = |z| e^{j\varphi}$$

From the definitions of complex variables, the next rules for the product and the ratio of two complex variables are derived:

$$|z_1 z_2| = |z_1||z_2|; \qquad \left|\frac{z_1}{z_2}\right| = \frac{|z_1|}{|z_2|}$$

$$\arg z_1 z_2 = \arg z_1 + \arg z_2; \qquad \arg \frac{z_1}{z_2} = \arg z_1 - \arg z_2$$

Further, for complex variables the common rules for integration and differentiation are valid. These rules are frequently used in the next chapters.

4.1.2 Complex notation of signals and transfer functions

A sinusoidal signal is fully characterized by its amplitude, frequency and phase. If such a sine wave signal passes through an electronic network, its amplitude and phase may change, but the frequency remains unchanged.

Complex variables   53

The amplitude $\hat{x}$ and phase φ of a sinusoidal signal $x(t) = \hat{x}\cos(\omega t + \varphi)$ show a noticeable similarity with the modulus and argument of a complex variable. A complex variable $X = |X|e^{j(\omega t + \varphi)}$ is represented in the complex plane as a rotating vector, with a length $|X|$ and angular speed ω (compare Figure 4.1). At the moments $t = \pm nT$, the argument of X equals φ. Hence, the modulus $|X|$ is equivalent to the amplitude $\hat{x}$; the argument $\arg X$ is equivalent to the phase φ. Further, the real part of X is equal to the time function $x(t)$.

To distinguish between complex variables and real or time variables, the first are written in capitals (X, V, I), the time variables in lower-case letters (x, v, i, respectively).

In Chapter 3, several transfer functions were defined as the ratio between output quantity and input quantity. Complex transfer functions are defined in the same way. For example, the complex voltage transfer function of a two-port network is denoted as $A_v = V_o/V_i$. The amplitude transfer follows directly from $|A_v|$ and the argument of A_v represents the phase difference between the input and output: $|A_v| = |V_o|/|V_i| = \hat{v}_o/\hat{v}_i$ and $\arg A_v = \arg V_o - \arg V_i = (\omega t + \varphi_o) - (\omega t + \varphi_i) = \varphi_o - \varphi_i$.

4.1.3 *Impedances*

The ratio of a complex voltage and a complex current is in general a complex quantity. The ratio V/I is called the impedance Z. The inverse ratio is the admittance: $Y = 1/Z = I/V$. The impedance can be considered as a complex resistance; the admittance as a complex conductance. We will now derive the impedance of a capacitance and a self-inductance.

The voltage–current relation of a self-inductance is $v = L\,di/dt$. A sinusoidal current can be represented as a complex current $I = |I|e^{j(\omega t + \varphi)}$. The complex voltage of the self-inductance is: $V = L\,dI/dt = L|I|j\omega e^{j(\omega t + \varphi)} = j\omega LI$. This is the complex relation between the voltage and the current of a self-inductance. The impedance of the self-inductance becomes $Z = V/I = j\omega L$.

The impedance of a capacitance is found in a similar way. In the time domain, the current through the capacitance is $i = C\,dv/dt$, so in complex notation $I = C\,dV/dt = j\omega CV$. Hence, the impedance of a capacitance is $Z = V/I = 1/j\omega C$.

From the foregoing, it follows that the impedances of a self-inductance and a capacitance have imaginary values. The impedance of a resistor is real. The impedance of a composition of network elements is, in general, a complex quantity. For the computation of these impedances, the same rules for the series and parallel combinations can be used as for networks with only resistances.

Example 4.1
The ratio of V and I in the network of Figure 4.2 is equal to:

$$Z = \frac{(R_1 + 1/j\omega C)R_2}{R_1 + 1/j\omega C + R_2} = R_2\frac{1 + j\omega R_1 C}{1 + j\omega(R_1 + R_2)C}$$

Mathematical tools

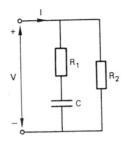

Figure 4.2 An example of a complex impedance.

The modulus of Z represents the ratio of the voltage amplitude and current amplitude: $|Z| = |V|/|I| = \hat{v}/\hat{\imath}$. The argument equals the phase difference between the sine-shaped voltage and current. The modulus of the impedance of a self-inductance is $|Z| = \omega L$, hence directly proportional to the frequency. The argument amounts to $\pi/2$. As $\arg V = \arg I + \arg Z = \arg I + \pi/2$, the current through a self-inductance lags the voltage across it by $\pi/2$ radians.

The modulus of the impedance of a capacitance is $|Z| = 1/\omega C$, hence inversely proportional to the frequency. The argument is $-\pi/2$, thus the current through a capacitance leads the voltage across it by $\pi/2$ radians. In composite networks, the phase difference is in general a function of frequency.

Example 4.2

The modulus $|Z|$ of the network from Figure 4.2 is:

$$|Z| = \frac{R_2|1 + j\omega R_1 C|}{|1 + j\omega(R_1 + R_2)C|} = R_2\sqrt{\frac{1 + \omega^2 R_1^2 C^2}{1 + \omega^2(R_1 + R_2)^2 C^2}}$$

For $\omega \to 0$, $|Z|$ approaches R_2. This can also be concluded directly from Figure 4.2, as for DC the capacitance behaves as an infinitely large resistance. In the case $\omega \to \infty$, the capacitance behaves as a short-circuit for the signals, hence:

$$|Z| = R_2\left(\frac{1/\omega^2 + R_1^2 C^2}{1/\omega^2 + (R_1 + R_2)^2 C^2}\right)^{1/2} \stackrel{\omega \to \infty}{=} \frac{R_1 R_2}{R_1 + R_2}$$

which is nothing else than the two resistances R_1 and R_2 in parallel. The phase difference between the current through the network and the voltage across it follows from:

$$\arg Z = \tan^{-1} \omega R_1 C - \tan^{-1} \omega(R_1 + R_2)C$$

Using the complex expressions for the impedances of a self-inductance and a capacitance, the transfer of two-port networks can be achieved rather quickly. From the complex transfer, the amplitude transfer and the phase difference can be derived immediately.

Example 4.3
The complex transfer of the network in Figure 4.3 is found directly, using the formula for the voltage divider network (Exercise 3.3):

$$H = \frac{V_o}{V_i} = \frac{R_2}{R_2 + R_1 + 1/j\omega C} = \frac{j\omega R_2 C}{1 + j\omega(R_1 + R_2)C}$$

The modulus and the argument are:

$$|H| = \frac{\omega R_2 C}{(1 + \omega^2(R_1 + R_2)^2 C^2)^{1/2}}$$

$$\arg H = \frac{\pi}{2} - \tan^{-1}\omega(R_1 + R_2)C$$

Both the amplitude transfer and the phase transfer are functions of frequency. In Chapter 6 we will introduce a simple method to make a plot of $|H|$ and $\arg H$ versus frequency, in order to get a quick insight in the frequency dependence of the transfer.

4.2 Laplace variables

The Laplace transform is a well-known method to solve linear differential equations. Using the Laplace transform, a linear differential equation of the order n is transferred into an algebraic equation of the order n. As electronic networks are characterized by linear differential equations, the Laplace transform may be a useful tool to describe the properties of such networks. The method is not restricted to sinusoidal signals, as complex variables are. Laplace variables are valid for (almost) arbitrary signal shapes.

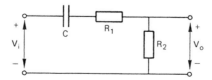

Figure 4.3 An example of a complex voltage transfer.

Mathematical tools

The Laplace transform can be considered as an extension of the Fourier transform (Section 2.2.2). The conditions for the existence of a Laplace transform are somewhat easier to fulfil (see Equation (2.7)). This aspect will be discussed briefly in Section 4.2.4.

4.2.1 *The Laplace transform*

The definition of the single-sided Laplace transform of a function $x(t)$ is:

$$X(p) = \mathbf{L}\{x(t)\} = \int_0^\infty x(t)\,e^{-pt}\,dt$$

Through this transformation, a function $x(t)$ is transformed into a function $X(p)$ in the Laplace domain. The Laplace operator p is also denoted by the letter s. In the double-sided Laplace transform, the integration range is from $-\infty$ to $+\infty$. Table 4.1 gives some time functions together with the corresponding Laplace functions. The time functions are supposed to be zero for $t<0$. For the transformation of other functions the next rules can be used. Let $\mathbf{L}\{x(t)\} = X(p)$, then:

$$\mathbf{L}\{ax(t)\} = aX(p) \tag{4.1}$$

$$\mathbf{L}\{x_1(t) + x_2(t)\} = X_1(p) + X_2(p) \tag{4.2}$$

$$\mathbf{L}\{e^{-at}x(t)\} = X(p+a) \tag{4.3}$$

$$\mathbf{L}\{x(t-\tau)\} = e^{-p\tau}X(p) \tag{4.4}$$

$$\mathbf{L}\left\{\frac{dx(t)}{dt}\right\} = pX(p) - x(0) \tag{4.5}$$

$$\mathbf{L}\left\{\int x(t)\,dt\right\} = \frac{1}{p}X(p) \tag{4.6}$$

$$\mathbf{L}\{x_1(t) * x_2(t)\} = X_1(p)X_2(p) \tag{4.7}$$

Successive repetition of the differentiation rule (Equation (4.5)) results in:

$$\mathbf{L}\left\{\frac{d^2x(t)}{dt^2}\right\} = p^2 X(p) - px(0) - x'(0) \tag{4.8}$$

with $x'(0)$ the first derivative of $x(t)$ for $t = 0$.

Table 4.1 Some time functions with their corresponding Laplace transforms

$x(t)$	$X(p)$	$x(t)$	$X(p)$
1	$\dfrac{1}{p}$	$t\cos\omega t$	$\dfrac{p^2-\omega^2}{(p^2+\omega^2)^2}$
$t^n\ (n\geq 0)$	$\dfrac{n!}{p^{n+1}}$	$t\sin\omega t$	$\dfrac{2p\omega}{(p^2+\omega^2)^2}$
e^{at}	$\dfrac{1}{p-a}$	$e^{-at}\cos\omega t$	$\dfrac{p+a}{(p+a)^2+\omega^2}$
$\cos\omega t$	$\dfrac{p}{p^2+\omega^2}$	$e^{-at}\sin\omega t$	$\dfrac{\omega}{(p+a)^2+\omega^2}$
$\sin\omega t$	$\dfrac{\omega}{p^2+\omega^2}$	$\delta(t)$	1

In the next section, we will see how the Laplace transform is used to solve network equations.

4.2.2 Solving differential equations with the Laplace transform

Chapter 3 showed that the relation between currents and voltages in an electric network is described with linear differential equations. To explain the use of the Laplace transform for solving such equations, we consider the network of Figure 4.4a. The relation between the output voltage v_o and the input voltage v_i is given as:

$$v_o + RC\frac{dv_o}{dt} = v_i$$

To solve this equation, we transform the time functions $v(t)$ into Laplace functions $V(p)$. Using the rules given above and rearranging terms we find:

$$V_o(p) + pRCV_o(p) - RCv_o(0) = V_i(p)$$

This is a linear algebraic equation from which $V_o(p)$ can easily be solved:

$$V_o(p) = \frac{V_i(p) + RCv_o(0)}{1 + pRC}$$

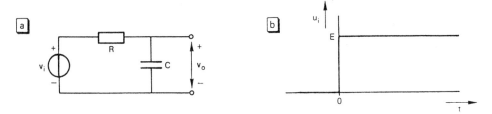

Figure 4.4 (a) An RC network with (b) a step input voltage.

If $v_i(t)$ is a known time function whose Laplace transform exists, $V_o(p)$ can be solved and by inverse transformation the output voltage $v_o(t)$ is finally found. This procedure is illustrated with the input voltage shown in Figure 4.4b, a step function with height E. The output voltage is called the step response of the network.

The Laplace transform of the input appears to be E/p (see Table 4.1). Suppose all voltages are zero for $t \leq 0$, then:

$$V_o(p) = E \frac{1}{p(1 + pRC)}$$

To find the inverse transform of this function, it must be split up into terms that are listed in Table 4.1. This can be achieved by splitting the right-hand side of the last equation into terms with denominators p and $1 + pRC$, respectively. This results in:

$$V_o(p) = E \left(\frac{1}{p} - \frac{1}{p + 1/RC} \right)$$

from which the output time function is found:

$$v_o(t) = E(1 - e^{-t/RC}), \quad t > 0$$

Figure 4.5 shows this time function. It appears that the tangent at the point $t = 0$ intersects the horizontal line $v_o = E$ (the end value or steady-state value) for $t = \tau = RC$. Using this property we can easily draw the step response, if the value of RC is known. The product RC is called the time constant of the network, a term that is applicable to each first-order system.

If the capacitor was charged at $t = 0$, then $V_o(p)$ contains an additional term $v_o(0)/(p + 1/RC)$, hence the expression for the output voltage given above must be extended with a term $v_o(t) e^{-t/RC}$.

If the input step starts at $t = t_1 > 0$ instead of at $t = 0$, the output voltage will be shifted over a time period of t_1 as well: in the expression for $v_o(t)$, t must be replaced by $t - t_1$. These two extra conditions, the initial charge and a time delay, are shown in

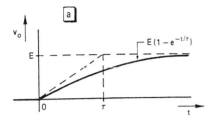
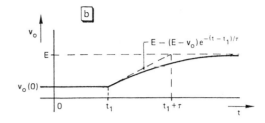

Figure 4.5 (a) Step response of the network from Figure 4.4a; (b) the step response of the same network, now with initial conditions and time delay.

Figure 4.5b. Notice that the tangent (now through the point $t = t_1$) still intersects the end value a time period τ after the input step.

4.2.3 Transfer functions and impedances in the p-domain

The relation between two signal quantities in a network is described generally by a differential equation of the order n:

$$a_n \frac{d^n y}{dt^n} + a_{n-1} \frac{d^{n-1} y}{dt^{n-1}} + \ldots + a_0 y = b_m \frac{d^m x}{dt^m}$$

$$+ b_{m-1} \frac{d^{m-1} x}{dt^{m-1}} + \ldots + b_0 x$$

Applying the Laplace transform and supposing zero initial conditions, this changes into:

$$a_n p^n Y + a_{n-1} p^{n-1} Y + \ldots + a_0 Y = b_m p^m X + b_{m-1} p^{m-1} X + \ldots + b_0 X$$

The ratio between X and Y is now:

$$\frac{Y}{X} = \frac{b_m p^m + b_{m-1} p^{m-1} + \ldots + b_0}{a_n p^n + a_{n-1} p^{n-1} + \ldots + a_0} \tag{4.9}$$

Equation (4.9) describes a transfer function $H(p)$ or an impedance $Z(p)$, depending on the dimensions of X and Y. The Fourier transform $H(\omega)$ in Section 4.1 gives a description of the system in the frequency domain (for sinusoidal signals). Likewise, the transfer function $H(p)$ describes the properties of the system in the p-domain, for arbitrary signals. This in parallel with the Fourier transform also holds for impedances: in the p-domain, the impedance of a capacitance is $1/pC$, that of a self-inductance is pL. The Fourier transform may be considered as a special case of the Laplace transform, namely $p = j\omega$, which means only sinusoidal functions.

Example 4.4
The transfer of the network depicted in Figure 4.4a can be written directly, without deriving the differential equation, as:

$$H(p) = \frac{V_o(p)}{V_i(p)} = \frac{1/pC}{R + 1/pC} = \frac{1}{1 + pRC}$$

The impedance of a resistor, a capacitance and a self-inductance all in series amount, in the p-domain, to $R + 1/pC + pL$. The impedance of a network composed of a self-inductance in parallel with a capacitance is

$$\frac{(1/pC)pL}{1/pC + pL} = \frac{pL}{1 + p^2 LC}$$

In Equation (4.9), the values of p for which the numerator is zero are called the zeros of the system. The values of p for which the denominator is zero are the poles of the system. The transfer function of the network in Figure 4.4 has only one pole, $p = -1/RC$; the impedance of the network consisting of a capacitance and a self-inductance in parallel (Example 4.4) has one zero for $p = 0$ and two imaginary poles $p = \pm j/\sqrt{LC}$.

The dynamic behaviour of the system is fully characterized by its poles and zeros. In some technical disciplines, in particular control theory, the description of systems is based on its poles and zeros.

4.2.4 Relation with the Fourier integral

The Fourier series is the expansion of a periodic signal into discrete, sinusoidal components. The derived Fourier integral may be interpreted as an expansion into a continuous package of sinusoidal components. The Laplace operator p is a complex variable, $p = \alpha + j\omega$. Therefore, the transform can also be written as:

$$\mathbf{L}\{x(t)\} = \int_0^\infty x(t) e^{-\alpha t} e^{-j\omega t} dt$$

This shows that the Laplace transform is equivalent to the Fourier transform (Section 2.6), except that the function $x(t)e^{-\alpha t}$ rather than the function $x(t)$ is transformed. A proper choice of α allows the transformation of functions that do not converge for $t \to \infty$, assuming the function $x(t)e^{-\alpha t}$ satisfies the condition (Equation 2.7). In this respect, the Laplace transform can be interpreted as the expansion into a continuous package of signals of the form $e^{-\alpha t}\sin \omega t$, which are exponentially decaying sine waves.

SUMMARY

Complex variables

- A complex variable z can be written as $z = \operatorname{Re} z + j \operatorname{Im} z = |z|(\cos\varphi + j\sin\varphi) = |z|e^{j\varphi}$, with $|z|$ the modulus and φ the argument of z.
- The modulus of z is $|z| = ((\operatorname{Re} z)^2 + (\operatorname{Im} z)^2)^{1/2}$, the argument of z is $\arg z = \varphi = \tan^{-1}(\operatorname{Im} z / \operatorname{Re} z)$.
- The complex notation for a sinusoidal voltage or current $\hat{x}\cos(\omega t + \varphi)$ is $X = |X|e^{j(\omega t + \varphi)}$.
- The impedance Z of a two-terminal element is defined as the ratio between the complex voltage V and the complex current I. The impedance of a capacitance is $1/j\omega C$, that of a self-inductance is $j\omega L$.
- The complex transfer H of a two-port network is the ratio between the complex output signal and the complex input signal. The modulus $|H|$ represents the amplitude transfer, the argument $\arg H$ represents the phase transfer.
- The complex notation is only valid for sinusoidal signals.

Laplace variables

- The Laplace transform is defined as $\mathbf{L}\{x(t)\} = \int_0^\infty x(t)e^{-pt}\,\mathrm{d}t = X(p)$.
- The Laplace transform of the derivative of a function $x(t)$ is $\mathbf{L}\{\mathrm{d}x/\mathrm{d}t\} = pX(p) - x(0)$. From this property, it follows that a linear differential equation can be transformed into a linear algebraic equation with the Laplace operator p as variable.
- The Laplace transform allows the computation of the properties of an electric network for arbitrary signals; the complex notation is only suitable for sine waves.
- Transfer functions and impedances can be described in the p- or Laplace domain. The impedance of a capacitance C and a self-inductance L is $1/pC$ and pL, respectively. The rules for the composition of networks in the ω-domain are also valid in the p-domain.
- The zeros of a system, described by a Laplace polynomial $T(p)/N(p)$ are those values of p for which $T(p) = 0$; the poles of this system are those values of p for which $N(p) = 0$.
- The Fourier integral can be interpreted as the expansion of a function into a continuous series of sinusoidal components. Likewise, the Laplace integral can be interpreted as the expansion of the function into a continuous series of exponentially decaying sinusoids.

Exercises

Complex variables

4.1 Find the impedance of each of the networks a–f given in the figure below. Calculate the impedance of network d for $\omega = 1/\sqrt{LC}$ and $R = 0$. Calculate the impedance of the network e for $\omega = 1/\sqrt{LC}$ and $R = \infty$.

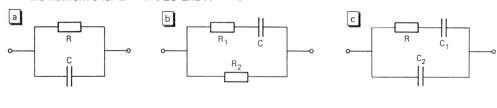

4.2 Find the complex transfer functions of the networks a–f depicted in the figure below.

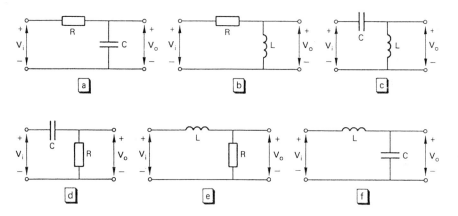

4.3 Find the complex transfer function of the network given in the figure below. At which condition with respect to R_1, R_2, C_1 and C_2 is the transfer independent of the frequency?

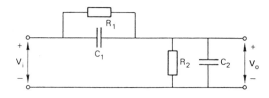

4.4 Find the transfer function of the bridge network given in the figure below. At which condition is $V_o = 0$, irrespective of V_i?

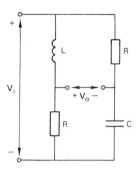

Laplace variables

4.5 Give the impedances in the p-domain of all the networks of Exercise 4.1. All currents and voltages are zero for $t < 0$.

4.6 Transform the complex transfer functions from Exercise 4.2 into the p-domain. Find the zeros and poles of these functions.

4.7 Calculate the output signal of the network from Exercise 4.2d, using the Laplace transform, for each of the following situations:
 (a) $v_i = 0$ for $t < 0$; $v_i = E$ for $t \geq 0$; C uncharged at $t = 0$;
 (b) $v_i = 0$ for $t < 0$; $v_i = E$ for $0 \leq t < t_1$; $v_i = 0$ for $t > t_1$, C uncharged at $t = 0$.
 Make a plot of the output voltage versus time, for both cases.

4.8 In the network shown in the figure below, the input voltage $v_i(t)$ is a sine wave. At $t = 0$ the input is connected to ground. At this moment, the voltage across the capacitance C is zero and the current through it is i_o. Calculate the output voltage $v_o(t)$ for three values of R:
 (a) $R = 400\,\Omega$;
 (b) $R = 120\,\Omega$;
 (c) $R = 200\,\Omega$.

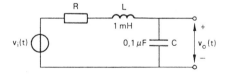

5 Models

A model of an electronic measurement system should account for those properties of the system that are of interest for the user. An extended model is not necessary and might be confusing. A restricted model does not give adequate information about the system behaviour and may cause a wrong interpretation of the measurement results.

The electronic properties of a measurement system can be modelled using a limited number of network elements (sources, impedances). The first part of this chapter deals with methods to obtain such models and how to use them. The second part of this chapter shows how noise and interference signals are modelled.

5.1 System models

An electronic circuit is modelled by a network of electronic elements. The model of a system with n external connections has (at least) n external terminals. In this chapter we will consider networks with two, three or four terminals, arranged in one or two signal ports. Particularly, we will discuss the influence on the signal transfer when connecting one system to another.

5.1.1 *Two-terminal networks*

In Chapters 3 and 4 we introduced two-terminal systems with passive elements. Such networks are equivalent if their impedances measured between the two terminals are the same. Such an equivalence also exists for systems containing active elements, like voltage sources and current sources. Two or more active two-terminal networks are equivalent if both the short-circuit current I_k and the open voltage V_o are equal (Figure 5.1). The ratio between the open voltage and the short-circuit current is the internal impedance or source impedance of the network: $Z_g = V_o/I_k$.

According to Thévenin's theorem, any active, linear two-terminal system can be fully characterized by one voltage source V_o and one impedance Z_g, connected in series (Figure 5.2a). An equivalent model consists of one current source I_k and one impedance Z_g (Figure 5.2b). This is called Norton's theorem.

The equivalence between both models can easily be verified. At open terminals, the current through Z_g in Figure 5.2a is zero, hence the voltage across the terminals is V_o. In Figure 5.2b, at open terminals all the current flows through Z_g, hence the

Figure 5.1 The determination of the open voltage and the short-circuit current of an active two-terminal network.

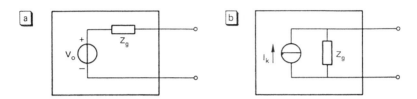

Figure 5.2 (a) Thévenin's equivalent of a two-terminal system; (b) Norton's equivalent.

output voltage equals $I_k \cdot Z_g = V_o$, that is the Thévenin voltage. Similarly, it can be shown that in both cases a current I_k flows through an external short-circuit.

Both models are fully equivalent. However, one model may be preferred over the other depending on the properties of the system that has to be modelled. Systems with a low value of the source impedance are preferably modelled with the voltage source model; systems that behave more like a current source (with a high value of Z_g) are better modelled by the current source model. In such cases, the source impedance characterizes the deviation from an ideal voltage source ($Z_g = 0$) or current source ($Z_g = \infty$), respectively.

There are other reasons for a particular choice. For instance a transducer whose output voltage is proportional to the measurement quantity is preferably modelled with a voltage source; a transducer that reflects the measurement quantity as a current is modelled with a current source model.

5.1.2 *Two-port networks*

Networks with three or four terminals generally have one port which is assigned as input port, and the other as output port (see also Section 3.1). In accordance with these functions, we define input quantities V_i and I_i, and output quantities V_o and I_o, with positive directions as indicated in Figure 5.3.

The ratio between input voltage and input current is called the input impedance $Z_i = V_i/I_i$ of the system. The output impedance is the ratio between the output voltage and output current: $Z_o = V_o/I_o$. There are many ways to model a two-port network. An obvious way is to express the output quantities in terms of the input quantities, because generally, the outputs are affected by the inputs. If the network

Figure 5.3 A two-port network with input and output quantities.

contains no independent internal sources, the system can be described by the next two equations:

$$V_o = AV_i + BI_i$$
$$I_o = CV_i + DI_i$$

where A, B, C and D are system parameters.

Although this is a rather logical choice to describe a system, other expressions are in use, in which both currents are expressed in both voltages or conversely:

$$V_i = A'I_i + B'I_o \quad \text{or} \quad I_i = A''V_i + B''V_o$$
$$V_o = C'I_i + D'I_o \quad \quad I_o = C''V_i + D''V_o$$

The advantage of these two relationships is that they can be transferred directly into an equivalent model (Figure 5.4). The system of Figure 5.3 is described by two voltage sources, one at the input and one at the output (Figure 5.4a). For Figure 5.4a, the system equations are:

$$V_i = Z_{11}I_i + Z_{12}I_o$$
$$V_o = Z_{21}I_i + Z_{22}I_o$$

The coefficients Z_{11}, Z_{12}, Z_{21} and Z_{22} have a clear physical meaning (impedances) and can be determined directly by measurements. The two sources in the network model are dependent sources; their values depend on other variables in the system.

In Figure 5.4b, both ports are modelled by a current source. The system equations are:

$$I_i = Y_{11}V_i - Y_{12}V_o$$
$$I_o = -Y_{21}V_i + Y_{22}V_o$$

The system parameters Y are convertible to the parameters Z of the model in Figure 5.4a. Another possible model is with one voltage source and one current source.

It should be noted that, in general, the input impedance previously defined is not equal to the impedance Z_{11}, neither is the output impedance Z_o equal to the

System models   67

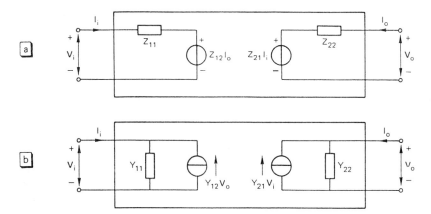

Figure 5.4 Various models for a two-port system: (a) with two voltage sources; (b) with two current sources.

impedance Z_{22} of the model. Both Z_i and Z_o depend on the impedances that may be connected to the output port or the input port. It follows from the system equations or the system models that the input impedance Z_i in Figure 5.4a is equal to Z_{11} only if the output current $I_o = 0$ (open output) and that $Z_o = Z_{22}$ only if the input current $I_i = 0$ (open input).

Example 5.1
A system whose model is depicted in Figure 5.5 is loaded with an impedance Z_L at the output. Its input impedance Z_i is defined as V_i/I_i:

$$Z_i = \frac{V_i}{I_i} = \frac{I_i Z_{11} + I_o Z_{12}}{I_i}$$

with

$$I_o = \frac{-Z_{21} I_i}{Z_{22} + Z_L}$$

From these equations it follows:

$$Z_i = Z_{11} - \frac{Z_{12} Z_{21}}{Z_{22} + Z_L}$$

Similarly, the output impedance of the same system, now with a source impedance connected at the input, is found to be:

$$Z_o = Z_{22} - \frac{Z_{12} Z_{21}}{Z_{11} + Z_g}$$

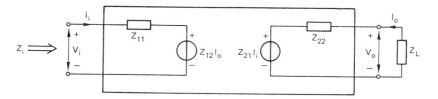

Figure 5.5 The input impedance of a system depends on the load at its output.

From Figure 5.4a, it appears that the output voltage is $V_o = Z_{21}I_i + Z_{22}I_o$, and thus is a function of both the input current and output current. A similar relation holds for the input voltage $V_i = Z_{11}I_i + Z_{12}I_o$. Apparently, there is a bidirectional signal transfer, from input to output and vice versa. In many systems, like amplifiers and measurement systems, only a unidirectional signal transfer (from input to output) is allowed. Most measurement systems are designed so that the reverse transfer is negligible. So, in the model, the source $Z_{12}I_o = 0$. A consequence of a unidirectional signal path is that the input impedance is independent of the load connected to the output and furthermore, the output impedance is independent of the source circuit connected to the input. Suppose in Figure 5.5 the impedance $Z_{12} = 0$, in that case, $Z_i = Z_{11}$ and $Z_o = Z_{22}$. Such a system is fully described by its input impedance, its output impedance and a third system parameter (Z_{21} in the case of Figure 5.5).

Example 5.2
The model of a voltage-to-current converter is depicted in Figure 5.6. Input and output have a common ground terminal; the input impedance is Z_i, the output impedance is Z_o. The current model at the output is chosen because the output should have the character of a current source. The value of the output current is S times the input voltage; S is the voltage-to-current sensitivity of the system.

There is no internal source at the input side of the model; apparently the input voltage and current are not influenced by the output signals; there is no internal feedback.

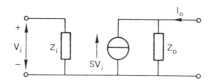

Figure 5.6 A model of a voltage-to-current converter.

5.1.3 *Matching*

In electronic systems, the currents and voltages are the major carriers of the information. A proper transfer of information requires a proper transfer of voltages or currents.

If a voltage is the information carrier, the voltage transfer from one system to the other should be as accurate as possible, without loss of information. Suppose a signal source with internal source impedance Z_g is connected to a system whose input impedance is Z_i (Figure 5.7). The actual input voltage is

$$V_i = \frac{Z_i}{Z_g + Z_i} V_g$$

So, the input voltage is always smaller than the source voltage. The deviation is minimized by reducing the ratio Z_g/Z_i. For $Z_g \ll Z_i$, the input voltage of the system is $V_i \approx (1 - Z_g/Z_i)V_i$, so the relative error is about $-Z_g/Z_i$. If both Z_g and Z_i are known, the measurement error can be corrected for this additional signal attenuation. Otherwise, the input impedance of a voltage-measuring system should be as high as possible, to minimize the load error. This is called the voltage matching of the systems.

If the current is the information carrier, the model of Figure 5.7b is preferred. The current I_i through the system's input is:

$$I_i = \frac{Z_g}{Z_i + Z_g} I_g$$

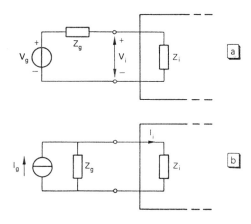

Figure 5.7 Matching between two systems: (a) voltage matching; (b) current matching.

Here as well, the signal is attenuated. If $Z_i \ll Z_g$, the input current is $I_i \approx (1 - Z_i/Z_g)I_g$, and the relative error is about $-Z_i/Z_g$. Again, to minimize this error, the input impedance of the measurement system must be as low as possible; this is called the current matching of the systems.

Summarizing these results:

- a voltage measurement system requires a high input impedance, to minimize loading errors;
- a current measurement system requires a low input impedance.

At perfect voltage or current matching, the signal transfer from the source to the input terminals of the measurement system is 1; the power transfer, however, is zero. This is true because when $Z_i = 0$, the input voltage is zero, and for $Z_i \to \infty$ the input current is zero. So, in both cases, the input power is zero as well. At first sight this seems to be an undesirable situation, because most output transducers require a lot of signal power to be activated. Fortunately, electronic components offer an almost infinite signal power amplification, so voltage or current matching does not necessarily result in a low power transfer.

Another consequence of current or voltage matching is the zero power supplied by the signal source. This is a great advantage especially when the input transducer cannot supply power by itself, or when no power may be extracted from the measurement object, for instance in order not to reduce the measurement accuracy.

Voltage or current matching is not always possible. In particular, at high frequencies it is difficult to realize a high input impedance. Any system has an input capacitance different from zero, originating from the input components as well as from the connector and the connecting wires. The impedance of a capacitance decreases with increasing frequency, so does the input impedance of the system.

Furthermore, high frequency signals may reflect at the interfaces between system parts. Such reflections may introduce standing waves between two points of the signal path, disturbing a proper propagation of the measurement signal.

Such effects can be avoided by applying another type of matching, called characteristic matching (Figure 5.8). The systems have a particular input and output impedance, which is the same for all the systems involved. This characteristic impedance R_k has a fixed, relatively low and real value, for instance 50 or 75 Ω.

The voltage, current and power signal transfer of a characteristic system differ essentially from that of the systems described so far. The voltage transfer V_i/V_g of the system in Figure 5.8a is just ½. To calculate the power transfer from one system to another, we refer to Figure 5.8b, showing a voltage source with source impedance R_g, loaded with an impedance R_i. The signal power that is supplied to the load is:

$$P_i = I_i^2 R_i = \frac{V_g^2 R_i}{(R_g + R_i)^2}$$

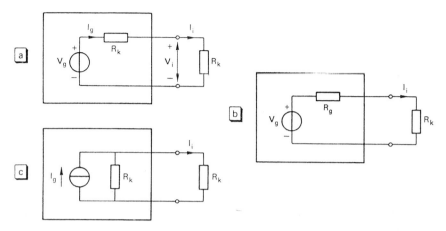

Figure 5.8 (a) Characteristic matching of the signal source; (b) the source supplies the maximum signal power to the system if $R_g = R_k$; (c) characteristic loading of a current source.

and depends on both R_i and R_g. To find the condition for maximum power transfer, we take the first derivative of P_i with respect to the variable R_i:

$$\frac{dP_i}{dR_i} = V_g^2 \frac{(R_g + R_i)^2 - 2R_i(R_g + R_i)}{(R_g + R_i)^4}$$

This is zero for $R_i = R_g$, so maximum power transfer occurs at a characteristic coupling of the two systems. Half of the available power is dissipated in the source resistance: $I_g^2 R_k = V_g^2/4R_k$. The other half is transferred to the load: $V_i \cdot I_i = V_g^2/4R_k$. For the current source in Figure 5.8c, the same conclusions can be drawn. This type of matching is also called power matching because the power transfer is maximized.

Many characteristic systems only have a characteristic input impedance if the system is loaded with R_k, and a characteristic output impedance if the system is connected to a source whose source resistance is just R_k (Figure 5.9). Such systems have bidirectional signal transport.

Figure 5.9 Some systems have only a characteristic input impedance if loaded with a characteristic impedance, and vice versa.

5.1.4 Decibel notation

Consider two systems, connected in series. Let system 1 be characterized by $x_{o1} = H_1 x_{i1}$ and system 2 by $x_{o2} = H_2 x_{i2}$. These systems are coupled such that $x_{o1} = x_{i2}$. The output of the total system equals $x_{o2} = H_1 H_2 x_{i1}$. Obviously, the total transfer of a number of systems connected in series is equal to the product of the individual transfers (taking into account the corrections for non-ideal matching).

In several technical disciplines (telecommunications, acoustics), it is usual to express the transfer as the logarithm of the ratio between output and input quantity. This simplifies the calculation of the transfer of cascaded systems. The logarithmic transfer of the total system is simply the sum of the individual transfers, according to the rule $\log(a \cdot b) = \log(a) + \log(b)$.

The logarithmic transfer is defined as $\log(P_o/P_i)$, base 10 (unit bel), or as $10 \cdot \log(P_o/P_i)$ (unit decibel or dB).

In Figure 5.10, the input power of the system is $P_i = v_i^2/R_i$ and the power supplied to the load amounts to $P_o = v_o^2/R_L$. So the power transfer from source to load is $10 \cdot \log(P_o/P_i) = 10 \cdot \log[(v_o/v_i)^2 R_i/R_L]$ dB, and hence depends on the load resistance, and the square of the voltage transfer. The power transfer of a characteristic system is always equal to the square of the voltage transfer, because $R_i = R_L$.

The logarithmic expression of a voltage transfer is defined as $20 \cdot \log(v_o/v_i)$. This definition does not account for the interfacing of the system.

Example 5.3
The resistance values in the system of Figure 5.10 are: $R_i = 100\,\text{k}\Omega$, $R_o = 0\,\Omega$, $R_L = 1\,\text{k}\Omega$; further, the gain of the amplifier is $A = 100$.
The power transfer appears to be equal to $10 \cdot \log(A^2 R_i/R_L) = 10 \cdot \log 10^6 = 60\,\text{dB}$.
The voltage transfer of the system itself is $v_o/v_i = A = 100$ or $20 \cdot \log A = 40\,\text{dB}$.
This is not the voltage transfer from source to load, which is v_o/v_g.

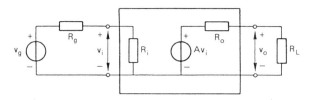

Figure 5.10 The power transfer of this system is proportional to the square of the voltage transfer.

5.2 Signal models

Any measurement system is influenced by interference signals, obscuring the measurement signals. Interference either originates from the system itself, or may enter the system from outside. Undoubtedly the designer has undertaken efforts to minimize noise, to avoid cross-over of auxiliary signals to the measurement signal and to prevent the induction of spurious signals from outside, for instance by proper shielding.

The designer and user of the system must take care in keeping unwanted signals outside the system.

Despite careful design, a completely interference-free operation can never be guaranteed. Internally generated interference (like noise) and the sensitivity to external interference should be specified by the manufacturer of the measurement system.

Interference signals introduce measurement errors. Such errors can be categorized in three groups: destructive errors, multiplicative errors and additive errors. Destructive errors lead to a total malfunctioning of the system which, of course, is not acceptable. Multiplicative and additive errors may be allowed, but up to a specified level. Multiplicative errors result from a deviation in the transfer of the system. The output of the system can be expressed as $x_o = H(1 + \varepsilon)x_i$, with H the nominal transfer. The output remains zero for zero input, but the transfer deviates from the specified value. The relative measurement error is ε; the absolute error, $H\varepsilon x_i$, depends on the input signal.

Multiplicative errors are caused by, for instance, drift of the component values (due to temperature and ageing) or non-linearity. In the rest of this section we will only consider additive errors, particularly noise.

5.2.1 Additive errors

Additive errors are observed when the output signal differs from zero at zero input:

$$x_o = Hx_i + x_n$$

with H the transfer of the system and x_n the error signal. This error signal accounts for all kinds of additive interference, like (internally generated) offset, cross-over of internal auxiliary signals or unwanted external signals entering the system and becoming mixed up with the measurement signal.

If x_n is a DC signal, we call it offset (Section 1.2):

$$x_o = Hx_i + x_{o, \text{off}}$$

or

$$x_o = H(x_i + x_{i, \text{off}})$$

where $x_{o,off}$ and $x_{i,off}$ are the output and input offset. The absolute error in the output signal does not depend on the input signal, as is the case with multiplicative errors; the relative error increases with decreasing input signal. Therefore, offset errors are always expressed in absolute terms.

The influence of all additive error signals can be described by only two additional signal sources at the input of the system (Figure 5.11). These sources are called equivalent error signal sources. The system itself is supposed to be free from errors; all system errors are concentrated into these two equivalent error signal sources v_n and i_n.

In Figure 5.11, the signal source is modelled by a Thévenin circuit, consisting of a voltage source v_g and a source resistance R_g. At open input terminals, the input signal of the system is $v_i = i_n R_i$, so the output signal is $v_o = H i_n R_i$. At short-circuited terminals, the output is $v_o = H v_n$. The total output signal is:

$$v_o = H(v_g + v_n + i_n R_g) \frac{R_i}{R_g + R_i}$$

The output signal depends on the source resistance R_g. When the measurement system is connected to a voltage source (low value of R_g), v_n is the dominating error at the output. Therefore, a voltage measurement system should have a low value of v_n. At current measurements, the system should have a low value of i_n, to minimize the contribution of $i_n R_g$ to the output error signal.

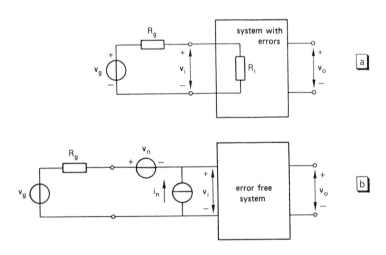

Figure 5.11 (a) A system with internal error sources; (b) modelling of the error signals by two external sources: the system is made error free.

Signal models

The error sources v_n and i_n can represent DC signals (for instance offset, drift) or AC signals, for instance noise. In the first case, the signal is expressed in its momentary value, while in the second case rms values are preferred. It is important to make this distinction; the momentary value of two (or more) voltage sources in series or current sources in parallel equals the sum of the individual values; the rms value does not follow this linear summing rule. According to the definition (Section 2.1.2), the rms value of two voltages $v_1(t)$ and $v_2(t)$ in series is:

$$v_{s,\text{rms}} = \left(\frac{1}{T}\int (v_1+v_2)^2\,dt\right)^{1/2}$$

$$= \left(\frac{1}{T}\int (v_1^2 + 2v_1v_2 + v_2^2)\,dt\right)^{1/2}$$

If the two signals are completely independent (uncorrelated), the average of $v_1 v_2$ is zero, hence:

$$v_{s,\text{rms}} = (v_{1,\text{rms}}^2 + v_{2,\text{rms}}^2)^{1/2}$$

This square summing rule is applied when several stochastically independent error signals are combined into one set of equivalent error sources v_n and i_n at the input of the system.

Example 5.4

Two systems I and II each have an equivalent error voltage source v_{n1} and v_{n2}, and an equivalent error current source i_{n1} and i_{n2} (Figure 5.12a). We will try to find the equivalent set of error sources v_{ns} and i_{ns} of the two systems in series (Figure 5.12b).

Both models in Figure 5.12 represent the same system, so the models should be completely equivalent. This equivalence allows us to calculate the error sources v_n and i_n.

We take v_2 of system II as a criterion for equivalence: this voltage must be the same for both models, for any value of the source resistance. Take two extreme cases: $R_g = \infty$ and $R_g = 0$ (open input terminals and short-circuited input terminals, respectively) and derive for both situations the output voltage v_2.

At open input terminals:

$$v_2 = i_{n1}R_iA_1 + i_{n2}R_o + v_{n2} = i_{ns}R_iA_1$$

so

$$i_{ns} = i_{n1} + \frac{v_{n2} + i_{n2}R_o}{A_1 R_i}$$

Models

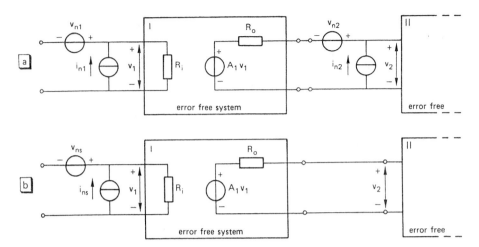

Figure 5.12 Two models for the error sources of two systems connected in series: (a) separate sources; (b) combined sources.

At short-circuited input terminals:

$$v_2 = v_{n1} A_1 + i_{n2} R_o + v_{n2} = v_{ns} A_1$$

so

$$v_{ns} = v_{n1} + \frac{v_{n2}}{A_1} + \frac{i_{n2} R_o}{A_1}$$

This calculation is only valid for the momentary values of the error signals. If the calculation is with respect to stochastic, uncorrelated error sources, then we must write:

$$i_{ns}^2 = i_{n1}^2 + \frac{v_{n2}^2 + i_{n2}^2 R_o^2}{A_1^2 R_i^2}$$

and

$$v_{ns}^2 = v_{n1}^2 + \frac{v_{n2}^2}{A_1^2} + \frac{i_{n2}^2 R_o^2}{A_1^2}$$

respectively.

From this example, it follows that a high gain of the first system is the most favourable situation with respect to the total system error; the errors from the second system contribute little to the total error. In other words, the additive error of a system is almost completely determined by the input components alone if these produce a high signal gain relative to the other system parts. In a proper design, the gain of the system is realized by its input components.

5.2.2 Noise

Any conductor or resistor exhibits thermal noise due to the thermal movements of the electrons in the material. The spectral power density (Section 2.1.3) amounts to $4kT$ (W/Hz) and is almost independent of frequency. Thermal noise behaves like white noise.

The noise power dissipated in a resistor with resistance R is $P = V^2/R = I^2 R$. Thermal noise can therefore be represented by a voltage source in series with the (noise-free) resistance, with value $v_n = \sqrt{4kTR}$, or a current source in parallel to the noise-free resistance, with value $i_n = \sqrt{4kT/R}$ (Figure 5.13).

Example 5.5
To extend the measurement range of a voltmeter to higher voltages, a voltage divider is connected to the input (Figure 5.14a). We need to know the equivalent error sources of the whole system (Figure 5.14b), in order to estimate the divider's contribution to the system noise. Take the input voltage v_i of the amplifier as a criterion for equivalence. The equivalent current source is found by calculating v_i at short-circuited terminals of the system:

$$v_i^2 = v_1^2 \left(\frac{R_2}{R_1 + R_2}\right)^2 + (i_2 + i_n)^2 \left(\frac{R_1 R_2}{R_1 + R_2}\right)^2 + v_n^2 = v_s^2 \left(\frac{R_2}{R_1 + R_2}\right)^2$$

from which follows:

$$v_s^2 = v_1^2 + \left(\frac{R_1 + R_2}{R_2}\right)^2 v_n^2 + R_1^2 (i_2 + i_n)^2$$

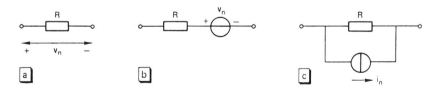

Figure 5.13 (a) Any resistor generates noise; (b) the resistance noise is represented by a voltage source or (c) a current source.

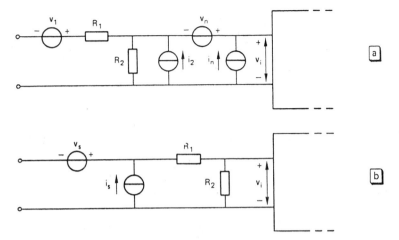

Figure 5.14 Modelling noise signals of a system with voltage divider (a) by separate error sources and (b) by combined error sources.

Similarly, the equivalent current source is calculated, at open inputs:

$$i_s^2 = i_2^2 + i_n^2 + \left(\frac{v_n}{R_2}\right)^2$$

These two equations clearly show the contribution of the voltage divider to the total noise error, relative to the noise of the system itself.

SUMMARY

System models

- Any two-terminal system is characterized by a voltage source V_o in series with a source impedance Z_g (Thévenin's theorem), or a current source I_k in parallel to a source impedance Z_g (Norton's theorem). V_o and I_k are the open voltage and short-circuiting current, respectively. The source resistance is $Z_g = V_o/I_k$.
- The input impedance of a system is the ratio between the (complex) input voltage and the input current; the output impedance is the ratio between the output voltage and the output current. Generally, the input impedance depends on the load impedance and the output impedance on the source impedance.
- Linear two-port systems can be characterized by a model with two sources (voltage source and/or current source) and two impedances. The model is mathematically described by two system equations.

- A voltage-measuring instrument requires a high input impedance Z_i; if the source impedance Z_g satisfies the inequality $Z_g \ll Z_i$, then the relative measurement error equals about $-Z_g/Z_i$.
- A current-measuring instrument requires a low input impedance Z_i; if the source impedance Z_g satisfies the inequality $Z_g \gg Z_i$, then the relative measurement error equals about $-Z_i/Z_g$.
- Characteristic matching is achieved when the output source impedance equals the input load resistance. The power transfer amounts to ½.
- The power transfer P_o/P_i expressed in decibels (dB) is $10 \cdot \log(P_o/P_i)$. The logarithmic voltage transfer is defined as $20 \cdot \log(V_o/V_i)$.

Signal models

- System errors can be classified into three categories: destructive, multiplicative and additive errors.
- Additive error signals originating from the system itself can be represented with two independent sources at the input of the system, an equivalent error voltage source and an equivalent error current source.
- The rms value of the sum of two uncorrelated signals is equal to the root of the summed squares: $v_{s,\text{rms}} = \sqrt{v_{1,\text{rms}}^2 + v_{2,\text{rms}}^2 + \ldots + v_{n,\text{rms}}^2}$.
- The spectral power of thermal noise is $4kT$ W/Hz. This noise can be represented by a voltage source in series with the noise-free resistance, $\sqrt{4kTR}$ [V/$\sqrt{\text{Hz}}$] or by a parallel current source with strength $\sqrt{4kT/R}$ [A/$\sqrt{\text{Hz}}$].
- The equivalent voltage error source of a system is found by equal outputs at short-circuited input terminals. The equivalent error current source is found by equal outputs at open input terminals.

Exercises

System models

5.1 Find the Thévenin equivalent circuits of the source circuits a–c given in the figure below.

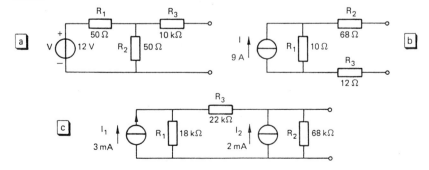

80 Models

5.2 In a list of specifications, the maximum figures for the rejection ratios are given as CMRR (at DC): 90 dB; CMRR (at 1 kHz): 80 dB; supply voltage rejection ratio SVRR: 110 dB. Specify these quantities in μV/V.

5.3 The current from a current source with source resistance 100 kΩ is measured with a current meter. The largest allowable error is 0.5%. What is the requirement for the input resistance of the current meter?

5.4 The source resistance of an unknown voltage source is measured as follows: first, its voltage is measured using a voltmeter with input resistance $R_i \geq 10$ MΩ. Next, the voltage is measured while the source is loaded with 10 kΩ. The successive measurement results are 9.6 and 8 V.

(a) Calculate the source resistance.
(b) What is the relative error in the first measurement, due to the load by R_i?

5.5 What is characteristic matching? What is a characteristic impedance? How much is the power transfer at characteristic coupling?

5.6 The figure below shows a model of a voltage amplifier. Find the voltage transfer $A_v = v_o/v_i$, the current transfer i_o/i_i and the power transfer $A_p = P_o/P_i$.

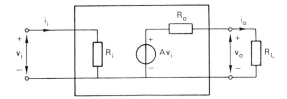

Signal models

5.7 The figure below shows the model of a voltage amplifier. R_L is a resistor with thermal noise voltage v_{nL}; all other components are error free. Find the equivalent error sources at the input terminals of the system, due only to the noise from R_L.

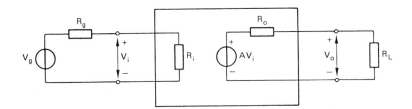

5.8 Refer to the model of Exercise 5.7. Now, only R_i and R_o genereate noise, with noise contributions i_{ni} and v_{no}, respectively. All other components are noise free. Determine the equivalent noise sources due to these two resistances.

5.9 Refer again to the figure in Exercise 5.7. Now, the offset voltage and offset current are represented by equivalent sources at the input, V_n and I_n. At 20°C, $V_n = 1\,\text{mV} \pm 10\,\mu\text{V/K}$; $I_n = 10\,\text{nA}$, which value doubles for each 10°C temperature rise. Further, $R_g = 10\,\text{k}\Omega$, $R_i = \infty$, $R_L = \infty$ and $A = 10$. Calculate the maximum output offset voltage:

(a) for $T = 20°C$,
(b) for a temperature range $0 < T < 50°C$.

5.10 The input power of a system is the product of its input current and input voltage. The signal-to-noise ratio S/N is the ratio between the input signal power P_s and the input noise power P_n. Prove that S/N is independent of the input resistance R_i.

6 Frequency diagrams

The complex transfer function of an electronic system generally depends on the frequency. The usual way to visualize the frequency dependence of the transfer function is by drawing a frequency diagram: both the modulus (amplitude transfer) and the argument (phase transfer) are plotted versus frequency. This set of characteristics is the Bode plot of the system, and will be discussed in the first part of this chapter. There are other ways to visualize the frequency dependence of a complex transfer function. One of them, the polar plot, represents the transfer in the complex plane. This method is discussed in the second part of the chapter.

6.1 Bode plots

6.1.1 *Systems of the first order*

A Bode plot consists of two diagrams: the amplitude characteristic and the phase characteristic. Both plots have a logarithmic frequency scale; the modulus (amplitude) scale is also logarithmic, the phase (argument) is plotted along a linear scale. Modulus and argument can be rather complicated functions of the frequency; nevertheless, the Bode plot can be drawn quite easily, using some approximations. The method will be explained by a simple network with one resistance and one capacitance (Figure 6.1).

The complex transfer function of this network is $H(\omega) = 1/(1 + j\omega\tau)$, with $\tau = RC$, the time constant of the network. The modulus of the transfer function equals $1/\sqrt{1 + \omega^2\tau^2}$. To make a plot of this function versus frequency, we will consider three particular conditions: very low frequencies ($\omega \ll 1/\tau$), very high frequencies ($\omega \gg 1/\tau$) and $\omega = 1/\tau$.

For low frequencies (relative to $1/\tau$), the modulus is about 1 (or 0 dB); there is no signal attenuation because the impedance of the capacitance at very low frequencies is

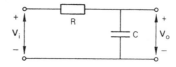

Figure 6.1 A first-order RC network.

high. For high frequencies (relative to $1/\tau$), the modulus $|H|$ can be approximated by $1/\omega\tau$: the modulus is inversely proportional to the frequency. With logarithmic scales, the characteristic is a falling straight line. Figure 6.2a shows these two parts, the asymptotes, of the amplitude characteristic. The asymptotes intersect at the point $\omega = 1/\tau$, the corner frequency of the characteristic. The amplitude characteristic can be approximated surprisingly well by these two asymptotes: $|H| = 1$ for $\omega\tau < 1$ and $|H| = 1/\omega\tau$ for $\omega\tau > 1$. The value of $|H|$ for $\omega\tau = 1$ equals $1/\sqrt{2}$ or $\tfrac{1}{2}\sqrt{2}$. Using decibel notation, this is equal to:

$$20 \log \tfrac{1}{2}\sqrt{2} = 20 \log(2)^{-1/2} = -10 \log 2 = -10 \cdot 0.301\,03 \approx -3 \text{ dB}$$

This explains why the frequency where the asymptotes intersect is also called the -3 dB frequency. The real characteristic can now be sketched easily along the asymptotes and through the -3 dB point. The deviation from the asymptotic approximation is only 1 dB for the frequencies $\omega\tau = \tfrac{1}{2}$ and $\omega\tau = 2$.

In the high frequency range, the modulus of the transfer is inversely proportional to frequency; the transfer halves at double the frequency (1 octave). The slope of the falling asymptote is a factor 2 decrease per octave or, using decibel notation, -6 dB/octave. Likewise, the transfer drops a factor 10 at a tenfold increase of the

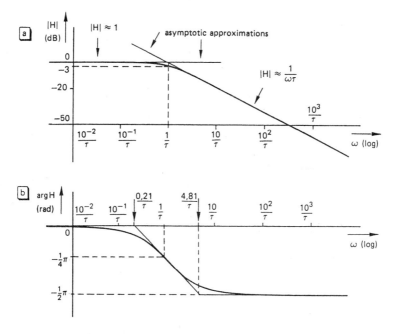

Figure 6.2 The Bode plot of the RC network of Figure 6.1; (a) amplitude transfer; (b) phase transfer.

frequency (decade); the slope of the characteristic is a factor 10 decrease per decade or -20 dB/decade.

The phase characteristic can be approximated in a similar way, by splitting up the frequency range into two parts. At low frequencies $\arg H$ approaches 0, and for high frequencies it approaches the value $-\pi/2$. These lines are two asymptotes of the phase characteristic. For $\omega\tau = 1$, the modulus equals $-\pi/4$, which appears to be the point of inflection of the actual characteristic. Now, the course of the curve can be roughly sketched. A better approximation is achieved when we know the direction of the tangent at the point of inflection. Its slope is found by differentiating $\arg H$ to $\log \omega$ (the vertical scale is linear, the frequency scale is logarithmic). The result is $-½ \ln 10$, and is used to find the intersections of this tangent with the asymptotes $\arg H = 0$ and $\arg H = -\pi/2$. These points are $\omega = 0.21/\tau$ and $\omega = 4.81/\tau$, hence about a factor 5 to the left and right of the point of inflection (Figure 6.2b). The actual phase transfer for $\omega\tau = 5$ and $\omega\tau = 1/5$ deviates only $(1/15)\pi$ rad from the approximations 0 and $-\pi/2$ at these points.

6.1.2 Systems of higher order

We will start with transfer functions that can be written as the product of first-order functions. The Bode plot of such a function is achieved simply by adding together the plots of the separate first-order functions.

Example 6.1

The transfer function $H = j\omega\tau_1/(1 + j\omega(\tau_2 + \tau_3) - \omega^2 \tau_2 \tau_3)$ can be written as the product of the functions $H_1 = j\omega\tau_1/(1 + j\omega\tau_2)$ and $H_2 = 1/(1 + j\omega\tau_3)$. The function H_2 has already been discussed. The characteristic of H_1 can be approximated in a similar way: for $\omega\tau_2 \ll 1$, $|H_1|$ can be approximated by $\omega\tau_1$, and for $\omega\tau_2 \gg 1$ the modulus of H_1 is about $\omega\tau_1/\sqrt{(\omega^2\tau_2^2)} = \tau_1/\tau_2$. Figure 6.3a shows the two individual first-order characteristics $|H_1|$ and $|H_2|$ with -3 dB points at $\omega = 1/\tau_2$ and $\omega = 1/\tau_3$, respectively. It is assumed that $\tau_1 < \tau_2 < \tau_3$. To find the characteristic of $|H|$, we recall that H is the product of both first-order functions and the modulus scale is logarithmic. So, the characteristic of $|H|$ is found by adding the characteristics of the two individual first-order characteristics. As can be seen in Figure 6.3a, within the frequency range $1/\tau_3 < \omega < 1/\tau_2$, $|H_2|$ decreases 6 dB/octave and $|H_1|$ increases with the same amount. Thus, in this interval, the total transfer remains constant.

In a similar way, the phase characteristic is found by adding the two individual phase characteristics of the first-order transfer functions; notice that the phase scale is linear, and that the phase of a series system is the sum of the phases of each single system.

This summing method offers a quick insight in the amplitude and phase characteristics even of rather complex systems. If factorization into first-order functions is not possible, this method cannot be used. The next step is to study the Bode plot of second-order systems, described by the expressions $H = 1/(1 + j\omega\tau_1 - \omega^2 \tau_2^2)$ or

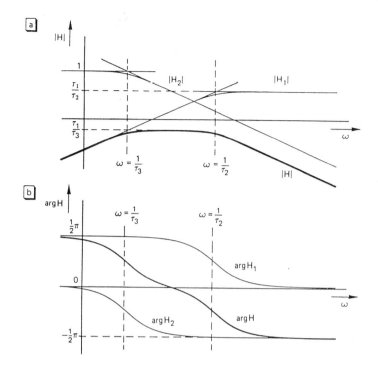

Figure 6.3 Bode plot of the transfer function H = H_1H_2. It is composed of the first-order Bode plots of H_1 and H_2; (a) amplitude characteristic; (b) phase characteristic.

$H = 1/(1 + 2j\omega z/\omega_0 - \omega/\omega_0^2)$. We will use the second equation with the parameters z and ω_0. Only when $z < 0$ is the function factorizable into two first-order transfer functions. Figure 6.4 shows the Bode plot for several values of $z < 1$.

The transfer shows overshoot for small values of the parameter z. By differentiation of $|H|$, it can easily be proved that overshoot occurs for $z < \frac{1}{2}\sqrt{2}$. This overshoot increases when z decreases. Therefore, the parameter z is called the damping ratio of the system. It is related to the parameter Q, the quality or relative bandwidth of the system: $Q = 1/2z$ (see Chapter 13).

From a further analysis of the second-order transfer function, the following construction rules are derived. At frequencies much lower than ω_0, the modulus is approximated by the line $|H| = 1$, and for frequencies much higher than ω_0 by the line $|H| = \omega_0^2/\omega^2$. The slope of the latter is -12 dB/octave or -40 dB/decade and is therefore twice as steep as in a first-order system. The transfer shows a maximum at the frequency $\omega = \omega_0\sqrt{(1 - 2z^2)}$ and amounts to $1/2z\sqrt{(1 - z_2)}$ ($z < \frac{1}{2}\sqrt{2}$). The frequency where the transfer has its peak value is called the resonance frequency. For very low values of z the resonance frequency is about ω_0; that is why ω_0 is called the undamped angular frequency.

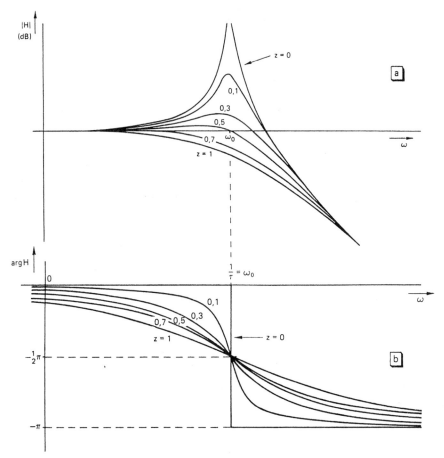

Figure 6.4 Bode plot of a second-order system, for various values of the damping ratio z; (a) amplitude characteristic; (b) phase characteristic.

The phase characteristic has the following properties. For very low frequencies $\arg H \approx 0$ and at very high frequencies $\arg H \approx -\pi$. At $\omega = \omega_0$, the phase transfer is just $-\pi/2$, irrespective of the parameter z. The slope of the tangent through that point (the point of inflection) is found by differentiating $\arg H$ to the variable $\log \omega$. The result is $-(1/z)\ln 10$. This tangent intersects the horizontal asymptotes of the phase characteristic at frequencies that are a factor $10^{\pi z/2\ln 10} = 4.81^z$ on both sides of ω_0.

6.2 Polar plots

The polar plot of a complex transfer function represents its modulus and argument in the complex plane with the frequency ω as a parameter. The method to derive polar plots will be illustrated by several examples.

6.2.1 First-order functions

The first example is the RC network from Figure 6.1, with $H = 1/(1 + j\omega\tau)$. For each value of ω, H is represented in the complex plane. There are two methods for doing so: either the calculation of $|H|$ and $\arg H$ (as for the Bode plot) or the calculation of $\operatorname{Re} H$ and $\operatorname{Im} H$. We will use the latter method.

The real and imaginary parts of an arbitrary complex polynomial can be found easily by multiplying numerator and denominator by the conjugate of the denominator (the conjugate of a complex variable $a + jb$ is $a - jb$). Hence:

$$H(\omega) = \frac{1}{1 + j\omega\tau} = \frac{1}{1 + j\omega\tau} \cdot \frac{1 - j\omega\tau}{1 - j\omega\tau} = \frac{1 - j\omega\tau}{1 + \omega^2\tau^2}$$

The denominator of the resulting expression is always real so the real and imaginary parts of H are achieved directly:

$$\operatorname{Re} H = \frac{1}{1 + \omega^2\tau^2}; \quad \operatorname{Im} H = \frac{-\omega\tau}{1 + \omega^2\tau^2}$$

For $\omega = 0$, the real part is 1 and the imaginary part is 0. For $\omega \to \infty$ $\operatorname{Re} H$ and $\operatorname{Im} H$ are both zero. In the special case $\omega = 1/\tau$, $\operatorname{Re} H = \frac{1}{2}$ and $\operatorname{Im} H = -\frac{1}{2}$. The polar plot is depicted in Figure 6.5.

From the polar plot, the Bode plot can be derived and vice versa. The polar plot in Figure 6.5 appears to be a semicircle. Let $\operatorname{Re} H = x$ and $\operatorname{Im} H = y$. Elimination of ω from the expressions for $\operatorname{Im} H$ and $\operatorname{Re} H$ gives:

$$x^2 + y^2 = x \quad \text{or} \quad \left(x - \frac{1}{2}\right)^2 + y^2 = \left(\frac{1}{2}\right)^2$$

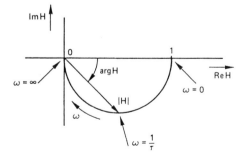

Figure 6.5 The polar plot of the function $H = 1/(1 + j\omega\tau)$; for $\omega\tau = 1$, $|H| = \frac{1}{2}\sqrt{2}$ and $\arg H = -\pi/4$.

This is just the equation for a circle with centre (½, 0) and radius ½. The polar plot is only half of this circle, because we consider only positive frequencies. It can be shown that the polar plot of any function $(a + jb\omega\tau)/(c + jd\omega\tau)$ is a (semi)circle; if c or d are zero, the circle degenerates into a straight line. The parameter ω moves along the plot in a clockwise direction.

We can sketch the polar plot of a lot of simple networks without calculating the real and imaginary parts. This is illustrated in the following examples (Figure 6.6).

Consider the network of Figure 6.6a. For $\omega = 0$, the transfer is zero (the capacitance blocks the signal transfer). The polar plot therefore starts at the origin. For $\omega \to \infty$, the transfer approaches the real value $R_2/(R_1 + R_2)$ (the capacitance acts as a short-circuit for signals); this is the end point of the semicircle, which can now be sketched directly bearing in mind the clockwise direction.

Finally, we discuss the polar plot of the complex impedance from Figure 6.6c. The plot starts at $\omega = 0$, for which Z has the real value R_2 (C has an infinite impedance) and ends at $\omega \to \infty$, where $Z = R_1 // R_2$ ($//$ means in parallel). The polar plot is a semicircle, depicted in Figure 6.6d.

6.2.2 Functions of higher orders

There is no simple recipe for the construction of the polar plot of an arbitrary complex function of order two and higher; the summing method as used with Bode plots is of little practical value. Nevertheless, there are some rules for the start and end points of the plot, and for the direction of the tangents at those points. Some of these rules will be given below.

Consider the general transfer function:

$$F(j\omega) = \frac{a_t(j\omega)^t + a_{t+1}(j\omega)^{t+1} + \ldots + a_{T-1}(j\omega)^{T-1} + a_T(j\omega)^T}{b_n(j\omega)^n + b_{n+1}(j\omega)^{n+1} + \ldots + b_{N-1}(j\omega)^{N-1} + b_N(j\omega)^N}$$

with t and T the lowest and highest power of $j\omega$ in the numerator and n and N the lowest and highest power of $j\omega$ in the denominator. None of the coefficients a_t, a_T, b_n and b_N is zero. Furthermore, for any physical system described by $F(j\omega)$, the order of the numerator never exceeds that of the denominator, $T \leq N$.

The starting point of the polar plot (at $\omega = 0$) is found from $\lim_{\omega \to 0} F(j\omega) = \lim_{\omega \to 0}(a_t/b_n)(j\omega)^{t-n}$, that is the approximation of $F(j\omega)$ by the lowest powers of numerator and denominator. Depending upon t and n, this approximation is 0 ($t > n$), a_t/b_n ($t = n$) or ∞ ($t < n$). The phase at the starting point equals $\arg(j\omega)^{t-n} = (t-n)\pi/2$; in other words, the starting point lies on either the real axis or the imaginary axis.

Similarly, the end point is found from $\lim_{\omega \to \infty} F(j\omega) = \lim_{\omega \to \infty}(a_T/b_N)(j\omega)^{T-N}$, hence the end point amounts to either a_T/b_N ($T = N$) or 0 ($T < N$). The phase at the end point is $\arg(j\omega)^{T-N} = (T-N)\pi/2$; in other words, the end point lies on either the real axis or the imaginary axis, like the starting point.

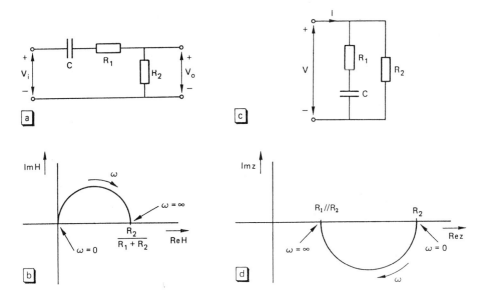

Figure 6.6 (a) A network with complex transfer function $H = V_o/V_i$; (b) the polar plot of H; (c) a network with complex impedance $Z = V/I$; (d) the polar plot of Z.

The tangents in the starting point and the end point are derived from $\lim_{\omega \to 0} dF/d\omega$ and $\lim_{\omega \to \infty} dF/d\omega$, respectively. The values of these limits depend on the coefficients a and b in $F(j\omega)$, but are always a multiple of $\pi/2$. This means that the tangents have either horizontal or vertical directions. Table 6.1 gives some results, without further proof. These results are illustrated in the following examples.

Example 6.2
The function $F(j\omega)$ is defined as:

$$F(j\omega) = \frac{j\omega\tau_1}{1 + j\omega\tau_2 + (j\omega\tau_3)^2}$$

starting point: $t = 1$; $n = 0$, so 0;
end point: $T = 1$; $N = 2$, so 0;
direction at the starting point: $(1 - 0)\frac{1}{2}\pi = \frac{1}{2}\pi$;
direction at the end point: $\pi + (1 - 2)\frac{1}{2}\pi = \frac{1}{2}\pi$.

The polar plot is depicted in Figure 6.7a. The intersection with the real axis is found from $\operatorname{Im} F = 0$, resulting in $\omega = 1/\tau_3$ and $\operatorname{Re} F = \tau_1/\tau_2$.

Table 6.1 Starting and end points of a polar plot and the direction of the tangents in these points: [a] $\frac{1}{2}\pi\,\text{sign}(a_{t+1}b_n - a_t b_{n+1})$; [b] $\frac{1}{2}\pi\,\text{sign}(a_T b_{N-1} - a_{T-1} b_N)$

	Starting point			End point	
	$t > n$	$t = n$	$t < n$	$T = N$	$T < N$
Value	0	a_t/b_n	∞	a_T/b_N	0
Tangent	$(t-n)\pi/2$	$\pm\pi/2$ [a]	$\pi + (t-n)\pi/2$	$\pm\pi/2$ [b]	$\pi + (T-N)\pi/2$

Example 6.3

The function $G(j\omega)$ is defined as:

$$G(j\omega) = \frac{1}{1 + j\omega\tau_1 + (j\omega\tau_2)^2 + (j\omega\tau_3)^3}$$

starting point: $t = n = 0$, hence $a_t/b_n = a_0/b_0 = 1$;
end point: $T = 0$; $N = 3$, hence 0;
direction at the starting point: $\text{sign}(0.0 - 1.1)\frac{1}{2}\pi = -\frac{1}{2}\pi$;
direction at the end point: $\pi + (0 - 3)\frac{1}{2}\pi = -\frac{1}{2}\pi$.

Figure 6.7b shows the polar plot. The intersections with the real and imaginary axes are found from $\text{Im}\,G = 0$ and $\text{Re}\,G = 0$, respectively: $G(\omega = 1/\tau_2) = -j/(\tau_1/\tau_2 - \tau_3^3/\tau_2^3)$ and $G(\omega = \sqrt{\tau_1/\tau_3^3}) = 1/(1 - \tau_1 \tau_2^2/\tau_3^3)$.

Example 6.4

The function $H(j\omega)$ is defined as:

$$H(j\omega) = \frac{1}{j\omega\tau_1 + (j\omega\tau_2)^2}$$

starting point: $t = 0$; $n = 0$, so ∞;
end point: $T = 0$; $N = 2$, so 0;
direction at the starting point: $\pi + (0 - 1)\frac{1}{2}\pi = \frac{1}{2}\pi$;
direction at the end point: $\pi + (0 - 2)\frac{1}{2}\pi = 0$.

A more accurate plot follows from further analysis. This can be done numerically (for instance using a computer program) or analytically:

$$\text{Re}\,H = -\frac{\tau_2^2/\tau_1^2}{1 + \omega^2 \tau_2^4/\tau_1^2} \quad \text{and} \quad \text{Im}\,H = -\frac{1/\omega\tau_1}{1 + \omega^2 \tau_2^4/\tau_1^2}$$

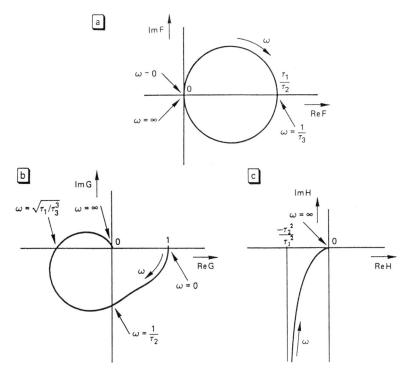

Figure 6.7 (a) Polar plot of $F = j\omega\tau_1/(1 + j\omega\tau_2 + (j\omega\tau_3)^2)$;
(b) Polar plot of $G = 1/(1 + j\omega\tau_1 + (j\omega\tau_2)^2 + (j\omega\tau_3)^3)$;
(c) Polar plot of $H = 1/(j\omega\tau_1 + (j\omega\tau_2)^2)$.

The real part is always negative, running from $-\tau_2^2/\tau_1^2$ ($\omega = 0$) to 0 ($\omega = \infty$); the domain of the imaginary part extends over all negative values. With these conditions, the polar plot can be drawn accurately.

For more complicated functions, it is not easy to give the polar plot; the Bode plot is easier to draw. A polar plot has the advantage of representing the frequency dependence of the transfer in a single diagram.

SUMMARY

Bode plots

- The Bode plot represents the modulus and the argument of a complex transfer function versus the frequency. The frequency and the modulus are plotted on logarithmic scales.

92 Frequency diagrams

- The Bode plot of a first-order function can easily be drawn using the asymptotic approximations for $\omega\tau \gg 1$ and $\omega\tau \ll 1$.
- The point $\omega\tau = 1$ in the Bode plot of a first-order system is the $-3\,\text{dB}$ frequency; the deviation from the asymptotic approximation amounts to $-3\,\text{dB}$.
- The Bode plot of a function consisting of the product of first-order functions is the sum of the individual characteristics of these first-order functions.
- The denominator of a second-order polynomial is written as $1 + 2j\omega z/\omega_0 - \omega^2/\omega_0^2$. In this expression, ω_0 is the undamped frequency and z the damping ratio. For $z < \tfrac{1}{2}\sqrt{2}$, the amplitude characteristic shows overshoot; its peak value increases with decreasing z.
- The frequency at which the amplitude transfer has its maximum value is called the resonance frequency of the system.

Polar plots

- A polar plot represents a complex function in the complex plane, with the frequency as a parameter.
- The polar plot of a first-order system describes a circle or semicircle in the complex plane. In particular cases, the circle degenerates into a straight line.
- The starting point and end point of a polar plot are found by taking the limit of the transfer function for $\omega = 0$ and $\omega \to \infty$, respectively.
- The direction of the tangent at the starting point is $\tfrac{1}{2}k_1\pi$, where k_1 is the difference between the lowest power of $j\omega$ in the numerator and in the denominator of the transfer function ($k_1 > 0$).
- The direction of the tangent at the end point is $\pi + \tfrac{1}{2}k_2\pi$, where k_2 is the difference between the highest power of $j\omega$ in the numerator and the denominator ($k_2 < 0$).
- At equal, lowest or highest powers of $j\omega$ in the numerator and denominator of the transfer function, the tangents at the starting point and the end point are oriented vertically.

Exercises

Bode plots

6.1 Make a sketch of the Bode plot (asymptotic approximation) of the networks a–c in the figure below.

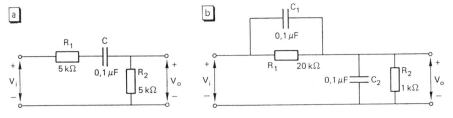

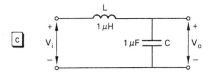

6.2 Find the following parameters of the network given in the figure below: the undamped frequency ω_0; the damping ratio z; the frequency for maximum amplitude transfer; the peak value of the amplitude transfer and the modulus of the transfer at $\omega = \omega_0$.

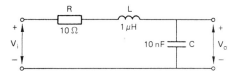

6.3 The complex transfer function of an all-pass network is given as:

$$H = \frac{1 - j\omega RC}{1 + j\omega RC}$$

Find $|H|$ and $\arg H$. Draw the Bode plot of H.

6.4 Make a sketch of the Bode plot of the following function in which $\tau = 10^{-3}$ s:

$$H = \frac{1}{(1 + j\omega\tau)^3}$$

6.5 Find the Bode plot of the transfer function:

$$H = \frac{1 + j\omega\tau_1}{(1 + j\omega\tau_2)(1 + j\omega\tau_3)}$$

where $\tau_1 = 0.1$ s, $\tau_2 = 0.01$ s and $\tau_3 = 0.001$ s.

Polar plots

6.6 Draw the polar plots of the networks a–c from Exercise 6.1.

6.7 Draw the polar plot of the transfer function from Exercise 6.3. Explain the name 'all-pass network'.

6.8 Draw the polar plot of the following function:

$$H = \frac{1 + j\omega\tau_1}{1 + j\omega\tau_2}$$

for three cases: $\tau_1 = 2\tau_2$; $\tau_1 = \tau_2$ and $\tau_1 = \tfrac{1}{2}\tau_2$.

7 Passive electronic components

So far we have considered only models (network elements) of electronic systems and components. In this chapter we will look at the actual electronic components and discuss their properties. Electronic components form the basic materials for electronic circuits and systems. They are divided into passive and active components. The active components need auxiliary power for operation. They offer the possibility of signal power amplification. Passive components do not require auxiliary energy for proper operation. They cannot perform signal power gain; on the contrary, signal processing is accompanied by power loss. In this chapter we will discuss such passive components as resistors, capacitors, inductors and transformers. The first part gives a description of these components as the elements for electronic circuits. The second part deals with special realizations for their application as a sensor.

7.1 Passive circuit components

7.1.1 *Resistors*

The ability of a material to transport charge carriers is called the conductivity σ of that material. The reciprocal of the conductivity is the resistivity ρ. The latter is defined as the ratio between the electric field strength E (V/m) and the resulting current density J (A/m^2). Hence: $E = \rho J$ and $\sigma = 1/\rho$. The conductivity of the common materials ranges from almost zero up to almost infinity. It is determined by the concentration of charge carriers and their specific mobility within the material. On the basis of the conductivity, materials are categorized into insulators, semiconductors or conductors (Figure 7.1). Conductivity is a material property independent of the dimensions, except for special cases in very thin films.

The resistance R (unit ohm, Ω) of a piece of material is defined as the ratio between the voltage V across two points of the material and the resulting current I (Ohm's Law). The resistance depends on the resistivity of the material as well as the dimensions. The reciprocal of the resistance is the conductance G (unit Ω^{-1}, sometimes siemens, S).

There is a tremendous variety of resistor types commercially available. They are categorized into fixed resistors, adjustable resistors and variable resistors. Important criteria when applying resistors are the resistance value or range, the inaccuracy (tolerance of the value), the temperature sensitivity (temperature coefficient) and the

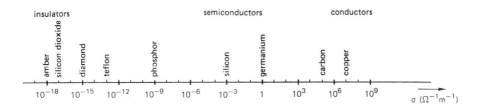

Figure 7.1 The conductivity of some materials.

maximum allowable temperature, voltage, current and power. These properties are determined by the material type and its construction. As materials we have carbon film, metal film and metal wire resistors.

Metal film and metal wire resistors have a much better stability than carbon film types, but the latter allow the realization of much higher resistance values. Metal wire resistors show a self-inductance and capacitance that cannot be neglected. The self-inductance can be minimized by special winding methods, applied in expensive, highly accurate resistors. Usually, the desired resistance value of a carbon or metal film resistor is achieved by cutting a spiral groove into a film around the cylindrical body. Table 7.1 shows the major properties of some commonly used resistor types.

Except for very special purposes, the resistance values are normalized in various series (International Electrotechnical Commission, 1952). The norm is the E-12 series: each decade is subdivided logarithmically into twelve parts. The subsequent values between 10 and 100 Ω are:

$$10, 12, 15, 18, 22, 27, 33, 39, 47, 56, 68, 82, 100$$

Resistors with narrower tolerances belong to other series, for instance the E-24 series (increasing with a factor $\sqrt[24]{10}$), E-48, E-96 and E-192. For further information on commercially available resistors, the reader is referred to the data books of the manufacturers.

Table 7.1 Some properties of various resistor types

	Range (Ω)	Tolerance (%)	Temperature coefficient ($10^{-6}\,\Omega/K$)	Maximum temperature (°C)	Maximum power (W)
Carbon film	$1 \ldots 10^7$	$5 \ldots 10$	$-500 \ldots +200$	155	$0.2 \ldots 1$
Metal film (NiCr)	$1 \ldots 10^7$	$0.1 \ldots 2$	50	175	$0.2 \ldots 1$
Wire-wound (NiCr)	$0.1 \ldots 5 \cdot 10^4$	$5 \ldots 10$	$-80 \ldots +140$	350	$1 \ldots 20$

7.1.2 Capacitors

A capacitor is a set of two conductors separated from each other by an insulating material called the dielectric. The capacitance C (unit farad, F) of this set is defined as the ratio between the charge Q on one of the conductors and the resulting voltage V across them: $C = Q/V$. At varying charge, the voltage varies too. The charge transport to or from the conductor per unit of time is the current I ($I = dQ/dt$), hence the relation between the current and the voltage of an ideal capacitor is $I = C\,dV/dt$.

The capacitance of a capacitor is determined by the geometry of the conductors and of the dielectric. The dielectric permittivity or dielectric constant is the ability of a material to be polarized. The dielectric constant of a material that cannot form electric dipoles is 1, by definition. Such a material behaves like a vacuum.

The capacitance of a capacitor consisting of two parallel flat plates a distance d apart, with surface area A, placed in a vacuum, equals $C = \varepsilon_0 A/d$ (this is an approximation and only valid when the surface dimensions are larger than d). In this expression, ε_0 is the absolute or natural permittivity or the permittivity of vacuum: $\varepsilon_0 = 8.85 \cdot 10^{-12}$ F/m. When the space between the conductors of a capacitor is filled with a material having a dielectric constant ε_r, then the capacitance increases by a factor ε_r: $C = \varepsilon_0 \varepsilon_r A/d$. A large capacitance requires a dielectric material with a large relative permittivity. Table 7.2 shows the relative permittivity of several materials.

There is an enormous variety of capacitor types. Capacitors are divided into fixed capacitors, adjustable capacitors (trimming capacitors) and variable capacitors. Common dielectric materials are air (for trimming capacitors), mica, ceramic materials, paper, plastic and electrolytic materials.

A capacitor never behaves as a pure capacitance; it shows essential anomalies. The major deviations are the dielectric loss, the temperature coefficient and the breakthrough voltage. We will discuss these items separately.

When an AC voltage is applied to the terminals of a capacitor, the dipoles in the dielectric must continuously change their direction. The resulting dissipation (heat loss) is called the dielectric loss. It is modelled by a resistance in parallel to the capacitance. A quality measure for the dielectric loss is the ratio between the current

Table 7.2 Dielectric constants of various materials

Material	ε_r	Material	ε_r
Vacuum	1	Ceramic, Al_2O_3	10
Air (0°C, 1 atm)	1.000576	Porcelain	6–8
Water, at 0°C	87.74	Titanate	15–12 000
at 20°C	80.10	Plastic, pvc	3–5
Glass, quartz	3.75	Teflon	2.1
Pyrex 7740	5.00	Nylon	3–4
Corning 8870	9.5	Rubber, Hevea	2.9
Mica	5–8	Silicone	3.12–3.30

through the loss resistance and the current through the capacitance; this ratio is called the loss angle δ, defined as $\tan \delta = |I_R|/|I_C| = 1/\omega RC$ (Figure 7.2).

The temperature dependence of the capacitance is caused by the expansion coefficient of the materials and the temperature sensitivity of the dielectric constant. There are capacitors with a built-in temperature compensation. The non-linearity of the capacitance is, in general, negligible.

A large value of the capacitance is achieved by a very thin dielectric. However, the electrical field strength increases with decreasing thickness, limiting the breakthrough voltage of the capacitor.

Very high capacitance values are achieved with electrolytic capacitors. Their dielectric consists of a very thin oxide layer, formed from the material of one of the conductors, usually aluminium. The surface of the conductor is first stained, resulting in pores in the metal surface, increasing the active area. Then the surface material is anodically oxidized. The resulting aluminium oxide can withstand a high electrical field strength (up to $700 \text{ V}/\mu\text{m}$), allowing a very thin layer without breakthrough occurring. In the wet aluminium capacitor, the counter electrode is connected to the oxide via a layer of paper impregnated with boric acid and a second layer of non-anodized aluminium. In the dry type, the paper is replaced by a fibrous material, glass, impregnated with manganese dioxide. Instead of aluminium, tantalum is also used with anodic tantalum oxide as the dielectric. Electrolytic capacitors are unipolar; they can only function correctly at the proper polarity of the voltage (indicated on the encapsulation).

The highest values of the capacitance that can be obtained are in the order of 1 F (electrolytic types); the lowest values are around 0.5 pF, with an inaccuracy of about $\pm 0.3 \text{ pF}$ (ceramic types). Accurate and stable capacitances are realized with plastic film (for instance polystyrene) or mica as the dielectric. Mica capacitors have a loss angle corresponding to $\tan \delta \approx 0.0002$, and are very stable: $10^{-6}/°\text{C}$. They can withstand a high electric field strength (60 kV/mm) and hence a high voltage (up to 5 kV).

7.1.3 *Inductors and transformers*

Inductors and transformers are components based on the phenomenon of induction: a varying magnetic field produces an electric voltage in a wire surrounding that magnetic field (Figure 7.3a). This effect is described by the magnetic induction B (unit tesla, T)

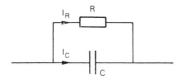

Figure 7.2 A model of a capacitor with dielectric losses.

98   Passive electronic components

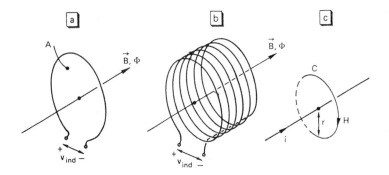

Figure 7.3 (a) A varying magnetic field produces an induction voltage in a conductor; (b) the induction voltage increases with the number of turns; (c) a current through a conducting wire generates a magnetic field.

and the magnetic flux Φ (unit weber, Wb). For a uniform magnetic induction field (the field has the same magnitude and direction within the space considered), the flux equals $\Phi = B \cdot A$, with A the surface area of the loop formed by the conductor (Figure 7.3a).

In vector notation, the magnetic flux is defined as $\Phi = \iint \mathbf{B} \, d\mathbf{A}$; this definition also holds for non-uniform fields. Both B and Φ are defined so the induced voltage satisfies the expression $v_{ind} = -d\Phi/dt$ (Faraday's Law of Induction): the induced voltage equals the rate of change of the magnetic flux. In a uniform magnetic field the induced voltage is $v_{ind} = -A \, dB/dt$.

The induced voltage is not influenced when short-circuiting the wire loop; a current equal to v_{ind}/R will flow, with R the resistance of the loop. The induced voltage increases with increasing rate of change of the flux (frequency), increasing loop area and increasing number of turns (Figure 7.3b). Each loop undergoes the same flux change and is connected in series, so $v_{ind} = -n \, d\Phi/dt$, with n the number of turns.

A varying magnetic field produces an induction voltage in a conductor; the reverse effect also exists: a current I through a conductor produces a magnetic field; its strength H (unit A/m) satisfies Biot and Savart's Law: $H = I/2\pi r$. This empiric law, which is only valid for a straight wire, shows that the magnetic field strength is inversely proportional to the distance r from the wire (Figure 7.3c). Another form of this law is Ampère's Law: the integral of the magnetic field strength over a closed contour equals the total current through the contour: $\oint_C \mathbf{H} \, d\mathbf{s} = \sum_{n=1}^{N} i_n$. This law shows that the magnetic field produced by a group of conducting wires is the sum of the fields from each individual wire.

Two conductors are coupled mutually: an electric current in one conductor produces a magnetic field that, varying with time, induces a voltage in the other conductor; this is called mutual inductance. Anyway, this effect also occurs in a single conductor: a varying current through the conductor produces a varying magnetic field that induces a voltage in the same conductor. This phenomenon is called self-inductance. The direction of the induced voltage is such that it tends to reduce its origin (Lenz's Law). The self-inductance depends strongly on the geometry of the conductor: the self-inductance increases when the conductor is coupled more closely to itself, such as in a coil.

We have introduced two magnetic quantities: the magnetic induction B and the magnetic field strength H. Because of the different definitions, these two quantities do not have the same dimensions, although they both describe a magnetic field property. They are coupled by the equation $B = \mu_0 H$, where $\mu_0 = 4\pi \cdot 10^{-7}$ V s/A m, the permeability of a vacuum. Obviously, the inductive coupling between two conductors (or between a conductor and itself) depends on the properties of the medium between the conductors. The property that accounts for the ability of the material to be magnetized (that is, magnetically polarized) is described by the relative permeability, μ_r. For a vacuum, the relative permeability is 1; the relation between B and H in a magnetizable medium is $B = \mu_0 \mu_r H$. A high value of μ_r means a higher flux for the same magnetic field strength. For most materials μ_r is about 1. Some electronic applications require materials with a high permeability (for instance supermalloy: $\mu_r = 250\,000$); the relative permeability of iron is about 7500. Magnetic materials are much less ideal than dielectric materials. The magnetic losses are high, the material shows magnetic saturation, resulting in a strong non-linearity and hysteresis.

The self-inductance L of a conductor is the ratio between the magnetic flux Φ and the current I through the conductor: $L = \Phi/I$. The relation between the AC current through the inductor and the voltage across it is $v = L\,di/dt$. We saw that the capacitance is increased by applying a dielectric with high permittivity. Likewise, the self-inductance of an inductor is increased by using materials with a high relative permeability. Inductors are constructed as coils (with or without a magnetic core) in various shapes (for instance solenoidal, toroidal).

Whenever possible, inductors are avoided in electronic circuits; they are expensive and bulky (especially when designed for low frequencies). They show non-linearity and hysteresis (due to the magnetic properties of the core material) as well as resistance (of the wires and due to magnetic losses) and capacitance (between the turns). Inductors are applied in filters for signals in the medium and higher frequency ranges.

A transformer is based on signal transfer by mutual inductance. In its most simple form, a transformer consists of two windings (the primary and the secondary winding) around a core of a material with high permeability. The ratio between the number of turns in the primary and secondary winding is the turn ratio n.

For an ideal transformer with a turn ratio $1:n$, the voltage transfer (from primary winding to secondary winding) is just n, and the current transfer is $1/n$. A proper coupling between the two windings is achieved by overlaying the two windings and by applying a toroidal core.

Transformers are mainly used in the power supply of a system, for down conversion of the mains voltage or up conversion in battery-operated instruments. They are also applied in special types of (de)modulators in the middle and high frequency range.

7.2 Sensor components

Most electric parameters of electronic components, such as resistance, capacitance, mutual inductance and self-inductance, depend on both the material properties and the dimensions of the component. This dependency allows us to construct special components that are essentially sensitive to a particular physical quantity. Such components are then suitable for use as a sensor, if they meet the usual requirements with respect to sensitivity and reproducibility. Some of these sensors use the sensitivity of material parameters to externally applied physical signals, for instance temperature, light intensity and mechanical pressure. Others have variable dimensions in order to be sensitive to quantities such as displacement and rotation. This section describes the basic construction and properties of sensors based on resistive, capacitive, inductive, piezoelectric and thermoelectric effects.

7.2.1 *Resistive sensors*

The resistance between the two terminals of a conductor equals $R = \rho F$, with ρ the resistivity of the material and F a geometric factor, determined by the shape and dimensions of the conductor. There are two groups of resistive transducers: one that is based on variations in ρ, and one that uses a variable geometry.

Temperature-sensitive resistors

The resistivity of a conductive material depends on the concentration of free charge carriers and their mobility. The mobility is a parameter that accounts for the ability of charge carriers to move more or less freely throughout the atom lattice; their movement is hampered constantly by collisions. Both concentration and mobility vary with temperature, at a rate that depends strongly on the material.

In intrinsic (or pure) semiconductors, the electrons are bound quite strongly to their atoms; only a very few have enough energy (at room temperature) to move freely. At increasing temperature more electrons will gain sufficient energy to be freed from their atom, so the concentration of free charge carriers increases with increasing temperature. As the temperature has much less effect on the mobility of the charge carriers, the resistivity of a semiconductor decreases with increasing temperature; its resistance has a negative temperature coefficient.

In metals, all available charge carriers can move freely throughout the lattice, even at room temperature. Increasing the temperature will not affect the concentration. However, at higher temperatures the lattice vibrations become stronger, increasing the chance of the electrons colliding and hampering a free movement throughout the material. Hence, the resistivity of a metal increases at higher temperature: its resistivity has a positive temperature coefficient.

The temperature coefficient of the resistivity is used to construct temperature sensors. Both metals and semiconductors are applied. They are called (metal) resistance thermometers and thermistors, respectively.

The construction of a resistance thermometer of high quality requires a material (metal) with a resistivity temperature coefficient that is stable and reproducible over a wide temperature range. Copper, nickel and platinum are suitable materials. Copper is useful in the range from −140 to 120°C, nickel from −180 to 320°C. By far the best material is platinum, due to a number of favourable properties. Platinum has a high melting point (1769°C), is chemically very stable, resistant against oxidation and available with high purity. The normalized temperature range of a platinum resistance thermometer runs from −180 up to 540°C, but it can be used to over 1000°C at reduced stability. Platinum resistance thermometers are used as international temperature standards for temperatures between the boiling point of oxygen (−182.97°C) and the melting point of antimony (680.5°C).

A platinum thermometer has a high linearity. Its temperature characteristic is given by:

$$R_T = R_0(1 + \alpha T + \beta T^2 + \ldots)$$

with R_0 the resistance at 0°C. The normalized value of α is $3.908\,02 \cdot 10^{-3}\,\text{K}^{-1}$ and that of β is $5.8020 \cdot 10^{-7}\,\text{K}^{-2}$, according to the European norm DIN-IEC 751. The temperature coefficient is therefore almost 0.4%/K. A common value for R_0 is $100\,\Omega$; such a temperature sensor is called a Pt-100.

The material of a thermistor (contraction of the words 'thermally sensitive resistor') should have a stable and reproducible temperature coefficient. Commonly used materials are sintered oxides from the iron group (chromium, manganese, nickel, cobalt, iron); these oxides are doped with elements of different valency to obtain a lower resistivity. Several other oxides are added to improve the reproducibility. Other materials that are used for thermistors are the semiconductors germanium, silicon, gallium arsenide and silicon carbide.

Thermistors cover a temperature range from −100 to 350°C. Their sensitivity is much larger than that of resistance thermometers. Furthermore, the size of thermistors can be very small, so they are applicable for temperature measurements in or on small objects. Compared to resistance thermometers, a thermistor is less stable in time and shows a much larger non-linearity.

As explained before, the resistance of most semiconductors has a negative temperature coefficient, as do most thermistors. That is why a thermistor is also called an NTC-thermistor or simply an NTC. The temperature characteristic of an NTC satisfies the next equation:

$$R_T = R_0 \exp B \left(\frac{1}{T} - \frac{1}{T_0} \right)$$

with R_0 the resistance at T_0 (0°C) and T the (absolute) temperature (in K). The temperature coefficient (or sensitivity) of an NTC is:

$$\alpha = \frac{1}{R} \frac{dR}{dT} = -\frac{B}{T^2}\,(\text{K}^{-1})$$

The parameter B is in the order of 2000–5000 K. For instance, at $B = 3600$ K and room temperature ($T = 300$ K), the sensitivity amounts to -4%/K. To obtain a stable sensitivity, thermistors are aged by a special heat treatment. A typical value of the stability after ageing is 0.2% per year.

Next to NTC thermistors there are PTC thermistors too. The temperature effect differs essentially from that of a thermistor. PTCs are made up of materials from the barium–strontium–lead–titanate complex, with a positive temperature coefficient over a rather restricted temperature range. Using various materials, a range from about -150 to 350°C is covered. Within the range of a positive temperature coefficient, the characteristic is approximated by $R_T = R_0 e^{BT}$, ($T_1 < T < T_2$). The sensitivity in that range is B (K^{-1}) and can be as high as 60%/K. PTC thermistors are rarely used for temperature measurements, because of the lack of reproducibility. They are mainly applied as safety components to prevent overheating at short-circuits or overloads.

Light-sensitive resistors

The resistivity of some materials depends on the intensity of incident light (photoresistive effect). A resistor of such a material is the light-dependent resistor (LDR) or photoresistor. A common material is cadmium sulphide. In the absence of light, the concentration of free charge carriers is low, hence the resistance of the LDR is high. When light falls on the material, free charge carriers are generated; the concentration increases and thus the resistance decreases with increasing intensity. The light sensitivity depends on the wavelength of the light, and is highest at about 680 nm (red light). Below 400 nm and above 850 nm, the LDR is not usable (Figure 7.4). Even in complete darkness, the resistance appears to be

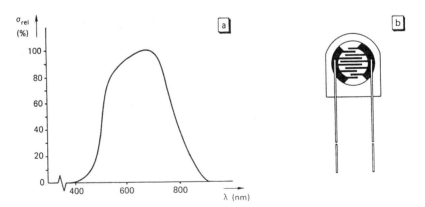

Figure 7.4 (a) Relative conductivity of an LDR versus wavelength, at constant light intensity; (b) example of an LDR.

finite; this is the dark resistance of the LDR, which can be more than 10 MΩ. The light resistance is usually defined as the resistance at an intensity of 1000 lux; it may vary from 30 to 300 Ω for different types. Photoresistors change their resistance value rather slowly; the response time from dark to light is about 10 ms; from light to dark, the resistance varies only about 200 kΩ/s.

Force-sensitive resistors

When an electric conductor, for instance a metal wire, is stressed, its resistance increases, because the diameter decreases and the length increases. Semiconducting materials also change their resistivity when subjected to a mechanical force (piezoresistive effect).

Strain gauges are components based on these effects. They are used to measure force, pressure, torque and (small) changes in length, in (for instance) mechanical constructions. Such strain gauges consist of a meander-shaped metal wire or foil (Figure 7.5a), fixed on an isolating flexible carrier (e.g. an epoxy).

A strain gauge undergoes the same strain as the construction to which it is glued (Figure 7.5b). The resistance change is proportional to the change in length. This proportionality factor or sensitivity is called the strain gauge factor or simply gauge factor. Metals always have a gauge factor of 2; semiconductor strain gauges have a much higher gauge factor, but are less stable and less linear.

Strain gauges are sensitive to temperature changes. To eliminate measurement errors due to temperature changes, temperature compensation is required. For further information on strain gauges the reader is referred to the data sheets of the manufacturers.

Resistive displacement sensors

A resistive displacement sensor commonly used is the potentiometer: a conductive path with a movable ruler (Figure 7.6). The conductor may consist of a spiral wire or a

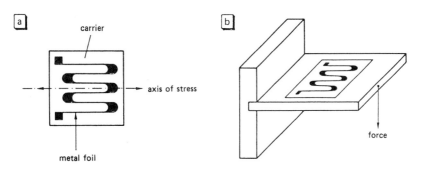

Figure 7.5 (a) An example of a strain gauge; (b) an application of a strain gauge for the measurement of bending.

104   Passive electronic components

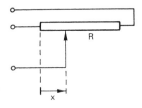

Figure 7.6 A linear resistive displacement sensor.

homogeneous path of carbon or a conductive polymer. Linear displacement sensors have a straight conductive path; angular displacement sensors have a circular (one turn) or helix-shaped path (multiturn potentiometer).

The major quality parameters are non-linearity (determined by the homogeneity of the conductive path), the resolution (for film types, infinite) and the temperature coefficient. There are linear potentiometers with a range from several mm up to 1 m, angular sensors up to ten revolutions and a non-linearity well below 0.1%. Potentiometers also exist with a prescribed non-linear relationship between displacement (rotation) and resistance change, for instance logarithmically. Such potentiometers perform some arithmetic signal processing directly.

7.2.2 Inductive sensors

Inductive sensors are based on a change in self-inductance L or mutual inductance M of a component, usually consisting of two parts that can move relative to each other. Figure 7.7a shows an inductor (coil) with a movable core; its self-inductance L varies with the position of the core. Due to leak fields through the winding and at the ends of the coil, the self-inductance varies only linearly with displacement Δx over a limited range (Figure 7.7b). Another disadvantage of this type of sensor is the necessity to measure an impedance (self-inductance) which is more difficult than a simple voltage or current.

The construction of Figure 7.8a is better in this respect. This sensor, called a linear variable differential transformer or, for short, LVDT, consists of one primary winding, two secondary windings connected in series but wound in an opposite direction, and a movable core. When the core is exactly in the centre of this symmetric construction, the voltages of the secondary windings are equal but opposite, so their sum is zero. A displacement of the core from the centre position results in an imbalance of the two secondary outputs; hence, the total output amplitude v_0 increases with displacement Δx; the phase of v_0 with respect to v_i is 0 or π, depending on the direction of the displacement.

LVDTs are made for displacements from several mm up to 1 m, with a non-linearity less than $\pm 0.025\%$ of the nominal displacement. The sensitivity of an LVDT is expressed as mV (output voltage) per mm (displacement) per V (input voltage), and ranges from 10 to about 200 mV/mm V, depending on the type.

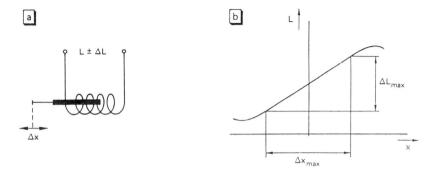

Figure 7.7 (a) A coil with movable core as a displacement sensor; (b) the self-inductance as a function of the core displacement.

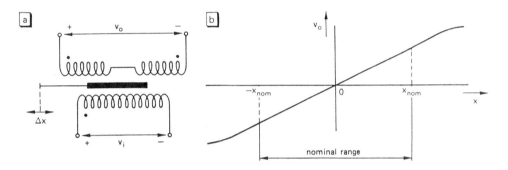

Figure 7.8 (a) Linear variable differential transformer; (b) output voltage amplitude versus displacement.

Inductive sensors with a movable core can also be used for the measurement of a force (with a spring) or acceleration (with a mass–spring system). In fact, all displacement sensors can be used to measure velocity or acceleration, by differentiating once or twice their output ($v = dx/dt$; $a = dv/dt$).

Figure 7.9a shows another type of inductive sensor. It consists of a coil with fixed core. The coil is connected to an AC voltage source, so it generates a varying magnetic field in front of the sensor head. Now if a conductor is close to the sensor, a voltage is induced in the conductor (Section 7.1.3). As there is no preferred current path in the object, currents will flow in arbitrary directions; these are called eddy currents. Due to mutual coupling, these currents induce in turn a voltage in the sensor coil, opposing the original voltage, thus reducing the self-inductance. The effect depends on the distance between coil and object, and can be used for the construction of a contactless displacement sensor. Such an eddy current displacement sensor is

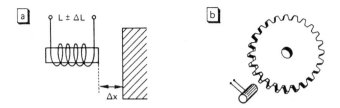

Figure 7.9 (a) Eddy-current sensor; (b) an example of an application of an eddy-current sensor as tachometer.

used for the measurement of (short) distances, and a number of other applications such as the measurement of the number of revolutions (tachometer), Figure 7.9b.

7.2.3 Capacitive sensors

The capacitance of a set of conductors is given as $C = \varepsilon_0 \varepsilon_r F$, with F a factor determined by the geometry of the conductors. For a capacitor consisting of two parallel flat plates, $F = A/d$, with A their surface area and d the distance between the plates. Capacitive sensors are based on changes in ε_r or in F. Some examples of the first possibility are:

- Capacitive temperature sensors, using the temperature dependence of ε_r; these are used for the measurement of temperatures close to absolute zero.
- Capacitive level sensors: a usually linear, tubular capacitor connected vertically in a vessel or tank; the dielectric varies with the level, as does the capacitance.
- Capacitive concentration sensors: the dielectric constant of materials like powders or grains depends on the concentration of a particular (dielectric) substance (for instance water).

Capacitive displacement sensors are based on the variation of the factor F. Figure 7.10 shows some examples of such sensors.

A balanced configuration (Figure 7.10d and e) is used where a high sensitivity is required; common changes (for instance due to temperature) are cancelled out. There exist cylindrical constructions (similar to the LVDT) that have extreme low non-linearity (0.01%) over the whole nominal range (±2 mm up to ±250 mm).

7.2.4 Thermoelectric sensors

Seebeck effect

Free charge carriers in different materials have different energy levels. When two different materials are connected to each other, the charge carriers at the junction will rearrange due to diffusion, resulting in a voltage difference across this junction. Of

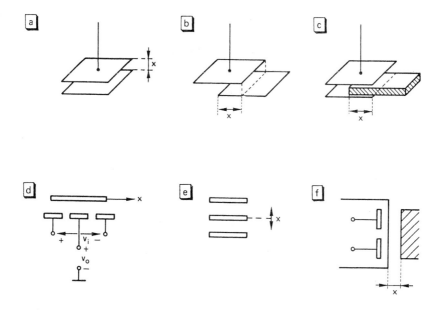

Figure 7.10 Examples of capacitive displacement sensors: (a) variable distance between the electrodes; (b) variable electrode surface area; (c) variable dielectric; (d) and (e) balanced configurations (differential capacitor); (f) as proximity sensor.

course, neutrality is maintained for the whole construction. The value of this junction potential depends on the type of materials and the temperature.

In a couple of two junctions of materials a and b in series (Figure 7.11), two voltages are generated at both junctions, but with opposite polarity; the voltage across the end points of the couple is zero, as long as the junction temperatures are equal. If the two junctions have different temperatures, the thermal voltages do not cancel, so there is a net voltage across the end points of the couple, which satisfies the expression:

$$V_{ab} = \beta_1(T_1 - T_2) + \tfrac{1}{2}\beta_2(T_1 - T_2)^2 + \ldots$$

This phenomenon is called the Seebeck effect, after the discoverer, Thomas Johann Seebeck (1770–1831). V_{ab} is the Seebeck voltage. The coefficients β depend on the materials, and somewhat on the temperature. The derivative of the Seebeck voltage with respect to the variable T_1 is:

$$\frac{\partial V_{ab}}{\partial T_1} = \beta_1 + \beta_2(T_1 - T_2) + \ldots = \alpha_{ab}$$

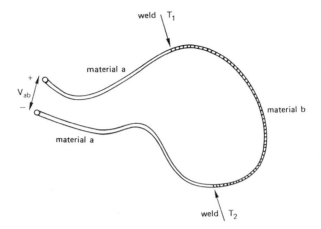

Figure 7.11 A thermocouple generates a DC voltage V_{ab} between its end points, if the junction temperatures are not equal.

and is called the Seebeck coefficient, which depends on the materials and the temperature as well. The Seebeck coefficient can always be written as the difference between two other coefficients: $\alpha_{ab} = \alpha_{ar} - \alpha_{br}$, with α_{ar} and α_{br} the Seebeck coefficients of the couples of materials a and r (a reference material) and b and r, respectively. Usually, the reference material is lead; sometimes copper or platinum.

The Seebeck effect is the basis for the thermocouple, a thermoelectric temperature sensor. One of the junctions, the reference junction or cold junction, is kept at a constant, well-known temperature (for instance 0°C); the other junction (the hot junction) is connected to the object whose temperature has to be measured. Actually, the thermocouple measures only a temperature difference, not an absolute temperature. Thermocouple materials should have a Seebeck coefficient that is high, to achieve high sensitivity, with a low temperature coefficient, to obtain high linearity and be stable in time (for a good long-term stability of the sensor).

Thermocouples cover a temperature range from almost 0 K to over 2800 K, and belong to the most reliable and accurate temperature sensors in the range from 630 to 1063°C. Table 7.3 shows some properties of various thermocouples.

The sensitivity of most couples from Table 7.3 depends on the temperature. For example, the sensitivity of the couple copper/constantan at 350°C has reached a value of 60 μV/K.

Altogether, metal thermocouples have a relatively low sensitivity. To obtain a sensor with a higher sensitivity, a number of couples are connected in series; all cold junctions are thermally connected to each other, as are all hot junctions. The sensitivity of such a thermopile is n times that of a single junction where n is the number of couples.

Table 7.3 Some properties of various thermocouples

Materials	Composition	Sensitivity at 0°C (μV/K)	Temperature range (°C)
Iron/constantan	Fe/60% Cu + 40% Ni	45	0–760
Copper/constantan	Cu/60% Cu + 40% Ni	35	−100–370
Chromel/alumel	90% Ni + 10% Cr/ 94% Ni + 2% Al + bal	40	0–1260
Platinum/platinum + rhodium	Pt/90% Pt + 10% Rh	5	0–1500

Peltier effect and Thomson effect

Experiments showed that when an electric current I flows through the junction of two materials a and b, there is a heat flow Φ_w from the environment towards the junction. The direction of the heat flow changes when reversing the current. The heat flow appears to be proportional to the current I:

$$\Phi_w = -\Pi_{ab}(T)I$$

with Π the Peltier coefficient, named after the discoverer, Jean Peltier (1785–1845). The Peltier coefficient depends only on the types of materials involved and the absolute temperature. In a series of two junctions, heat flows from one junction to the other (Figure 7.12), raising the temperature at one end of the connection and lowering it at the other end. So, the Peltier effect can be applied for cooling.

A Peltier element is a cooling device based on the Peltier effect. To achieve a reasonable cooling power, many cold junctions are connected thermally in parallel. The hot junctions are connected to a heat sink to reduce heating of the hot face in favour of a lower temperature at the cold face; after all, the Peltier effect generates a

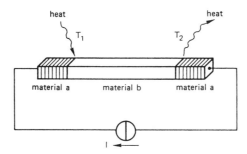

Figure 7.12 Basic construction of a Peltier element: when a current flows through a couple of junctions, one of these junctions heats up, the other cools down.

temperature difference. The temperature difference across a Peltier element is limited by dissipation, due to the Peltier current itself. The maximum difference is determined by the Peltier coefficient, the electrical resistance and the thermal resistance between the cold and hot faces of the element. Theoretically, the maximum temperature difference is about 65°C (at room temperature). This value is obtained with selected materials like doped bismuth telluride (Bi_2Te_3). Cooling to lower temperatures is achieved by piling up Peltier elements. However, the maximum temperature difference is much less than that of the sum of the single elements.

From the theory of thermodynamics, it follows that for a junction of materials a and b, the heat production is $\Pi_{ab}(T)I$ on the one hand, and $\alpha_{ab}(T)IT$ on the other hand. So, there is a relation between the Peltier and Seebeck coefficients:

$$\Pi_{ab}(T) = T\alpha_{ab}(T)$$

with T the absolute temperature of the junction.

Heat exchange with the environment also occurs when a current I flows through a homogeneous conductor that has a temperature gradient. This is the Thomson effect, after William Thomson (Lord Kelvin), 1824–1907. The heat flow appears to be proportional to the current I and the temperature gradient in the conductor:

$$\Phi_w = \mu(T)I\frac{dT}{dx}$$

with $\mu(T)$ the Thomson coefficient of the material. Theoretically, μ is equal to $T\,d\alpha/dT$.

Usually, in a thermocouple all three effects, the Seebeck effect, the Peltier effect and the Thomson effect, occur simultaneously. To avoid self-heating or self-cooling of the thermocouple, it must be connected to a measurement circuit with high input resistance; the current through the junctions is then kept to a minimum.

7.2.5 *Piezoelectric sensors*

Some materials show electrical polarization when subjected to a mechanical force; this is called the piezoelectric effect. The polarization originates from a deformation of the molecules, and results in a measurable surface charge or, via the relation $Q = CV$, a voltage across the material. The relation between force and charge is highly linear, up to several thousands of volts.

The sensitivity of piezoelectric materials is expressed in terms of the piezoelectric charge constant d (C/N or m/V).

A material with natural piezoelectricity is quartz (SiO_2); many ceramic materials and some polymers can be made piezoelectric by poling at high temperatures (above the Curie temperature). Piezoelectric materials are used for the construction of piezoelectric force sensors and accelerometers. In the latter, a well-defined mass (seismic mass) is connected to the piezoelectric crystal of the sensor.

Sensor components 111

A piezoelectric accelerometer (Figure 7.13a) has a strong peak in the amplitude characteristic (Figure 7.13b), due to the mass–spring system of the construction. Another important shortcoming of a piezoelectric sensor is the inability to measure static forces and accelerations; the charge, generated by a constant force, will gradually leak away via the high resistance of the piezoelectric crystal, or along the surface of the edges.

Table 7.4 shows the range and the sensitivity of piezoelectric sensors as force sensors and accelerometers. The various values depend strongly on the size and construction of the sensor.

Table 7.4 Range and sensitivity of piezoelectric force sensors and accelerometers

	Force sensor	Accelerometer
Range	10^2–10^6 N	$5 \cdot 10^3$–10^6 m s^{-2}
Sensitivity	2–4 pC/N	0.1–50 pC/m s^{-2}
Resonance frequency		1–100 kHz

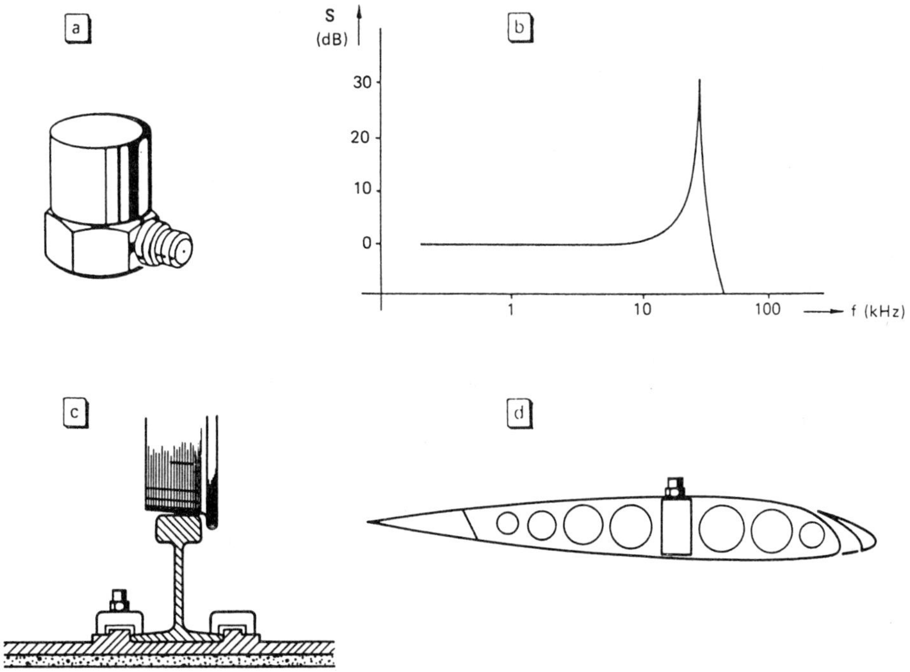

Figure 7.13 (a) An example of a piezoelectric accelerometer; (b) the sensitivity of a piezoelectric accelerometer shows a pronounced peak in the frequency characteristic; (c) and (d) two application examples of piezoelectric accelerometers as a vibration sensor: on a rail and on the model of an airplane wing.

The piezoelectric effect is reversible; this makes the material suitable for the construction of actuators, for instance acoustic generators. Recently, linear motors also have been designed using the piezoelectric effect.

SUMMARY

Passive circuit components

- The resistivity or specific resistance ρ of a material is the ratio between the electric field strength E and the resulting current density J: $E = \rho J$; the reciprocal of the resistivity is the conductivity, $\sigma = 1/\rho$.
- Ohm's Law for a resistor: $V = RI$.
- The capacitance of a set of two conductors is the ratio between the charge Q and the voltage V: $Q = CV$. For two parallel plates with surface area A and distance d from each other, the capacitance is $C = \varepsilon_0 A/d$ (in vacuum).
- The dielectric constant or relative permittivity ε_r of a material is a measure of its electric polarizability (the formation of electrical dipoles). For a vacuum, $\varepsilon_r = 1$; a capacitor with a dielectric material has a capacitance that is ε_r as much as in vacuum.
- A useful measure of a capacitor is the loss angle δ, defined as $\tan \delta = I_R/I_C$, with I_R and I_C the resistive and capacitive currents through the capacitor, respectively.
- The voltage induced by a varying magnetic field is $V_{ind} = -d\Phi/dt$, with Φ the magnetic flux, defined in turn as the integral of the magnetic induction B: $\Phi = \iint \bar{B} \cdot d\bar{A}$.
- The magnetic field strength H produced by a current through a straight wire at distance r is $H = I/2\pi r$.
- The magnetic permeability of a material is a measure of its magnetic polarizability (the formation of magnetic dipoles), and is defined via the relation $B = \mu_0 \mu_r H$. For vacuum, $\mu_r = 1$.
- The self-inductance of a coil (inductor) is the ratio between the produced flux Φ and the current through the inductor: $\Phi = LI$.
- The voltage transfer of an ideal transformer is equal to the ratio of the numbers of primary and secondary turns, n; the current transfer amounts to $1/n$.

Sensor components

- To measure physical quantities and parameters, special resistors are designed whose resistance varies in a reproducible way with those physical quantities, such as temperature, light intensity, relative displacement or angular displacement.

- Resistance thermometers are based on the positive temperature coefficient of the resistance of a metal. A common type of sensor is the Pt-100; this sensor has a resistance of 100.00 Ω at 0°C, and a sensitivity of about 0.39 Ω/K.
- Thermistors have a negative temperature coefficient; their resistance decreases exponentially with increasing temperature.
- Strain gauges are resistive sensors whose resistance varies upon applying a mechanical force. Temperature compensation is obligatory.
- An LVDT (linear variable differential transformer) is a cylindrical transformer with movable core; the voltage transfer is proportional to the linear displacement of the core.
- Capacitive sensors are based on the variation of their active dimensions or the dielectric properties. The first is used for the measurement of linear and angular displacement, the second for measuring, for instance, level, moisture content in a substance, or thickness.
- Thermocouples are temperature sensors based on the Seebeck effect. With thermocouples, temperature differences can be measured accurately, over a wide range.
- The Peltier effect is another thermoelectric effect that can be used for cooling.
- The piezoelectric effect is used for the construction of force sensors and accelerometers. Static measurements are not possible. The frequency characteristic of a piezoelectric accelerometer shows a pronounced peak, due to mechanical resonance.

Exercises

Passive circuit components

7.1 Give the relation between the current I and the voltage V of an (ideal) capacitor with capacitance C; the same question for the charge Q and the voltage V.

7.2 What is the loss angle of a capacitor?

7.3 Give the relation between the current I and the voltage V of an (ideal) coil with self-inductance L; the same question for the flux Φ and the current I.

7.4 What are the relations between the magnetic induction B, the magnetic flux Φ and the magnetic field strength H? Also give the dimensions of these three quantities.

7.5 Show that the impedance R_2 between the terminals of the secondary winding of the ideal transformer is equal to $n^2 R_1$ (see figure below).

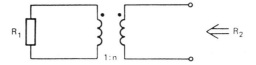

Sensor components

7.6 The transfer characteristic of a platinum resistance is given as $R(T) = R_0(1 + \alpha T + \beta T^2)$, with $R_0 = 100\ \Omega$, $\alpha = 3.9 \cdot 10^{-3}\,\text{K}^{-1}$ and $\beta = -5.8 \cdot 10^{-7}\,\text{K}^{-2}$. Calculate the maximum non-linearity error, referred to the line $R(T) = R_0(1 + \alpha T)$, over a temperature range from -50 to $100°C$. Express the error in % and in °C.

7.7 Both the connecting wires of the Pt-100 sensor from Exercise 7.6 have a resistance r. What is the maximum value of r, to keep the measurement error due to this resistance below an equivalent value of $0.1°C$?

7.8 A chromel–alumel thermocouple is used to measure the temperature of an object with a required inaccuracy of less than $0.5°C$. Find the maximum allowable input offset voltage of the voltage amplifier.

7.9 A strain gauge measurement circuit consists of four strain gauges, connected to a circular bar in such a manner that the resistances vary upon torsion as indicated in the figure below. The required inaccuracy of the system is a torsion corresponding to a relative resistance change of $\Delta R/R = 10^{-6}$. Find the minimum CMRR of the amplifier A.

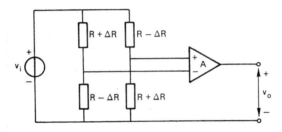

7.10 Give the Thévenin equivalent circuit of a thermocouple, taking into account the resistance of the connecting wires; also give the Norton equivalent of a piezoelectric force sensor, taking into account the capacitance and the leakage resistance of the crystal.

7.11 A linear potentiometric displacement sensor has a range from 0 to 20 mm and a total resistance of $R = 800\ \Omega$ (see the figure below). The sensor is connected to an ideal voltage source and loaded with a measurement instrument with input resistance $R_L = 100\ \text{k}\Omega$. What is the maximum non-linearity error, and at which displacement?

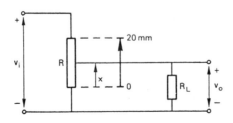

8 Passive filters

An electronic filter enables the separation of signals on account of a particular property, usually the frequency of the signal components. In this context, an electronic filter can be considered as a system with a predescribed frequency response. Filters have a wide application area. Some examples are:

- Improvement of the signal-to-noise ratio; if the frequency range of a measurement signal differs from that of the interference or noise signals, the latter two can be removed from the measurement signal by filtering.
- Improvement of the dynamic properties of a control system; circuits with a particular frequency response are connected to a control system in order to meet specific requirements for stability and other criteria.
- Instruments for signal analysis; many frequency-selective measurement instruments contain special filters, for instance spectrum analysers and network analysers.

In this chapter we will discuss filters composed only of passive components (resistors, capacitors, inductors); active filters are discussed in Chapter 13.

The advantages of passive filters are:

- high linearity (passive components are highly linear);
- wide voltage and current range;
- no power supply required.

The disadvantages are:

- the required filter properties cannot always be combined with other requirements, for instance with respect to the input and output impedances;
- inductors for applications at low frequencies are rather bulky; furthermore, an ideal self-inductance is difficult to realize;
- not all kinds of filter characteristics can be made with resistors and capacitors alone.

There are four main filter types (Figure 8.1): low-pass, high-pass, band-pass and band-reject (or notch) filters. Signal components with frequencies lying within the pass band are transferred properly; other frequency components are attenuated as much as possible.

116 Passive filters

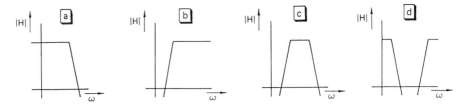

Figure 8.1 Amplitude transfer characteristic of four main filter types: (a) low-pass; (b) high-pass; (c) band-pass and (d) notch filter.

Filters with an ideal frequency characteristic, that is a flat pass-band up to the cut-off frequency and zero transfer beyond it, do not exist. The selectivity can be increased by increasing the order of the network (hence the number of components). The first part of this chapter deals with simple, inductorless filters of up to the second order. The second part deals with filters of a higher order.

8.1 First- and second-order *RC* filters

8.1.1 *Low-pass first-order* RC *filter*

Figure 8.2 shows the circuit diagram and the amplitude characteristic (frequency response of the amplitude) of a low-pass filter comprising only one resistance and one capacitance. The transfer characteristic is discussed earlier (Section 6.1.1). The amplitude transfer is about 1 for frequencies up to $1/2\pi\tau$ Hz ($\tau = RC$). Signals with a higher frequency are attenuated; the transfer decreases with a factor of 2 as the frequency is doubled. The cut-off frequency of this *RC* filter is equal to the -3 dB frequency.

The frequency characteristic represents the response to sinusoidal input voltages. Let us look at the response to another signal, a stepwise change of the input (Figure

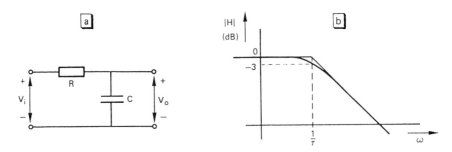

Figure 8.2 (a) A low-pass RC filter of the first order; (b) with corresponding amplitude transfer diagram.

8.3a). When the input voltage makes a positive step upwards, the capacitor will still be uncharged, so its voltage is zero. The current through the resistance R is, at that moment, equal to $I = E/R$, where E is the height of the input step. The capacitor is now charged with this current. Gradually, the voltage across the capacitor increases, while the charge current decreases until, finally, the charge current drops to zero. In steady state, the output voltage equals the input voltage E.

A more precise calculation of the step response of this filter requires solving the system differential equation (Section 4.2). The solution appears to be $v_o = E(1 - e^{-t/\tau})$.

If the input voltage jumps back to zero again, the discharge current is, in the very beginning, equal to E/R. This current decreases gradually until the capacitance is fully discharged and the output voltage is again zero (Figure 8.3b). The solution of the differential equation for this situation is derived in Section 4.2: $v_o = E e^{-(t-t_0)/\tau}$, an exponentially decaying voltage. The 'speed' of the filter is characterized by the parameter τ, which is the time constant of the system. It is apparent that the time constant of this filter is the reciprocal of the -3 dB frequency. The step response can be drawn quite easily, when we realize that the tangent at the starting point intersects the end value at a time $t = \tau$ after the input step.

Now we can compare the system behaviour as described in the time domain and in the frequency domain. For high frequencies, the transfer is approximated by $1/j\omega\tau$ (Section 6.1); this corresponds with integration in the time domain (Section 4.2). From Figure 8.3, it can be seen that a periodic rectangular input voltage results in a

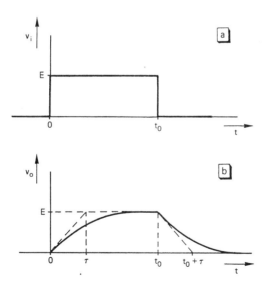

Figure 8.3 (a) Step voltage at the input of an RC low-pass filter; (b) corresponding step response.

triangular-shaped output voltage, in particular for high input frequency. For this reason, the RC network is called an integrating network, although it has only integrating properties for frequencies much higher than $1/\tau$.

Example 8.1

A sine-shaped measurement signal of 1 Hz is masked by an interference signal with equal amplitude and a frequency of 16 Hz (Figure 8.4a). This composite signal is applied to the low-pass filter of Figure 8.4b, in order to improve its signal-to-noise ratio. When the cut-off frequency (or -3 dB frequency) is 4 Hz ($\tau = 1/8\pi$ s), the amplitude transfer at 1 Hz appears to be 0.97 (using the theory of Chapters 4 and 6), whereas the amplitude transfer at 16 Hz is only 0.24 (Figure 8.4c).

To achieve better suppression of the interference signal at 16 Hz, we may shift the -3 dB frequency of the filter away from the signal frequency, for instance to 2 Hz. In that case, the amplitude transfer for 1 and 16 Hz is 0.89 and 0.12, respectively (Figure 8.4d). A better discrimination between the measurement signal and the interference signal is virtually impossible, because a further decrease of the -3 dB frequency will reduce the measurement signal itself. For a corner frequency at exactly 1 Hz, the transfers are 0.71 (or -3 dB) and 0.06, respectively. The best improvement of the signal-to-noise (signal-to-interference) ratio in this example is just a factor 16, because the frequency ratio is 16, and the slope of the filter characteristic is -6 dB/octave.

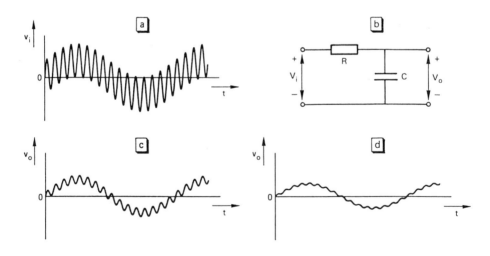

Figure 8.4 (a) The input signal v_i and (b) the filter from Example 8.1; (c) the output signal v_o for $\tau = 1/8\pi$ s; (d) the output for $\tau = 1/2\pi$ s.

We will now show the influence of non-ideal matching on the filter properties. Suppose the filter of Figure 8.2 is connected to a voltage source with source resistance R_g, and loaded with a load resistance R_L (Figure 8.5).

The complex transfer function of the total circuit is:

$$\frac{V_o}{V_g} = \frac{R_L}{R_L + R_g + R} \cdot \frac{1}{1 + j\omega R_p C}$$

with

$$R_p = \frac{R_L (R_g + R)}{R_L + R_g + R}$$

The transfer in the pass band is lowered and the cut-off frequency is also dependent on the load resistance. A higher source resistance and a higher load resistance will shift the cut-off frequency to lower frequencies. When applying an RC filter in a system, the influence of the source and load resistance must not be overlooked.

8.1.2 High-pass first-order RC filter

Figure 8.6 shows the circuit diagram and the corresponding amplitude transfer of a first-order RC high-pass filter. This filter consists of a single resistance and capacitance. Its transfer function is $V_o/V_i = j\omega\tau/(1+j\omega\tau)$, with $\tau = RC$. Signal components with a frequency that exceeds $1/2\pi\tau$ are transferred unattenuated; at lower frequencies the transfer decreases with 6 dB/octave.

The step response of this filter can be derived in a similar way as described in the preceding section. The exact expression for the step response can be found by solving the differential equation of the network: $v_o = f(v_i)$. The solution appears to be: $v_o = E\,e^{-t/\tau}$, with E the height of the step (Figure 8.7).

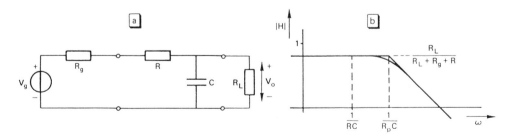

Figure 8.5 (a) A low-pass RC filter connected to a source with source resistance R_g and loaded with a load resistance R_L; (b) the corresponding transfer characteristic.

120 Passive filters

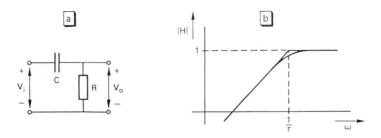

Figure 8.6 (a) A high-pass RC filter of the first order; (b) the corresponding amplitude transfer.

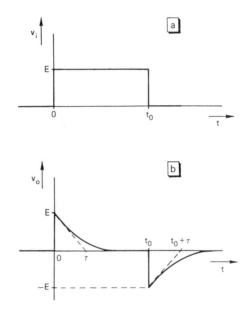

Figure 8.7 (a) A step voltage at the input of a high-pass RC filter; (b) the corresponding step response.

For low frequencies, the transfer can be approximated by $j\omega\tau$, corresponding to a differentiation in the time domain. Therefore, this network is called a differentiating network, although its differentiating properties only occur at low frequencies ($\omega \ll 1/\tau$).

Example 8.2
The signal from Figure 8.4a (drawn again in Figure 8.8a) is now connected to the input of the high-pass filter of Figure 8.8b. We consider the signal with frequency

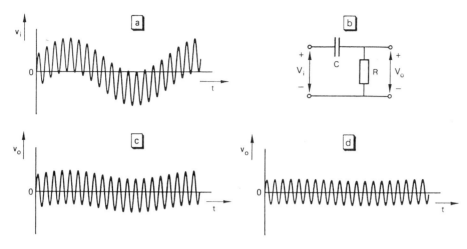

Figure 8.8 (a) The input signal v_i and (b) the filter circuit from Example 8.2; (c) the output voltage v_o at $\tau = 1/8\pi$ s; (d) the output for $\tau = 1/32\pi$ s.

16 Hz as the measurement signal and the component at 1 Hz as the interference signal. At a cut-off frequency of 4 Hz, the amplitude transfer at 1 Hz is about 0.24, and for a frequency of 16 Hz it is about 0.97 (Figure 8.8c).

To suppress further the interference signal of 1 Hz, the cut-off frequency of the filter must be shifted to a higher value. Suppose the -3 dB frequency is 8 Hz; in that case, the amplitude transfer at 1 Hz is only 0.12, and that at 16 Hz is almost 0.89. For the same reason as mentioned in Example 8.1, it is not possible to reduce the interference signal without reducing the measurement signal at the same time. Figure 8.8d shows a favourable situation: a cut-off frequency at 16 Hz, resulting in an amplitude transfer of 0.06 at 1 Hz and 0.71 (-3 dB) at 16 Hz, respectively.

The frequency response of a high-pass RC filter changes when it is connected to a source impedance and a load impedance. We consider the situation of a source resistance and a load capacitance (Figure 8.9).

The complex transfer of the total system is:

$$\frac{V_o}{V_g} = \frac{j\omega RC}{1 + j\omega(RC + R_g C + RC_L) - \omega^2 R_g RC_L C}$$

The exact position of the corner frequencies is found by writing the denominator in a form $(1 + j\omega\tau_1)(1 + j\omega\tau_2)$ and solving the equations $\tau_1 \tau_2 = R_g RC_L C$ and $\tau_1 + \tau_2 = RC + R_g C + RC_L$. The corner frequencies are $1/\tau_1$ and $1/\tau_2$ (Section 6.1.2). Their approximate values can be found more easily using the conditions $R_g \ll R$ and

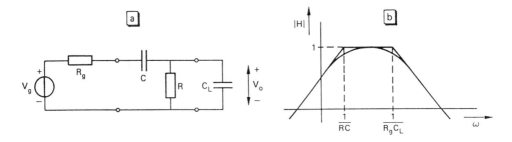

Figure 8.9 (a) A high-pass RC filter with source resistance and load capacitance; (b) due to the capacitive load, the character of the filter is changed.

$C_L \ll C$, conditions for optimal voltage matching. In that case, the denominator can be factorized into $(1 + j\omega RC)(1 + j\omega R_g C_L)$, from which follows the corner frequencies $1/RC$ (the original frequency of the high-pass filter) and $1/R_g C_L$ (an additional corner frequency due to the source and load). This example again illustrates the importance of taking the source and load impedances into account.

8.1.3 Band-pass filters

A simple way to obtain a band-pass filter is to connect a low-pass filter and a high-pass filter in series (Figure 8.10).

Due to mutual loading, the transfer is not equal to the product of the two sections individually. It follows:

$$H = \frac{V_o}{V_i} = \frac{j\omega\tau_2}{1 + j\omega(\tau_1 + \tau_2 + a\tau_2) - \omega^2 \tau_1 \tau_2}$$

with $\tau_1 = R_1 C_1$, $\tau_2 = R_2 C_2$ and $a = R_1/R_2$. The ratio τ_2/τ_1 determines the width of the pass band (the bandwidth of the filter). The transfer has a maximum at $\omega = 1/\sqrt{\tau_1 \tau_2}$, for which $|H| = 1/(1 + a + \tau_1/\tau_2)$. Because the denominator contains the term $(j\omega)^2$, the filter is said to be of the second order. The slope of the amplitude characteristic is ±6 dB/octave; its selectivity is rather poor, for two close frequencies cannot be clearly separated by this filter.

A band-pass filter with a similar amplitude transfer can be made in various ways with the same number of components, for instance by changing the order of the sections. Figure 8.11 shows some possibilities.

The complex transfer of the filter in Figure 8.11a is:

$$H = \frac{j\omega\tau_1}{1 + j\omega(\tau_1 + \tau_2 + a\tau_2) - \omega^2 \tau_1 \tau_2}$$

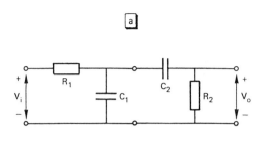
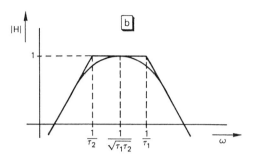

Figure 8.10 (a) A band-pass RC filter composed of a low-pass filter and a high-pass filter in series; (b) the corresponding amplitude transfer characteristic.

with $\tau_1 > \tau_2$; the transfer of the circuit in Figure 8.11b is:

$$H = \frac{j\omega\tau_1/a}{1 + j\omega(\tau_1 + \tau_2 + \tau_1/a) - \omega^2 \tau_1 \tau_2}$$

In both cases, $\tau_1 = R_1 C_1$, $\tau_2 = R_2 C_2$ and $a = R_1/R_2$. The filter of Figure 8.11b has the advantage that both a capacitive load and a resistive load can easily be combined with C_2 and R_2, respectively. They do not introduce additional corner frequencies to the characteristics. The same holds for a source resistance or capacitance in series with the input. Of course, they may change the position of the corner frequencies, but do not change the shape of the characteristic.

8.1.4 Notch filters

There are many types of notch filters composed of only resistors and capacitors. A rather common type is the symmetric double-T filter (or bridged-T filter), depicted in Figure 8.12.

At perfect voltage matching (zero source resistance and infinite load resistance), the complex transfer function is:

$$\frac{V_o}{V_i} = \frac{j\beta}{4 + j\beta}$$

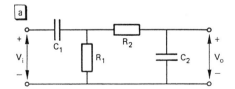
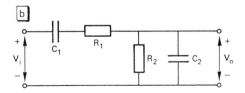

Figure 8.11 Some examples of a second-order RC band-pass filter.

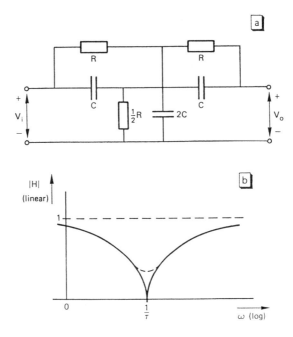

Figure 8.12 (a) The symmetric double-T notch filter; (b) the corresponding amplitude transfer characteristic.

with $\beta = \omega/\omega_o - \omega_o/\omega$, and $\omega_o = 1/RC$. This transfer is zero for $\beta = 0$, or $\omega = 1/RC$. So this filter completely suppresses a signal with frequency ω_o. This is only true if the circuit component values exactly satisfy the ratios as given in the figure. Even a small deviation results in a finite transfer at $\omega = 1/RC$ (dashed curve in Figure 8.12).

8.2 Filters of higher order

The selectivity of first-order filters is rather poor; the slope of the amplitude transfer characteristic is not more than 6 dB/octave or 20 dB/decade. The same holds for the second-order band-pass filter from the preceding section. It is very well possible to realize second-order band-pass filters with a much higher selectivity, using the principle of resonance. An example is the combination of a capacitance and an inductance. Resonance can also be achieved without inductances, using active circuits. Such filter types, which are discussed in Chapter 13, have a high selectivity but only for a single frequency: their bandwidth is very narrow.

Filter characteristics with both an extended flat pass-band and steep slopes are realized by increasing the order of the networks. Band-pass filters can be composed of a low-pass and a high-pass filter; further, low-pass and high-pass filters show a close resemblance, so this section is restricted to only low-pass RC filters.

8.2.1 Cascading first-order RC filters

Consider a number of n first-order RC low-pass filters with equal time constants τ and connected in series. The transfer function of this system can be written as $H = 1/(1+j\omega\tau)^n$. Here it is supposed that the sections do not load each other (this effect will be discussed later). For frequencies much higher than $1/\tau$, the modulus of the transfer is approximated by $|H| = 1/(\omega\tau)^n$; this means that the transfer decreases with a factor 2^n when doubling the frequency; in other words, the slope is $6n$ dB/octave. The attenuation at $\omega = \omega_o$ amounts to -3 dB for a single section, so $-3n$ dB for a filter of order n. The attenuation within the pass band is considerable.

Example 8.3
Figure 8.13 shows a filter consisting of three low-pass RC sections, each having a cut-off frequency $\omega = 1/RC$. The impedance of the components in the successive sections is increased by a factor of 10, starting from the source side. Hence, the effect of mutual loading can be neglected, and the transfer can be approximated by $1/(1+j\omega\tau)^3$.

At the cut-off frequency ω_c of the filter, the transfer is $(\tfrac{1}{2}\sqrt{2})^3 \approx 0.35$, and so a substantial attenuation. The amplitude transfer of a first-order low-pass filter at a frequency $\omega_c/10$ is about 0.995, and so is close to unity. For the third-order filter in this example, this value is $(0.995)^3 \approx 0.985$, which is 1.5% lower than the ideal case. If the filter was designed with equally valued resistances and capacitances, then the situation would be even worse. The transfer in that case is $H = 1/(1 + 6j\omega\tau - 5\omega^2\tau^2 - j\omega^3\tau^3)$; its modulus for $\omega = \omega_c$ is 0.156, and for $\omega = \omega_c/10$, it is only 0.89 (see Figure 8.14, curve e).

8.2.2 Approximations of the ideal characteristics

It appears clearly from Example 8.3 that the coefficients of the terms $j\omega$, $(j\omega)^2$ and $(j\omega)^3$ have a large influence on the shape of the transfer characteristic. One could try to find the proper coefficients for the best approximation to the ideal characteristic, at a

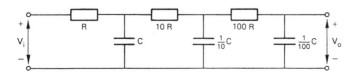

Figure 8.13 A third-order low-pass filter composed of three cascaded first-order RC sections.

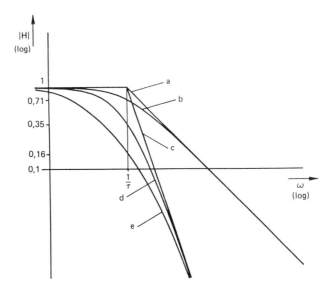

Figure 8.14 The amplitude transfer characteristic of: (a) the asymptotic approximation of a first-order low-pass filter; (b) a real first-order filter; (c) the asymptotic approximation of a third-order filter; (d) the filter from Figure 8.13; (e) the same filter, now with equal resistances and equal capacitances.

given order of the filter. In finding those optimal values, one can follow several criteria, resulting in different characteristics. Three such approximations will be discussed here.

Butterworth filter

The criterion is a maximally flat characteristic in the pass band, that is up to the desired cut-off frequency $\omega = \omega_c$. The shape of the characteristic beyond the cut-off frequency is not considered. The transfer function of order n, satisfying this criterion, appears to be:

$$|H| = \frac{1}{\sqrt{1 + (\omega/\omega_c)^{2n}}}$$

In terms of poles and zeros (Section 6.2), the poles of the complex transfer function are equidistantly positioned on the unity circle in the complex plane.

Chebyshev filter

The criterion is a maximally steep slope from the cut-off frequency; the shape in the pass band is not considered here. It appears that satisfying the condition for the slope, the pass band shows a number of oscillations. This number increases with increasing order. To find the proper values of the filter components, the designer of a Chebyshev filter usually makes use of special tables.

Bessel filter

The Bessel filter has an optimized step response. The criterion is a linear phase transfer up to the cut-off frequency. The step response shows no overshoot. Here as well, the filter designer uses tables to find the proper component values.

Figure 8.15 shows a qualitative comparison between the amplitude transfer characteristics and the step responses of the three types.

None of the filter types discussed here can be realized with resistances and capacitances alone. It is necessary to use either inductors or active components.

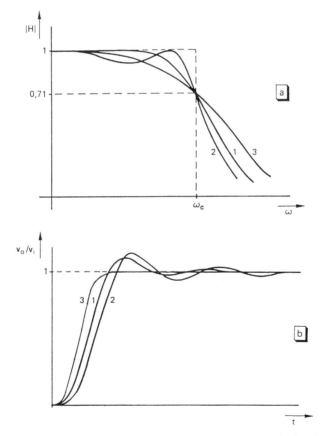

Figure 8.15 (a) The amplitude transfer and (b) the step response of (1) a Butterworth filter, (2) a Chebyshev filter and (3) a Bessel filter, all of the fourth order.

128 Passive filters

So far we have considered only the amplitude transfer of the filter. However, the phase transfer can also be of importance, in particular for filters of a high order. The higher the order, the larger the phase shift (Section 6.1.2). There is a relation between the amplitude transfer and the phase transfer of a linear network: the Bode relation. So, having finished a filter design on the basis of a specified amplitude transfer, it is necessary to check the resulting phase transfer. This may be of paramount importance, in particular in feedback systems, where an improper phase characteristic may endanger the system stability.

SUMMARY

First- and second-order RC filters

- There are four basic filter types: low-pass, high-pass, band-pass and band-reject (or notch) filters.
- The amplitude transfer of a first-order low-pass RC filter (integrating network) with time constant τ is approximated by its asymptotes 1 (0 dB) for $\omega\tau \ll 1$ and $1/\omega\tau$ (−6 dB/octave) for $\omega\tau \gg 1$. The phase transfer runs from 0 via $-\pi/4$ at $\omega\tau = 1$ to $-\pi/2$.
- The amplitude transfer of a first-order high-pass RC filter (differentiating network) with time constant τ is approximated by its asymptotes $\omega\tau$ (6 dB/octave) for $\omega\tau \ll 1$ and 1 (0 dB) for $\omega\tau \gg 1$. The phase runs from $\pi/2$ via $\pi/4$ (at $\omega\tau = 1$) to 0.
- When applying a low-pass or high-pass RC filter, the influence of the source impedance and the load impedance must be taken into account.
- A band-pass filter, composed of a low-pass and a high-pass RC filter, has an amplitude transfer that has slopes of ± 6 dB/octave. The selectivity is poor.
- A double-T or bridged-T is an example of an RC notch filter. Under certain conditions, the transfer is zero for just one frequency.

Filters of higher order

- The amplitude transfer of a filter of order n, composed of n cascaded, perfectly matched sections, has a slope of 2^n (or $6n$ dB) per octave. The attenuation at the cut-off frequency is $(\frac{1}{2}\sqrt{2})^n$ or $-3n$ dB.
- A (low-pass or high-pass) Butterworth filter has a maximally flat transfer in the pass band, at given order.
- A Chebyshev filter has a maximally steep slope from the cut-off frequency at a given order. Its pass band transfer is oscillatory, the oscillations being more pronounced at higher order.
- A Bessel filter has an optimized step response; its phase transfer is linear up to the cut-off frequency.

Exercises

First- and second-order RC filters

8.1 The time constant of a passive first-order low-pass filter is $\tau = 1$ ms. Find the modulus and the argument of the transfer at the following frequencies:
(a) $\omega = 10^2$ rad/s;
(b) $\omega = 10^3$ rad/s;
(c) $\omega = 10^4$ rad/s.

8.2 A passive first-order high-pass filter has a time constant $\tau = 0.1$ s. For which frequency is the transfer equal to:
(a) 1;
(b) 0.1;
(c) 0.01?

8.3 A measurement signal with bandwidth 0–1 Hz is interfered by a sinusoidal signal of 2 kHz. One tries to attenuate the interference signal by a factor of 100, using a first-order filter. The measurement signal itself should not be attenuated more than 3%. Calculate the limits of the filter time constant.

8.4 A periodic, triangular signal contains a third harmonic component whose amplitude is a factor 9 below that of the fundamental (even harmonics fail). This signal is applied to a low-pass filter, to reshape the signal into a sine wave. Find the resulting distortion of the output sine wave (the ratio of the amplitudes of the third harmonic and the fundamental), for a filter of order:
(a) 1;
(b) 2;
(c) 3.

8.5 A measurement signal of 10 Hz and an interference signal with the same amplitude and frequency 50 Hz must be separated; the interference signal must be suppressed at least a factor of 100 compared to the measurement signal. Find the minimum order to meet this requirement.

Filters of higher order

8.6 Find the type of filter characteristic (low-pass, high-pass, band-pass, notch) for the filters a–c depicted in the figure below.

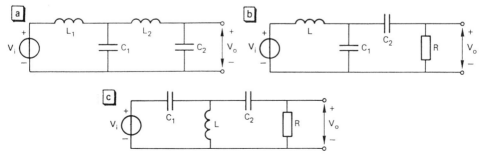

8.7 The modulus of the transfer function of an nth-order low-pass Butterworth filter is $|H| = [1 + (\omega/\omega_c)^{2n}]^{-1/2}$. Calculate the amplitude transfer of this filter for the frequency $\omega = \frac{1}{2}\omega_c$, and express this transfer in dB for:
(a) $n = 2$;
(b) $n = 3$.

8.8 Given a third-order LC filter with source resistance and load resistance (shown in the figure below), calculate the modulus of the complex transfer function, and show that this is a Butterworth filter.

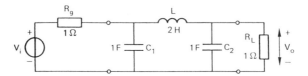

8.9 The inductor in the filter given in the figure below has a value $L = 1$ mH. Find the values of C and R, so that the -3 dB frequency is 10^5 rad/s, and the characteristic satisfies the Butterworth condition.

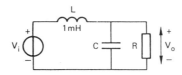

9 PN diodes

Semiconductors play a key role in electronics; all active components are based on the properties of semiconducting materials; there are some exceptions such as vacuum tubes, used for special applications. In particular the element silicon is extensively used, because it allows the construction of many electronic components integrated on to a small piece of this material. Other semiconducting materials currently used are germanium (the first transistor was made from this material, now used for high frequency signal processing) and gallium arsenide (for optical components). The first part of this chapter deals with the basic concepts of semiconductors, in particular silicon. Further, the operation and characteristics of a particular semiconductor device, the pn diode, is discussed. The second part of this chapter illustrates the application of such pn diodes in a number of signal processing circuits.

9.1 Properties of pn diodes

9.1.1 *Operation of pn diodes*

Pure silicon (also called intrinsic silicon) has a rather low conductivity at room temperature: the concentration of free charge carriers (Section 7.1) is small. By a special treatment, it is possible to increase the charge carrier concentration up to a predefined level, resulting in an accurately known conductivity. This is achieved by adding impurities to the silicon crystal, a process called doping.

Silicon is a crystalline material, consisting of a rectangular symmetric lattice of atoms. A silicon atom has four electrons in its valence band, each contributing to a covalent bond to one of the four neighbouring atoms. Very few electrons have enough energy to escape from this bond. Once escaped, they leave an empty place, called a hole.

In a crystal doped with atoms that have five electrons in their valence band, four of those are used to form the covalent bonds with four neighbouring silicon atoms; the fifth is only very weakly bound, and therefore able to move freely through the lattice. Notice that these electrons do not leave a hole. They do, however, contribute to the concentration of free charge carriers in the material and are therefore called free electrons. The material responsible for this donation of free carriers is called a donor and the resulting doped silicon is said to be n-type silicon, because the electrons have negative charge. At room temperature, almost all impurity atoms give rise to an additional free electron, so the free charge carrier concentration equals the donor concentration.

When silicon is doped with atoms that have only three electrons in the valence band, all three will contribute to the bonding with neighbouring silicon atoms; there is even one link missing, a hole that can easily be filled up by a free electron, if available. The material that causes this easy acceptance of free electrons is called an acceptor; the resulting doped silicon is p-type silicon, because there is a shortage of electrons (equivalent to holes). There are approximately as many holes as there are impurity atoms. We might think that the conductivity of p-type silicon is even lower than that of intrinsic material. This is not the case. On the contrary; suppose a piece of p-type silicon is connected to a battery. Electrons at the negative pole of the battery drop into the surplus of holes, and, due to the electric field produced by the battery, they move from hole to hole towards the positive pole. Obviously, this type of conduction differs essentially from the conduction by free electrons. Any time an electron moves to another hole, it leaves a new hole, resulting in a movement of holes in the opposite direction. To distinguish between conduction in n-type and p-type silicon, we consider holes as positive charge carriers, free to move through the silicon. The mobility of electrons differs somewhat from that of holes, so the conductivity of n-type silicon differs somewhat from that of p-type, even when equally doped and at the same temperature.

Suppose we connect a piece of p-type silicon to a piece of n-type silicon. This pn junction, which has particular properties, plays a key role in almost all electronic components. In modern technology, the junction is not made by simply putting two materials together, as in the early days of the pn junction, but merely by partially doping n-type material with acceptor atoms or p-type material with donor atoms. We will now discuss the main properties of a pn junction.

Consider the junction of p- and n-type just at the moment of connecting. The p-type contains a high concentration of holes, while in the n-type there are hardly any holes. Due to this sharp gradient, holes will diffuse from the p-type to the n-type material. For the same reason, electrons drift from the n-type to the p-type region. However, when an electron meets a hole, they recombine and nothing is left. The result of this recombination is a thin layer on both sides of the junction that contains neither free electrons nor free holes; it is depleted of free charge carriers, and for this reason this is called the depletion region or depletion layer (Figure 9.1).

The depletion region is no longer a neutral region; when the electrons have drifted away, positive ions remain in the n-type, while negative ions remain in the p-type. There are positive and negative space charge regions on both sides of the junction. Such a dipole space charge is accompanied by an electric field, pointing from the positive to the negative region, that is from the n-type to the p-type side. This electric field prevents a further diffusion of electrons and holes; at a certain width of the depletion layer there is an equilibrium between the electric field strength and the tendency of the free charge carriers to reduce any concentration gradient.

The width of the depletion layer depends on the concentration of the free charge

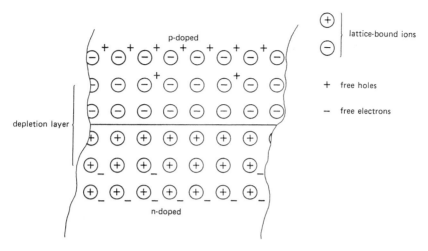

Figure 9.1 On both sides of a pn junction there is a region without free charge carriers, the depletion region.

carriers in the original situation; further, the width can be influenced by an external electric field. Suppose we connect the n-side of the junction to the positive pole of a battery and the p-side to the negative pole. This external field has the same direction as the internal field across the junction.

Moreover, as the depletion region has a very low conductivity (lack of free charge carriers), the external field appears almost completely across the junction itself, reinforcing the original field. We have seen that there is an equilibrium with respect to the depletion width and the electric field. Applying an external field (that is added to the internal field), a new equilibrium must exist, now at a wider depletion layer.

For an external field with opposite polarity, the reverse holds; the internal field decreases, corresponding to a decrease of the depletion layer width.

As the depletion layer is almost free from free charge carriers, it behaves as an insulator: no current can pass the junction. However, when the external electric field is further increased until it fully compensates the internal field, then a current can flow through the material, according to the (rather low) conductivity of the doped silicon. We can now draw the voltage–current characteristic of the pn junction (Figure 9.2). Disregarding the leftmost part of the characteristic (at V_z), we see that current can flow in only one direction. Therefore, this element is called a diode, referring to the similar property of the vacuum diode. Furthermore, current starts flowing from only a particular positive value of the external voltage.

The theoretical relationship between the current through a diode and the voltage across it is given by Equation (9.1):

$$I = I_0(e^{qV/kT} - 1) \tag{9.1}$$

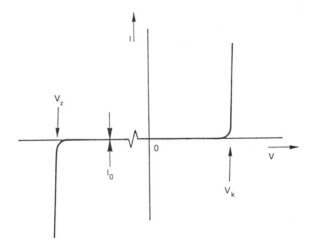

Figure 9.2 Voltage–current characteristic of a pn diode. V_z is the Zener voltage, I_0 the leakage current or reverse current, V_k the threshold voltage.

with q the electron charge ($1.6 \cdot 10^{-19}$ C), k Boltzmann's constant ($1.38 \cdot 10^{-23}$ J/K) and T the temperature in K. At room temperature ($T = 300$ K), the factor q/kT equals about 40 V^{-1}. For a negative voltage (reverse voltage), qV/kT is small compared to 1, hence the current is about $-I_0$, called the reverse current or leakage current of the diode. For positive voltages (forward voltage), the current can be approximated by $I \approx I_0 e^{qV/kT}$: the current increases exponentially with increasing forward voltage by an amount e (≈ 2.78) per 1/40 V = 25 mV. Due to this strong rise of current with voltage, we can approximately state that the diode conducts above a certain voltage (V_k in Figure 9.2), the threshold voltage. Below that voltage the current is almost zero. V_k depends on the material; for silicon it has a value between 0.5 and 0.8 V. In practice, a value of 0.6 V is used as a rough indication. The reverse current I_0 depends on the material and strongly on the temperature. Silicon diodes for general use have a leakage current below 10^{-10} A, which increases exponentially with temperature. For this reason the diode current also increases exponentially with temperature, by about a factor of 2 per 6–7°C. Conversely, at constant current, the voltage across the diode decreases about 2.5 mV as the temperature increases by 1°C.

From Figure 9.2 and Equation (9.1), it appears that a pn diode is a non-linear element; it is not possible to describe a diode in terms of a single resistance value. To characterize the diode in simpler terms as with the exponential relation of Equation (9.1), we will introduce the differential resistance, defined as $r_d = dV/dI$. From Equation (9.1), it follows that $V = (kT/q)\ln(I/I_0)$, hence the differential resistance of the diode is kT/qI, which is inversely proportional to the current I. As at room temperature $kT/q = 1/40$, the differential resistance of the diode is about 25 Ω for a current $I = 1$ mA. This value is independent of the construction, and is the same for all diodes. Using this rule of thumb, the differential resistance at an arbitrary current can be found easily.

The reciprocal value of the differential resistance is the (differential) conductance: $g_d = 1/r_d$, which corresponds to the slope of the tangent at a point of the characteristic of Figure 9.2. The conductance at 1 mA is 40 mA/V, and is directly proportional to the current.

Other important characteristics of a diode are the maximum allowable current (in forward direction) and voltage (in reverse direction). The maximum current is about 10 mA for very small types and may be several kA for power diodes. The maximum reverse voltage ranges from a few volts up to several tens of kilovolts. At maximum reverse voltage, the reverse current increases sharply (see left part of Figure 9.2), due to a breakdown mechanism. One such mechanism is Zener breakdown, an effect that is employed in special diodes, called Zener diodes, which can withstand breakdown. In the breakdown region these diodes have a very low differential resistance (the slope of the I–V characteristic is very steep); the voltage hardly changes at a varying current. This property is used for voltage stabilization. For applications that require a very stable Zener voltage, diodes are constructed with an additional diode for temperature compensation; the net temperature coefficient of the Zener voltage can be as low as $10^{-5}/°C$. Zener diodes are constructed for voltages ranging from about 6 V up to several hundred volts. Figure 9.3 shows the symbols of a normal diode and a Zener diode.

9.1.2 *Photodiodes*

The leakage current (reverse current) of a diode originates from the thermal generation of free charge carriers (electron–hole pairs). Charge carriers produced within the depletion layer of the diode will drift away due to the electric field; electrons drift to the n-side, holes to the p-side of the junction. Both currents contribute to an external leakage current, I_0. The generation rate of such electron–hole pairs depends on the energy of the charge carriers; the more energy, the more electron–hole pairs are produced. It is possible to generate extra electron–hole pairs by adding optical energy; if light is allowed to fall on the junction, free charge carriers are created, increasing the leakage current of the diode. Diodes designed to employ this effect (light-sensitive diodes or photodiodes) have a reverse current that is almost proportional to the intensity of the incident light.

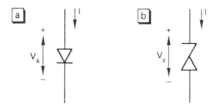

Figure 9.3 Symbols for (a) a diode; (b) a Zener diode.

The major characteristics of a photodiode are:

- spectral response, expressed as ampere per watt or ampere per lumen (Figure 9.4a). Silicon photodiodes have a maximum response for light with a wavelength of about 800 nm;
- dark current, the reverse current in the absence of light. As can be expected from the nature of the reverse current, the dark current of a photodiode increases strongly with increasing temperature; usually it is large compared to the reverse current of normal diodes, and ranges from several nA to μA, depending on the surface area of the device;
- quantum efficiency, the ratio between the number of optically generated electron–hole pairs and the number of incident photons; this efficiency is greater than 90% at the peak wavelength.

Photodiodes have a faster response to light variations compared to photoresistors (Section 7.2.1): they can follow light pulses with a frequency of several megahertz, and are, therefore, suitable for applications in glass fiber transmission (Figure 9.5a).

9.1.3 *Light-emitting diodes (LEDs)*

Some semiconductor materials generate photons when the forward current of such materials through a pn junction exceeds a certain value. Such a diode is called a light-emitting diode or LED. Silicon is not a suitable material for this effect; gallium arsenide, usually with small amounts of phosphorus, aluminium or indium, is a better semiconductor in this respect. The colour of an LED is determined by the composition of the semiconductor. The spectrum of available LEDs ranges from infrared to blue.

The major characteristics of an LED are:

- The peak wavelength: dependent on the material type, between 500 (blue) and 950 nm (infrared).
- Polar emissivity diagram: there are LEDs producing very narrow beams (usually by applying a built-in lens) or wider beams (Figure 9.4b); narrow-beam LEDs are suitable for the coupling to glass fibers (Figure 9.5a).
- The maximum allowable current, also determining the maximum intensity: the peak current is in the order of several hundred milliamps, and the corresponding forward voltage is 1–2 V.

LEDs are constructed in various encapsulations, with or without a lens, with only one element or a whole array. They are widely used as alphanumeric display. Figure 9.5b shows a simple type. There are also LEDs that emit two colours, depending on the direction of the current; actually this device consists of two independent LEDs in a single encapsulation, connected in antiparallel.

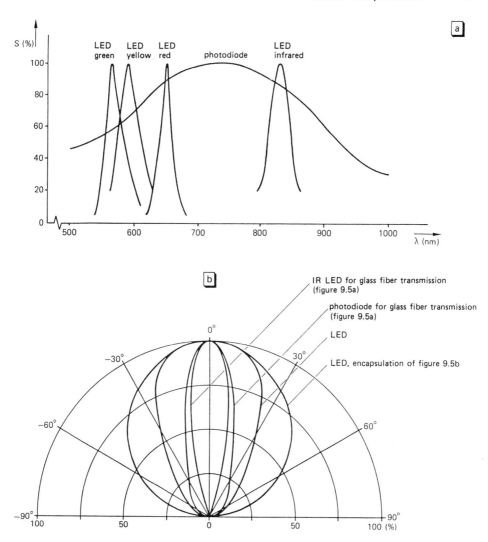

Figure 9.4 (a) Spectral response S and (b) polar sensitivity diagram of a photodiode and several light-emitting diodes; the shape of the polar diagram depends on the encapsulation (with or without built-in lens) of the device.

9.2 Circuits with pn diodes

PN diodes can be used for non-linear signal processing. There are two possibilities. The first is based on the exponential relationship between the voltage and the current, resulting in circuits for exponential and logarithmic signal converters and for analog

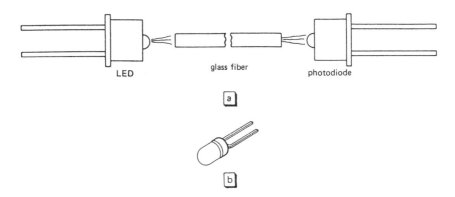

Figure 9.5 (a) An LED and a photodiode applied for glass fiber communication links; (b) common-type LED.

multipliers. We will discuss such circuits in Chapter 13. Looking to the diode characteristic of Figure 9.2, it appears that for voltages below V_k the diode behaves like a very large resistance; when conducting, the (differential) resistance is low (25 Ω around the point 1 mA). Further, the voltage across the diode remains almost V_k irrespective of the forward current. Similarly, for negative currents, the voltage equals the Zener voltage V_z. So, a diode behaves as an electronic switch; when conducting, the switch is closed, with a series voltage equal to V_k in the forward mode and V_z in the reverse mode (only for Zener diodes). If the voltage is below V_k, the switch is off and the current is zero. This rough approximation is used in this section to analyse a number of circuits commonly used with pn diodes.

9.2.1 Limiters

A limiter or clipper is a circuit that has a prescribed limited output voltage. Figure 9.6a shows a circuit for the limitation up to a maximum voltage; Figure 9.6b gives the corresponding transfer characteristic.

As long as the input voltage $v_i < V_k$, the current through the diode is zero, hence the output voltage $v_o = v_i$. When v_i reaches V_k, the diode becomes forward biased, conducts, and its voltage remains V_k. The diode current equals $(v_i - V_k)/R$, so with R the input current can be limited as well.

When the connections of the diode are reversed, the output voltage is limited to a minimum value, equal to $-V_k$. By connecting two diodes in antiparallel (Figure 9.6c), v_o is limited between $-V_k$ and $+V_k$ (Figure 9.6d). Such a limiter circuit is often connected across the input of a sensitive measurement system, for overload protection.

Limiting to another voltage other than V_k is achieved using Zener diodes (Figure 9.7).

Circuits with pn diodes   139

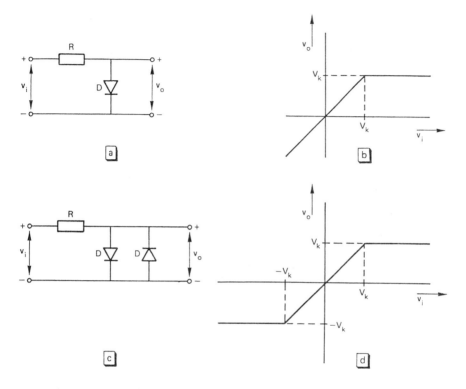

Figure 9.6 (a) A limiter for the limitation to a maximum voltage; (b) corresponding transfer characteristic; (c) limiter for both positive and negative maximum voltages; (d) corresponding transfer characteristic.

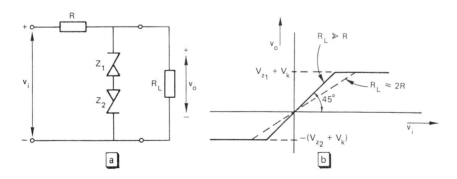

Figure 9.7 (a) A limiter with Zener diodes; (b) the transfer characteristic for two values of the load resistance R_L; $V_{z1} > V_{z2}$.

140 PN diodes

The maximum output voltage is that for which both diodes are conducting. The voltage across Z_1 is the Zener voltage V_{Z1} and that across Z_2 is the normal voltage drop V_k. The minimum voltage occurs when both diodes are conducting in the reverse direction; the voltage across Z_1 is the forward-biased voltage V_k and the voltage over Z_2 is its Zener voltage V_{Z2}. The output limits are not affected by a load resistance (Figure 9.7b). Between the two output limits (when both diodes are reverse biased), the transfer of the circuit is $v_o = v_i R_L/(R + R_L)$, the formula for the normal voltage divider circuit.

9.2.2 Peak detectors

A peak detector is a circuit that creates an output DC voltage equal to the peak value of a periodic input voltage. Figure 9.8a shows a peak detector circuit in its most simple form: a capacitor and a diode.

To explain the operation of this circuit, we start with an uncharged capacitor and consider the upgoing part of the input voltage (Figure 9.8b). At $t = 0$, the voltage across the diode is zero; the diode is reverse biased. The output v_o is also zero and stays zero until the input voltage v_i reaches V_k. Then the diode turns into the state of conduction which means that a current can flow and the voltage remains V_k. Hence, the output keeps track with the input, with a constant difference of V_k. The capacitor is charged by a (positive) current through the diode. As the current cannot flow in the inverse direction, the capacitor cannot be discharged. So, when the input voltage drops, having reached its peak value, the output voltage will remain constant. Its value equals the peak value of the input minus V_k. This situation is maintained until the input voltage exceeds $v_o + V_k$. At only a (slightly) higher value of the next peak, the diode will conduct and the capacitor is charged up to the new peak value, as is depicted in Figure 9.8c for a triangular input voltage.

The peak detector of Figure 9.8a can only detect the absolute maximum occurring. When we want the circuit to respond also to the peak value of a gradually decreasing amplitude, the capacitor must partly discharge between two successive peaks. This is achieved by connecting a resistor across the diode or the capacitor (Figure 9.9a). Now, even at reverse-biased diode, the capacitance discharges, resulting in a (small) decrease of the output voltage (Figure 9.9b). The capacitor is charged again at each new maximum that exceeds $v_o + V_k$. Thus, the circuit can follow the peak value of a periodic signal with a gradually decreasing amplitude as well. The price of this simple measure is a small ripple in the output signal, even at constant amplitude. To estimate an appropriate resistance value for a minimum output ripple, we assume a linear discharge curve (instead of a negative exponential curve). At $t = 0, T, 2T, \ldots$, v_o equals $\hat{v}_i - V_k$ and falls with a rate $\hat{v}_i/\tau$ (V/s). The output ripple is:

$$\Delta v = (\hat{v}_i - V_k) - \left[(\hat{v}_i - V_k)\left(\frac{\tau - T}{\tau}\right) \right] = (\hat{v}_i - V_k)\frac{T}{\tau}$$

Circuits with pn diodes 141

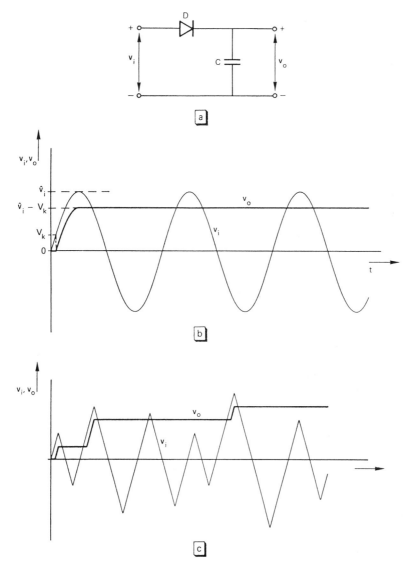

Figure 9.8 (a) Diode peak detector for positive peak values; (b) output voltage v_o for a sinusoidal input voltage v_i; (c) output voltage v_o for a triangular input voltage v_i.

The true value is somewhat smaller than this approximation, because v_o rises again from a moment slightly before $t = T, 2T, \ldots$. To minimize the ripple, τ must be large. To keep track of a decreasing amplitude, τ must be small. So, a compromise has to be found between ripple amplitude and the response time to decreasing amplitudes.

142 PN diodes

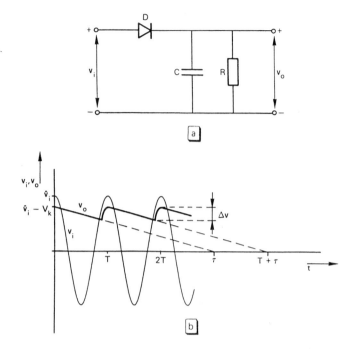

Figure 9.9 (a) Peak detector with discharge resistance, to allow detection of slowly decreasing amplitude; (b) determination of the ripple voltage.

The output voltage of a peak detector is always about V_k below the actual top value. When changing the polarity of the diode in Figure 9.8a, the output voltage responds to the negative of the peak value: $v_o = -\hat{v}_i + V_k$. An AC voltage whose amplitude is below V_k (about 0.6 V) cannot be detected with this simple circuit.

9.2.3 Clamp circuits

A clamping circuit shifts the average level of an AC signal up or down such that the top has a fixed value. Figure 9.10 shows the most simple circuit configuration and an example of a sinusoidal input and output voltage.

At $t = 0$, the capacitor is assumed to be uncharged. Obviously, the output voltage can never exceed the value V_k. Further, the capacitor can never be discharged at a diode voltage below V_k. With these conditions in mind, the operation of the circuit is clear. At increasing input v_i, the diode remains reverse biased; no current flows, hence $v_o = v_i$. As soon as v_i reaches V_k, the diode will conduct and behaves like a short-circuit: $v_o = V_k$. This situation is maintained until the input has reached its maximum. In the meantime, the capacitor is charged up to a voltage $v_c = \hat{v}_i - V_k$, but cannot discharge. Hence, the output follows the input, with a constant difference of just V_k, as long as the output does not exceed the value V_k.

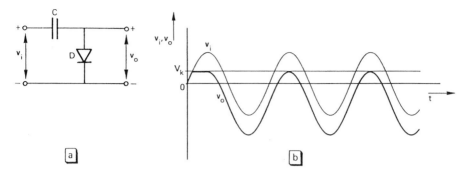

Figure 9.10 (a) Clamping circuit; (b) response to a sinusoidal input signal

As the capacitor can only be charged and not discharged, the circuit only operates well for constant or increasing signal amplitude. This drawback is eliminated by connecting a resistor across the diode or the capacitor, to allow some discharge of the capacitor between two successive maxima of the input signal (compare the peak detector of Section 9.2.2).

An arbitrary clamping level is obtained using a voltage source in series with the diode. The clamping level becomes $V + V_k$. Changing the polarity of the diode results in clamping on the negative peaks of the input. Figure 9.11 shows a circuit that clamps the negative peaks at a level $V - V_k$, and that responds to slowly decreasing amplitudes as well, as depicted in Figure 9.11b. Clamping circuits are used to set a proper DC level to AC signals. It can be combined with a peak detector (Figure 9.8a) to create a peak-to-peak detector; the output equals the peak-to-peak value of the input minus $2V_k$.

9.2.4 *DC voltage sources*

Most electronic systems require a more or less stable power supply voltage, sometimes a very stable and accurately known reference voltage. The voltage should not be affected by the current through it and so behave as an ideal voltage source. For high-stability requirements, special integrated circuits are available that give a stable output voltage for various output currents and over a wide power range. In situations where the requirements are less severe, simpler methods can be applied. In this section we introduce two DC voltage sources: the first uses an unstabilized DC voltage as a primary power source, the second an AC power source.

Voltage stabilizer with Zener diodes

As explained in Section 9.1, the Zener voltage is almost independent of the reverse current through the diode. This property is employed for the realization of a simple but rather stable voltage source (Figure 9.12).

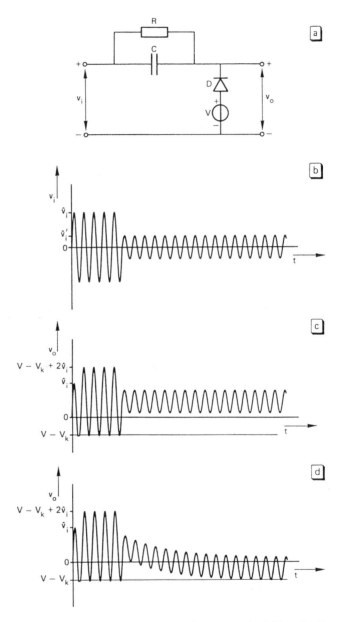

Figure 9.11 (a) Clamp circuit with discharge resistance and additional voltage source; (b) input signal with stepwise change of the amplitude by a factor of 3; (c) resulting output voltage for $R = \infty$; (d) output voltage for finite value of R.

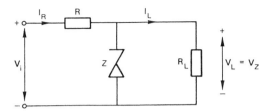

Figure 9.12 Voltage stabilization using a Zener diode.

The input of this circuit is a voltage V_i that may vary over quite a wide range. The diode is biased by a reverse current, whose value is set by resistor R. Obviously, under the condition $V_i > V_Z$, the output voltage V_L equals V_Z, which is almost independent of the current through the diode. When the circuit is loaded with R_L, a current flows through the load equal to $I_L = V_Z/R_L$. As long as $I_R > I_L$, enough current remains for the Zener diode to maintain its Zener voltage. The output is constant, irrespective of the load resistance. For this application, Zener diodes are available with a maximum allowable power from 0.3 to 50 W.

Graetz bridge

The bridge circuit of Graetz (Figure 9.13) is extensively used in systems that require a DC supply voltage, but where only AC power is available.

First, the primary AC voltage is reduced to a proper value v_i by the transformer.

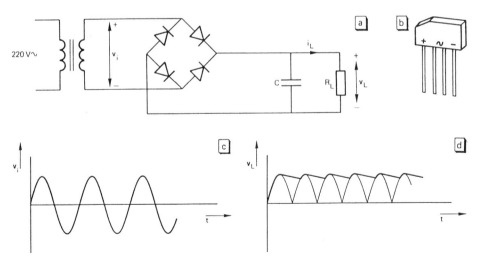

Figure 9.13 (a) Simple power supply circuit, consisting of a transformer, a diode bridge and a capacitor; (b) rectifier bridge in a single housing; (c) input voltage of the bridge circuit; (d) rectified output; the ripple depends on the load current and the capacitance C.

Due to the four diodes, the current I_L through R_L can only flow in the direction indicated in Figure 9.13a. The polarity of the output voltage v_L is always positive, irrespective of the polarity of v_i (Figure 9.13d, thin line). Due to this property the diode bridge is also called a double-sided rectifier; a single-sided rectifier simply clips the negative halves of the sine wave. The capacitor C forms, together with the diodes of the bridge, a peak detector (Section 9.2.2). The ripple in the output signal v_L (Figure 9.13d, bold line) is caused by discharge of the capacitor through the load resistance. To keep this ripple below a specified value (even at high load currents), a large value of C is required.

Rectifying bridges are available as a single component (Figure 9.13b), for a wide range of the maximum allowable current and power.

SUMMARY

Properties of pn diodes

- The conductivity of n-doped silicon is mainly determined by the concentration of free electrons; in p-doped silicon the conductivity is mainly determined by the concentration of holes.
- The relation between the current through a pn junction and the voltage is given by $I = I_0(e^{qV/kT} - 1)$. For $V < 0$, $I = I_0$, this is the leakage current or reverse current of the diode. This current is highly temperature dependent.
- A silicon diode becomes conductive at a forward voltage of about 0.6 V. At a constant current, the temperature sensitivity of the forward voltage is roughly -2.5 mV/K.
- The differential resistance of a pn diode is $r_d = kT/qI$; r_d is about 25 Ω at $I = 1$ mA.
- Zener diodes show an abrupt increase of the reverse current at the Zener voltage. They are used as voltage stabilizers and voltage reference sources.
- The leakage current of light-sensitive diodes or photodiodes is proportional to the intensity of the incident light. For proper operation the diode should be reverse biased.
- Light-emitting diodes or LEDs emit a light beam whose intensity is roughly proportional to the forward current. The colour or wavelength of the light is determined by the composition of the semiconductor material.

Circuits with pn diodes

- A pn diode can be used as an electronic switch; for $V_d < V_k$, its resistance is very high; when conducting, the resistance is low, and the voltage across the diode is about 0.6 V.
- Limiters with diodes use the property that the diode voltage is limited to about $V_k = 0.6$ V or (with Zener diodes) to the Zener voltage V_Z.

- With the diode–capacitor circuit of Figure 9.8, the peak value of a periodic signal can be measured. The output voltage is 0.6 V below the peak voltage. The response to decreasing amplitudes is improved by an additional resistor; this introduces a ripple voltage at the output.
- With the diode–capacitor circuit of Figure 9.10, the top of a periodic signal is shifted to a fixed value, irrespective of the signal amplitude. The response to slowly increasing amplitude is improved by an additional resistance.
- With a Zener diode, a reasonably stable voltage source can be realized.
- The Graetz bridge is a double-sided rectifier; with a relatively large capacitor this rectified voltage is converted into a DC voltage with a small ripple.

Exercises

Properties of pn diodes

9.1 What is the theoretical relationship between the current through a pn diode and the voltage across it?

9.2 Give the approximate value of the differential resistance of a pn diode at 1 mA, 0.5 mA and 1 μA. Also give the values of the conductance.

9.3 What is the change in the diode voltage at constant current, due to a temperature increase of 10°C?

9.4 What is the change in the diode voltage at constant temperature, due to an increase in current by a factor of 10?

9.5 To obtain the value of the series resistance r_s of a diode, the voltage is measured at two different currents: 0.1 and 10 mA. The results are 600 and 735 mV, respectively. Find r_s.

Circuits with pn diodes

9.6 Draw the transfer function (output voltage versus input voltage) of each of the circuits a–d depicted in the figure below. The diode has an ideal V–I characteristic: $V_k = 0.5$ V.

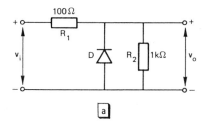
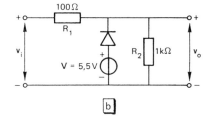

148 PN diodes

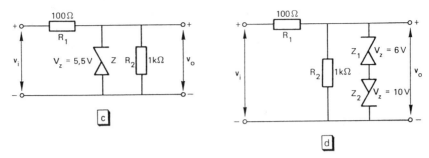

9.7 Find the output voltage of the peak detector given below, for sinusoidal input voltages with amplitude 6, 1.5 and 0.4 V and zero average value. The diode has ideal properties: $V_k = 0.6$ V.

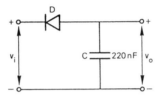

9.8 Refer to the circuit of the preceding exercise. $C = 0.1\ \mu F$; the input voltage is a sine wave with frequency 5 kHz. A resistor R is connected in parallel to C. Find the minimum value of R for which the ripple (peak-to-peak value) is less than 1% of the amplitude of v_i. The diode voltage V_k can be ignored with respect to the input amplitude.

9.9 Find the average value of v_o for each of the next circuits a–c in the figure below, for an input amplitude of 5 V. Assume $V_k = 0.6$ V.

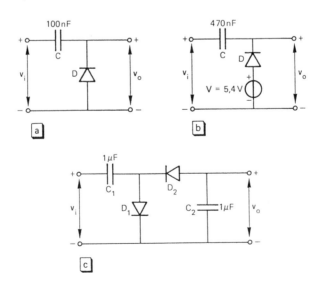

Exercises 149

9.10 The next figure shows a clamp circuit for clamping the negative tops of a signal to an adjustable voltage V. What is the range of V for a proper operation of this circuit? The input signal is a sinusoidal voltage with average zero and peak value $\hat{v}_i$; $V_k = 0.6$ V.

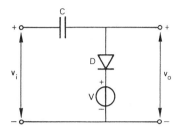

9.11 Make a sketch of the output signal of a Graetz bridge, loaded with a resistor of 20 Ω, and connected to:
(a) an AC voltage of 50 Hz, rms value 10 V;
(b) a DC voltage of +10 V;
(c) a DC voltage of −10 V.
What is the ripple voltage for these three cases? $V_k = 0.6$ V.

9.12 The same questions as in Exercise 9.11 above, but now a capacitance of 1500 μF is connected to the bridge.

9.13 Find the minimum value of R_L in the circuit shown in the figure below, for which the output voltage remains just 5.6 V.

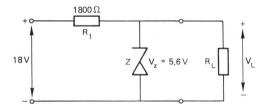

10 Bipolar transistors

A bipolar transistor consists of three layers of alternately p-type and n-type silicon. By making the middle layer of these three very thin, this component offers the possibility of amplifying electronic signals. The first part of this chapter starts with an explanation of the operating principle and proceeds with a description of the transistor's behaviour as an active electronic component. The second part of this chapter gives some examples of amplifier circuits with bipolar transistors.

10.1 Circuits of bipolar transistors

10.1.1 *Construction and characteristics*

Figure 10.1 shows a schematic structure of a bipolar transistor. In accordance with the successive materials, two types of transistors exist: npn transistors and pnp transistors. The same figure names the respective parts (the corresponding terminals have identical names) and shows the circuit symbol of both types.

As a first step, we should consider the transistor as a series of two pn junctions (or pn diodes) with one part in common. One of the diodes is connected to a reverse voltage, the other to a forward voltage. The common part is called the base of the transistor, the adjacent part of the reverse-biased diode is the collector and the adjacent part of the forward-biased diode is the emitter. Due to the forward voltage of the base–emitter diode, a current flows through that junction. For an npn transistor, this current is composed of electrons, flowing from the emitter to the base region. Under normal conditions, these electrons will leave the transistor via the base connection; furthermore, no current will flow through the base–collector junction because of the reverse voltage.

The situation changes significantly when the width of the base region is made very thin, as indicated in Figure 10.1. The greater part of the electrons that enter the base will survive a journey through the base and enter the collector region; once arrived there, they are pulled into the collector due to the electric field across the base–collector junction. Only a fraction of the electrons do not succeed in reaching the collector and leave the transistor via the base terminal. So, in an npn transistor, with properly connected voltages, electrons flow from emitter to collector, corresponding to a physical current from collector to emitter, thereby passing the forward-biased base–emitter junction and the reverse-biased base–collector junction.

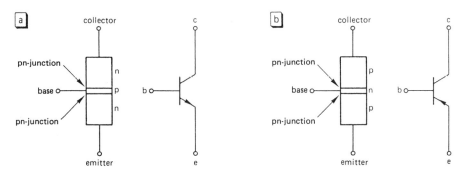

Figure 10.1 (a) Schematic structure and the symbol of an npn transistor; (b) the same for a pnp transistor.

A pnp transistor operates exactly the same; here holes flow from the emitter via the thin base region towards the collector. The polarity of the (physical) current is positive from emitter to collector, because holes are positive charge carriers.

We have seen that only a fraction of the total current flows through the base. This fraction is an important parameter of the transistor; it determines the current gain β of the transistor, defined as the ratio between the collector current I_C and the base current I_B, so: $I_C = \beta I_B$. The current gain depends strongly on the width of the base region, and ranges from 100 to 300 for low power transistors up to 1000 for special types (super β transistors). High power transistors have a much lower β, 20 or even less.

The current through the base–emitter diode (the emitter current I_E) still satisfies the diode equation (9.1). As the collector current is almost equal to the emitter current (the difference is only the small base current), the collector current satisfies Equation (9.1) as well: $I_C \approx I_0 e^{qV_{BE}/kT}$. So, the collector current is determined by the base–emitter voltage V_{BE}; it is independent of the base–collector voltage V_{BC}, under the condition of a reverse-biased base–collector junction. In other words, the collector behaves as a current source; the collector current does not depend on the collector voltage. More specifically, it is a voltage-controlled current source, because the current is directly related to the base–emitter voltage. Further, it is also a current amplifier with current gain β, because the collector current satisfies the relation $I_C = \beta I_B$.

It is important to distinguish between these two properties of a bipolar transistor: its output signal (the collector current) can be controlled by either an input voltage (V_{BE}) or an input current (I_B).

Figure 10.2 shows typical characteristics of a bipolar transistor: in (a) the relation between the collector current and the base–emitter voltage (at a fixed base–collector voltage) and in (b) the collector current versus the collector voltage (relative to the emitter voltage) with V_{BE} as a parameter.

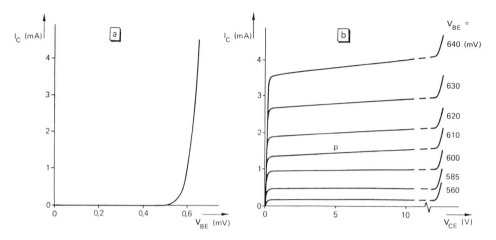

Figure 10.2 Typical characteristics of a bipolar transistor; (a) I_c versus V_{BE} at constant V_{BC}; (b) I_C versus V_{CE} for various values of V_{BE}. Collector breakdown occurs when the reverse voltage of the base–collector junction is increased too far.

Obviously, the actual characteristic differs in several aspects from the behaviour described above. First, the collector current (Figure 10.2a) increases by less than we should suppose from the exponential relation of the pn diode, due to the resistance of the emitter and base materials. The collector current (Figure 10.2b) is not fully independent of the collector voltage, due to the so-called Early effect. Further, but not visible in Figure 10.2, the current gain varies somewhat with the collector current; β decreases at very low and very high currents. Finally, a leakage current flows through the reverse-biased base–collector diode.

Other important parameters of a bipolar transistor are the maximum reverse voltage for the collector junction, the maximum forward voltage for the emitter junction and the maximum power $I_C \cdot V_{CE}$. The maximum power ranges from roughly 300 mW for the smaller types to over 100 W for power transistors with forced cooling. Figure 10.3 shows some transistors in various encapsulations.

Figure 10.3 Some examples of various transistors; (a) and (b) low power transistors (up to about 400 mW); (c) medium power (about 10 W); (d) high power transistor (about 100 W).

10.1.2 Signal amplification

To employ a bipolar transistor for linear signal amplification, it must be biased correctly, which means that all voltages between the terminals and consequently all currents through the terminals should have a proper value. This state is called the bias point or operating point of the transistor, for instance point P in Figure 10.2b. To bias the transistor, auxiliary power supply sources are required, such as batteries. A proper biasing is the first condition for adequate operation as a signal amplifier. The currents and voltages of the transistor must satisfy the equations:

$$I_E = I_0 e^{qV_{BE}/kT}$$
$$I_B = I_C/\beta \tag{10.1}$$
$$I_B + I_C = I_E$$

Once biased, the signals (voltages or currents) are superimposed on the bias voltage and current; they are considered as imposed fluctuations around the bias point. The signal can be applied to the transistor, for instance by varying the base–emitter voltage or the base current. Whichever variation is made, Equation (10.1) is still valid. So, when varying one of the transistor currents or voltages, all other quantities vary. To avoid non-linearity, the fluctuations are kept relatively small. The description of this small-signal behaviour is usually given in terms of the small-signal equations, found by differentiating Equation (10.1) at the bias point:

$$i_e = \frac{qv_{be}}{kT} I_E = \frac{v_{be}}{r_e} = g v_{be}$$

$$i_b = \frac{i_c}{\beta} \tag{10.2}$$

$$i_b + i_c = i_e$$

To distinguish between bias quantities and small-signal quantities, the former are written in capitals, the latter in lower case: i_e stands for dI_E or ΔI_E, v_{be} for dV_{BE}, etc. The parameter g (A/V or mA/V) is the transconductance of the transistor, and represents the sensitivity of the collector current to changes in the base–emitter voltage. Another notation is its reciprocal value, r_e, the differential emitter resistance (comparable to the differential resistance r_d of a diode).

The transistor's ability to amplify signal power follows from the next consideration. Assume the input terminal is the base, the output terminal is the collector. Suppose the input signal is the (change in the) base–emitter voltage v_{be}. Further, the collector is connected to a resistor R, so that the voltage across R equals $i_c R$. The input signal power is $p_i = i_b v_{be}$, the output signal power is $p_o = i_c^2 R$. Hence, the power transfer equals $p_o/p_i = i_c^2 R / i_b v_{be} = \beta g R$, a value that can easily be much larger than 1.

154 Bipolar transistors

Equation (10.2) describes the signal behaviour, irrespective of the biasing equation (10.1). To analyse transistor circuits, we use a transistor model based only on the small-signal equations. An example of such a model is depicted in Figure 10.4a, the T-equivalent circuit. By applying Kirchhoff's rules, it is easily verified that this model corresponds to Equation (10.2).

The model can be extended for a more precise description of the transistor. For example, the model depicted in Figure 10.4b accounts for the base resistance r_b; it is the resistance of the silicon between the base terminal and the internal base contact denoted as b'. The corresponding circuit equations are $i_e r_e = v_{b'e}$ and $i_b r_b = v_{b'b}$.

It is also possible to add capacitances to the model, to make it valid also for high signal frequencies. The next section illustrates the use of transistor models for the analysis of some basic electronic circuits with bipolar transistors.

10.2 Circuits with bipolar transistors

In the preceding section we saw that a transistor can only operate properly when it is biased correctly and that signals are merely fluctuations around the bias point. In this section we discuss a number of basic transistor circuits, by analysing both the biasing and the small-signal behaviour.

10.2.1 *Voltage-to-current converter*

Figure 10.5 shows how to use a bipolar transistor for a voltage-to-current converter; the input voltage V_i is converted to an output current I_o.

First we look at the bias of the circuit (input signal voltage zero). Biasing is performed by two auxiliary voltage sources V^+ (positive) and V^- (negative). The value of V^+ is such that for all possible values of V_i, the base–collector voltage remains positive (to maintain that junction's reverse bias). The output current I_o of this circuit is identical to the collector current I_C, which is almost equal to the emitter current I_E: $I_E = (V_E - V^-)/R_E$. As $V_B = V_i = 0$, the emitter voltage V_E is about -0.6 V (a forward-biased diode junction).

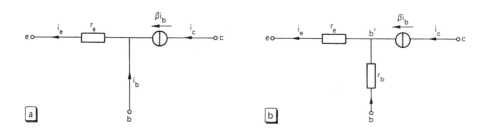

Figure 10.4 (a) A small-signal model of a bipolar transistor; (b) extended model, accounting for the base resistance.

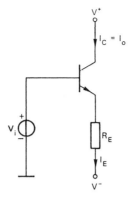

Figure 10.5 A voltage-to-current converter.

When the input voltage V_i changes, the emitter voltage V_E changes; the difference V_{BE} remains almost unchanged; hence I_E varies, as does I_o.

Example 10.1
Suppose $V^+ = 10$ V; $V^- = -10$ V and $R_E = 4.7$ kΩ. At zero input voltage, V_E is about -0.6 V, hence $I_E = (-0.6 + 10)/4700 = 2$ mA; this is also the output current I_o. For an input voltage of $V_i = 1$ V, V_E becomes $1 - 0.6 = 0.4$ V, and thus $I_E = 10.4/4700 = 2.2$ mA. For $V_i = -1$ V, $V_E = -1.6$ V and $I_E = 8.4/4700 = 1.8$ mA. Apparently, the voltage-to-current transfer of this circuit is about $1/R_E = 0.21$ mA/V.

So far we have assumed a constant base–emitter voltage of 0.6 V. However, as the current through the transistor changes, the base–emitter voltage also changes a little. To estimate the significance of this effect, we must make a more precise analysis. To that end, the transistor in Figure 10.5 is replaced by the model of Figure 10.4a, resulting in the circuit of Figure 10.6a.

The voltages V^+ and V^- of the power sources are constant, and have a very low internal resistance; the fluctuation of these voltages is zero, and so they can be connected to ground in the small-signal model. The signal model is redrawn in Figure 10.6b. The transistor equation (10.2) is still valid. Using this equation, the output (signal) current i_o can be calculated as a function of the input signal voltage v_i, for instance:

$$v_i = (i_b + \beta i_b)(r_e + R_E)$$
$$i_o = \beta i_b$$

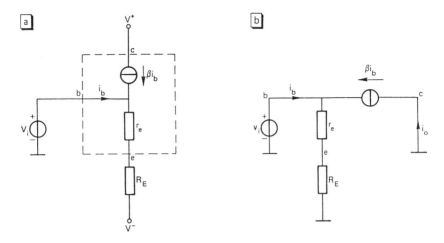

Figure 10.6 (a) A model of the voltage-to-current converter of Figure 10.5, where the transistor is replaced by the model of Figure 10.4a; (b) the corresponding small-signal model. Voltages from the auxiliary power supplies are constant, hence have zero value in the small-signal model.

from which follows:

$$\frac{i_o}{v_i} = \frac{\beta}{(1+\beta)(r_e + R_E)}$$

This expression accounts for a finite current gain β and a non-zero emitter differential resistance. For $\beta \to \infty$ and $r_e \to 0$, the transfer is $1/R_E$, a value found earlier.

Example 10.2
Suppose the transistor in Example 10.1 has a current gain $\beta = 100$; r_e is 12.5 Ω (because $I_E = 2$ mA). The transfer is $2.10 \cdot 10^{-4}$ A/V, which is almost equal to the approximated value $1/R_E = 2.13 \cdot 10^{-4}$.

Apparently, for a quick analysis of the circuit, we may consider V_{BE} being constant (0.6 V), β infinite and r_e small compared to R_E.

10.2.2 *Voltage amplifier stage with base-current bias*
The base of the transistor in the amplifier circuit shown in Figure 10.7 serves as the input terminal; the output signal is identical to the collector voltage.

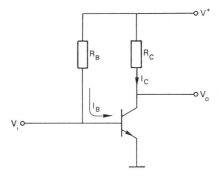

Figure 10.7 A voltage amplifier stage with base-current biasing; the bias currents are I_C and I_B.

The bias is performed by V^+ and a resistor R_B. The emitter is connected to ground, thus $V_E = 0$ and $V_B = 0.6$ V. The base current which flows through R_B equals $(V^+ - 0.6)/R_B$. The collector current is β times the base current. The output bias voltage (V_C) is $V^+ - I_C R_C$.

Example 10.3
Given: $V^+ = 15$ V; $\beta = 100$. Find the value for R_B for which the bias collector current is 1 mA, and find a proper value of R_C.

The base current should be 1 mA/100 = 10 µA: $I_B = (15 - 0.6)/R_B = 10^{-5}$, or $R_B = 1.4$ MΩ. The output voltage cannot exceed V^+ and may not drop below V_i to prevent the base–collector junction from being forward biased. To obtain a maximum range for output variations, set V_o halfway between the outermost values, about 7.5 V. This is achieved for $R_C = (15 - 7.5)/1 \cdot 10^{-3} = 7.5$ kΩ.

The input and output terminals have the emitter in common; such a circuit is called a common emitter circuit or CE circuit.

If the input of this amplifier is connected directly to a voltage source, the bias point would change significantly. The same happens when connecting a load to the output (for instance a resistance to ground). To avoid this problem, capacitors are connected in series with the input and the output terminals (coupling capacitors), see C_1 and C_2 in Figure 10.8. The capacitance of these components is so large that they act as a short-circuit for signals.

To calculate the voltage transfer of this amplifier, we make a model of the circuit (Figure 10.9a). This time we use the model of Figure 10.4b, to account for the internal base resistance. To simplify the analysis, we make the following assumptions: both C_1 and C_2 may be regarded as short-circuits for the signal frequencies; the source resistance $R_g = 0$ (ideal voltage source) and $R_L \gg R_C$ (neglecting the load). The model reduces to that of Figure 10.9b.

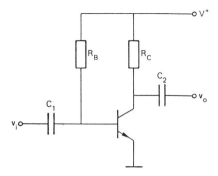

Figure 10.8 The voltage amplifier from Figure 10.7, extended with capacitors to connect source and load without affecting the bias.

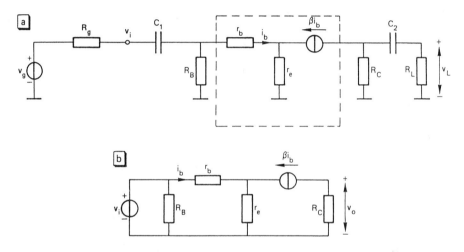

Figure 10.9 (a) Model of the circuit from Figure 10.8; (b) simplified model: $R_g = 0$ (hence $v_i = v_g$), $R_L \gg R_C$ (hence $v_L = v_o$) and the couple capacitors behave as short-circuits.

The voltage transfer $A_v = v_o/v_i$ is found from the next equations:

$$v_o = -\beta i_b R_C$$
$$v_i = i_b r_b + (i_b + \beta i_b) r_e$$

so:

$$A = \frac{v_o}{v_i} = -\frac{\beta R_C}{r_b + (1+\beta) r_e}$$

For most transistors, $\beta \gg 1$ and $r_b \ll \beta r_e$, hence $A \approx -R_C/r_e$. The minus sign is essential: at increasing input voltage, the collector current increases and the collector voltage (output) decreases with respect to the bias point.

If we want to take into account the effect of the load resistance R_L, in the model of Figure 10.9b, the resistance R_C is simply replaced by $R_C//R_L$ (the two resistors in parallel). The voltage gain now is roughly $-(R_C//R_L)/r_e$.

The approximations made in the previous analysis are allowed in most cases: the discrete resistance values (E12 series, Section 7.1.1), their tolerances and the tolerances of the transistor parameters make it useless to carry out a more precise analysis.

The biasing of the transistor depends fully on the parameter β. Unfortunately, β varies from transistor to transistor (even of the same type) and it is temperature dependent. Therefore, the biasing and (via r_e also the voltage gain) is not stable and not reproducible. To obtain a more accurate voltage transfer, irrespective of the transistor parameters, other biasing methods must be used, as illustrated in the following sections.

10.2.3 *Voltage amplifier stage with base-voltage bias*

In the amplifier stage shown in Figure 10.10, the base voltage is fixed with a voltage divider circuit $R_1 - R_2$ across the supply voltage V^+. Further, a resistor R_E is inserted between the emitter terminal and ground.

The voltage across R_E is $V_B - 0.6\,\text{V}$, resulting in a collector current $I_C \approx I_E = (V_B - 0.6)/R_E$. The coupling of the input and output voltages is done

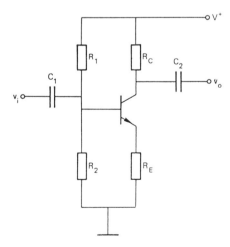

Figure 10.10 A voltage amplifier circuit with base-voltage biasing. The collector bias current is almost independent of the transistor parameters.

through couple capacitors, to prevent the bias from being affected by source and load circuits.

A further analysis of the amplifier comprises the following steps: biasing, signal voltage transfer, input resistance and output resistance.

Biasing

The resistance values R_1 and R_2 determine the base voltage: $V_B = V^+ R_2/(R_1 + R_2)$, where the bias current is neglected. This is only allowed when the resistance values of the voltage divider are not too high; the current through R_2 must be large compared to I_B. As $V_{BE} \approx 0.6$ V, $V_E = V_B - 0.6$ V. The current through R_E is V_E/R_E and equals the collector current. The (bias) output voltage is fixed at $V^+ - I_C R_C$, which should always be larger than V_B.

Small-signal voltage gain

The voltage gain is calculated using a small-signal model of the circuit (Figure 10.11). The coupling capacitors are considered as short-circuits.

The voltage gain is found using (for instance) the equations:

$$v_i = i_b r_b + (i_b + \beta i_b)(r_e + R_E)$$
$$v_o = -\beta i_b R_C \text{ (without load, } i_o = 0)$$

so that

$$A = v_o/v_i = \frac{-\beta R_C}{r_b + (1+\beta)(r_e + R_E)} \approx \frac{-R_C}{r_e + R_E} \approx \frac{-R_C}{R_E}$$

Input resistance

First, we put $R_1 // R_2 = R_p$, for simplicity. Elimination of i_b from the following equations:

$$v_i = (i_i - i_b) R_p$$
$$v_i = i_b r_b + (i_b + \beta i_b)(r_e + R_E)$$

results in:

$$v_i/i_i = R_p \frac{r_b + (1+\beta)(r_e + R_E)}{R_p + r_b + (1+\beta)(r_e + R_E)} = R_p // \{r_b + (1+\beta)(r_e + R_E)\}$$

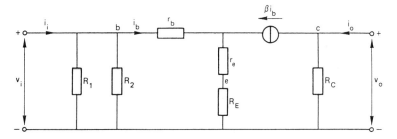

Figure 10.11 A model of the voltage amplifier circuit from Figure 10.10.

The input resistance of the circuit is equal to a resistance with value $r_b + (1+\beta)(r_e + R_E) \approx \beta R_E$ in parallel with both biasing resistors: $r_i = \beta R_E // R_1 // R_2$.

Output resistance

The output resistance at short-circuited input terminals is found from the equations:

$$i_o = \beta i_b + v_o/R_C$$
$$v_i = i_b r_b + (1+\beta) i_b (r_e + R_E) = 0$$

From the last equation, it follows that $i_b = 0$, hence $r_o = v_o/i_o = R_C$, which is simply the value of the collector resistance.

Resistor R_E has a favourable effect on the biasing, which is now almost independent of the transistor parameters. However, the gain factor is reduced to $-R_C/(R_E + r_e)$, compared to $-R_C/r_e$ when R_E would be zero. To combine a stable bias point and high voltage gain, R_E is decoupled by a capacitor C_E parallel to R_E (Figure 10.12a). This capacitor does not affect the (DC) bias of the circuit; for AC signals it behaves as a short-circuit.

The voltage transfer and the input and output impedances of this new circuit can easily be calculated by replacing R_E in the previous expressions with $Z_E = R_E/(1 + j\omega R_E C_E)$. The (complex) transfer function is now:

$$V_o/V_i = \frac{-R_C}{r_e + R_E/(1 + j\omega R_E C_E)} = \frac{-R_C}{r_e + R_E} \cdot \frac{1 + j\omega R_E C_E}{1 + j\omega R_E C_E r_e/(R_E + r_e)}$$

$$\approx \frac{-R_C}{R_E} \cdot \frac{1 + j\omega R_E C_E}{1 + j\omega r_e C_E}$$

For signals with a high frequency ($\omega \gg 1/r_e C_E$), the voltage transfer is about $V_o/V_i = -R_C/r_e$, the desired high value. Figure 10.12b shows the amplitude transfer as a function of the frequency.

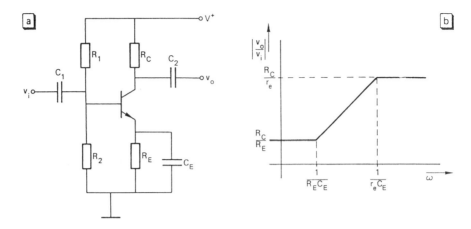

Figure 10.12 (a) A voltage amplifier stage with decoupled emitter resistance; (b) corresponding amplitude transfer characteristic.

Example 10.4

Given a power supply of 15 V, find proper values for the components in Figure 10.12a, such that the voltage gain is $A = -100$, for signal frequencies from 60 Hz.

Bias: let I_E be (for instance) 0.5 mA; this means $r_e = 1/40 I_E = 50\ \Omega$ and $R_C = A \cdot r_e = 100 \cdot 50 = 5\ \text{k}\Omega$. The voltage drop across R_C is $I_C \cdot R_C \approx I_E \cdot R_C = 2.5\ \text{V}$, so the collector voltage is $15 - 2.5 = 12.5$ V. The voltage may vary around this value, but never drops below the base voltage; so make V_B equal to (for instance) 7.5 V, hence $V_E = 6.9$ V and $R_E = 6.9/0.5 \cdot 10^{-3} = 13.8\ \text{k}\Omega$. The resistances R_1 and R_2 must be equal in order to reach $V_B = 7.5$ V. Take 100 kΩ: high enough for a reasonable input resistance (that should be as high as possible), low enough to neglect the influence of the base current on the bias point.

Choice of C_E: the lowest frequency f_L must satisfy the inequality $2\pi f_L r_e C_E \gg 1$, so $C_E \gg 1/(2\pi \cdot 60 \cdot 50) \approx 53\ \mu\text{F}$.

10.2.4 *Emitter follower*

We noticed that the voltage between the base and the emitter is almost constant; when the base voltage varies, the emitter voltage varies by the same amount. This property is used in a circuit that is denoted as emitter follower (Figure 10.13). The base is the input of the circuit and the emitter acts as an output. Although the voltage transfer of an emitter follower is only one, it has useful other properties. This becomes clear when we calculate the input and output resistance of an emitter follower. In the model (Figure 10.13b), the bias resistors R_1 and R_2 are omitted, to simplify the calculations. As they are in parallel with the input, they do not affect the voltage transfer and the output impedance.

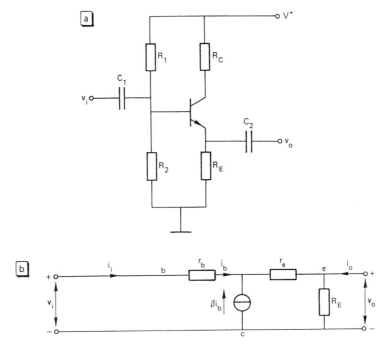

Figure 10.13 (a) An emitter follower; the voltage gain is almost 1; (b) a model of the emitter follower.

Voltage transfer (without load resistor, $i_o = 0$): elimination of i_b from the two equations:

$$v_i = i_b r_b + (i_b + \beta i_b)(r_e + R_E)$$
$$v_o = (i_b + \beta i_b) R_E$$

results in:

$$\frac{v_o}{v_i} = \frac{(1+\beta) R_E}{r_b + (1+\beta)(r_e + R_E)} \approx \frac{R_E}{r_e + R_E}$$

For R_E large compared to r_e, the transfer is almost 1.

Input resistance (when unloaded, $i_o = 0$): elimination of i_b from the following equations:

$$v_i = i_b r_b + (i_b + \beta i_b)(r_e + R_E)$$
$$i_i = i_b$$

164 Bipolar transistors

results in:

$$v_i/i_i = r_b + (1+\beta)(r_e + R_E) \approx \beta(r_e + R_E)$$

The resistances R_1 and R_2 are in parallel with the input, so the total input resistance is $r_i = \beta(r_e + R_E)//R_1//R_2$.

Output resistance (at short-circuited input, $v_i = 0$): elimination of i_b from the equations:

$$0 = i_b r_b + (i_b + \beta i_b) r_e + v_o$$
$$v_o = (i_b + \beta i_b + i_o) R_E$$

finally results in:

$$r_o = v_o/v_i = \frac{R_E[r_e + r_b/(1+\beta)]}{R_E + r_e + r_b/(1+\beta)} = R_E // \left(r_e + \frac{r_b}{1+\beta}\right) \approx r_e$$

From this analysis it appears that an emitter follower has a high input resistance (roughly β times R_E) and a low output resistance (about r_e). The emitter follower is therefore suitable as a buffer amplifier stage between two voltage transfer circuits, for minimizing load effects.

Example 10.5
Let the bias current of an emitter follower be 1 mA. Other transistor parameters are $\beta = 200$ and $r_b = 100\,\Omega$. R_E is 10 kΩ. For this situation, the voltage transfer is approximately $10^4/(25 + 10^4) \approx 0.9975$; the input resistance is $200(10^4 + 25) \approx 2$ MΩ and the output resistance is 25 Ω. The factor $r_b/(1+\beta)$ can be neglected compared to r_e.

10.2.5 *Differential amplifier stage*

A serious disadvantage of all the circuits with coupling capacitors discussed so far is that they cannot handle DC signals. The coupling capacitors separate the bias quantities (DC) from the signal quantities (AC). In DC amplifiers, coupling capacitors cannot be used; hence, the input voltage source and the load become part of the bias. Even when we succeed in solving this problem, there is still another one: the temperature. We know that at constant (bias) current, the base–emitter voltage varies with temperature (-2.5 mV/K). Such slow changes are not distinguished from slowly changing input signals.

Most of these problems are solved with the circuit of Figure 10.14. The basic idea is compensation of the temperature sensitivity of the transistor by a second, identical transistor. We will first discuss the biasing of this amplifier stage and then its signal behaviour.

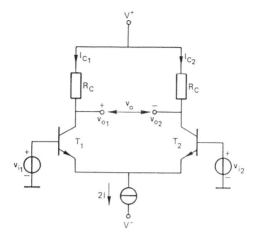

Figure 10.14 A differential amplifier stage: only the difference between v_{i1} and v_{i2} is amplified.

Bias

The base terminals of the transistors T_1 and T_2 act as input terminals of the amplifier; with respect to the biasing, both base contacts are at ground potential. The bias current is supplied by a current source with current $2I$ (for instance, the circuit of Figure 10.5 with a fixed input voltage). The base–emitter voltages of both transistors are equal, so their currents are the same: $I_{C1} = I_{C2}$. The collector voltage is set by V^+ and R_C: $V_C = V^+ - I_C R_C$. Of course this voltage must be higher than the highest occurring base voltage.

Signal properties

As long as $v_{i1} = v_{i2}$, the base–emitter voltages of both transistors remain equal, hence their collector currents remain I each. The amplifier is not sensitive to equal input signals or common mode signals. However, for $v_{i1} = -v_{i2}$, the collector currents of T_1 and T_2 will change but their sum remains $2I$. For positive v_{i1}, the collector current through T_1 increases with v_{i1}/r_e, whereas the collector current of T_2 decreases with the same amount: $-v_{i2}/r_e$. Hence, the output voltages (collector voltages) of T_1 and T_2 are $v_{o1} = -v_{i1}R_C/r_e$ and $v_{o2} = v_{i2}R_C/r_e$, respectively. A differential voltage $v_d = v_{i1} - v_{i2}$ between both inputs causes a differential output $v_o = v_{o1} - v_{o2} = -R_C v_d/r_e$.

The transfer for differential voltages is $-R_C/r_e$, just as for the normal CE amplifier stage. The transfer for common input signals appears to be zero. Due to asymmetry of the circuit (unequal transistor parameters), the common mode transfer may slightly differ from zero. The ratio between the differential transfer and the common mode transfer is the common mode rejection ratio or CMRR of the differential amplifier (Section 1.2).

This circuit, also known as long-tailed pair, forms the basic circuit for almost all types of operational amplifiers (Chapter 12).

SUMMARY

Properties of bipolar transistors

- There are two types of bipolar transistors: pnp and npn transistors.
- The three parts of a bipolar transistor are emitter, base and collector; the currents through the corresponding terminals are denoted as I_E, I_B and I_C.
- An essential condition for the proper operation of a bipolar transistor is a very narrow base region.
- The current gain of a bipolar transistor is $\beta = I_C/I_B$. Typical values of β are between 100 and 300. $I_E \approx I_C$.
- The collector current satisfies the equation $I_C \approx I_0 e^{qV_{BE}/kT}$ (like that of the pn diode). If the base–collector junction is reverse biased, I_C is almost independent of the collector voltage.
- The bipolar transistor functions as a voltage-controlled current source (from V_{BE} to I_C) or a current amplifier with current gain β.
- A bipolar transistor is biased with fixed DC voltages and currents. Signals are treated as fluctuations around the bias point.
- The change in the collector current, i_c, depends only on the change in base–emitter voltage: $i_c = g \cdot v_{be}$, with $g = 1/r_e$ the transconductance of the transistor and r_e the emitter differential resistance: $r_e = kT/qI_C$. Hence $r_e \approx 25\,\Omega$ at $I_C = 1\,\text{mA}$ (compare with r_d of a pn diode).

Circuits with bipolar transistors

- The base–emitter junction of a bipolar transistor is forward biased; the base–collector junction is reverse biased.
- The analysis of an electronic circuit with bipolar transistors can be split up into two parts: the biasing and the small-signal behaviour.
- To analyse the small-signal behaviour of a transistor circuit, a transistor model is used where all fixed voltages (for instance, from the power source) are zero.
- To avoid the source and load affecting the bias point, they are coupled to the circuit via couple capacitors. Consequently, this sets a lower limit to the signal frequency.
- The voltage transfer of a CE stage with emitter resistance R_E and collector resistance R_C is roughly $-R_C/R_E$. With a decoupled emitter resistance, the

transfer becomes approximately $-R_C/r_e$. The low frequency cut-off point is about $\omega = 1/C_E r_e$.
- An emitter follower has a voltage transfer of one, a high input resistance (about βR_E) and a low output resistance (about r_e). The circuit is used as a voltage buffer.
- The first part of a differential amplifier is designed for a low offset and drift and a high common mode rejection ratio. The latter is limited by the asymmetry of the components.

Exercises

Properties of bipolar transistors

10.1 Give the relation between the collector current I_C and the base–emitter voltage V_{BE} of a bipolar transistor, used in a linear amplifier circuit. Mention the conditions for V_{BE} and V_{BC} for a proper operation.

10.2 The current gain of a bipolar transistor is $\beta = 200$. Find the base current and the collector current, for an emitter current I_E of 0.8 mA.

10.3 What phenomenon is described by the Early effect?

10.4 In the circuit shown in the figure below, $V_{BE} = 0.6$ V. Find the output voltage V_o. The base current can be ignored.

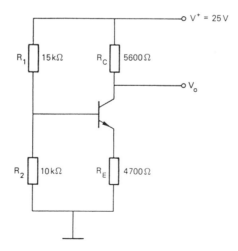

168 Bipolar transistors

10.5 In the following figure, $V_{BE} = 0.6$ V for the npn transistor, and $V_{EB} = -V_{BE} = 0.6$ V for the pnp transistor. Find the output voltage V_o. Hint: first calculate I_{E1}, then I_T and finally I_{E2}.

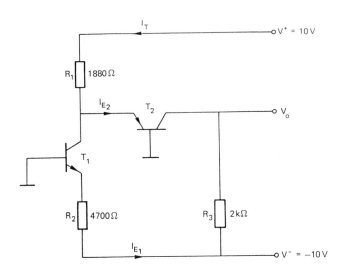

Circuits with bipolar transistors

10.6 Given the next three voltage amplifier stages a–c in the figure below, for all transistors $\beta = \infty$, $r_b = 0\,\Omega$ and $V_{BE} = 0.6$ V (for the pnp: -0.6 V). Calculate the emitter differential resistance of each of the transistors.

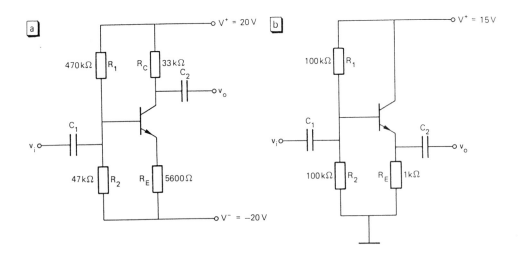

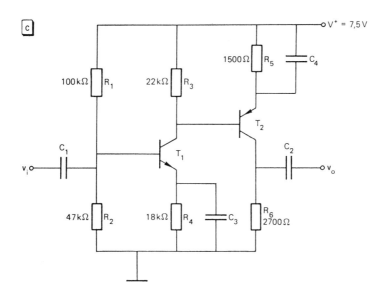

10.7 Refer to the preceding exercise. Find the voltage transfer of each of these circuits. The coupling capacitors may be considered as a short-circuit; furthermore, $r_e \ll R_E$.

10.8 The voltage transfer v_o/v_i in circuit 10.6b is not exactly equal to one, due to $r_e \neq 0$. What is the deviation from 1?

10.9 Due to a finite value of β and the fact that $r_b \neq 0$, the voltage transfer v_o/v_i of circuit 10.6b deviates somewhat from 1. Find that deviation, for $\beta = 100$ and $r_b = 100\,\Omega$.

10.10 Calculate the input resistance and the output resistance of circuit (b) as shown in Exercise 10.6, taking into account the components R_1, R_2 and the transistor parameters β and r_b: $\beta = 100$ and $r_b = 100\,\Omega$. Approximations up to $\pm 5\%$ are allowed.

10.11 In circuit (c) of Exercise 10.6, the output of the first stage (the collector of the npn transistor) is loaded with the input of the second stage (the base of the pnp transistor). Calculate the reduction in the voltage transfer of the whole circuit, due to this load, compared to the value found in Exercise 10.7. Take $\beta = 100$.

10.12 Calculate the emitter differential resistance r_e of the transistor in circuit 10.6b, assuming $\beta = 100$.

11 Field-effect transistors

A field-effect transistor or FET is an active electronic component, suitable for linear signal amplification, similar to the bipolar transistor. Other possible functions of the field-effect transistor are a voltage-controlled resistance and an electronic switch. Although the physical mechanism of an FET differs completely from that of a bipolar transistor, it has a remarkable resemblance to it, when used as a signal-amplifying component; the FET is also biased with DC voltages and the signals are treated as fluctuations around the bias point. There are two main kinds of FETs: the junction field-effect transistor or JFET and the metal oxide semiconductor field-effect transistor or MOSFET. We will first explain the principle of operation and the main properties of an FET. In the second part of the chapter, some linear amplifier circuits will be described in which FETs are used as active components.

11.1 Properties of field-effect transistors

The phrase field-effect refers to the possibility of affecting the conductivity of a semiconductor material by an electric field. A JFET utilizes the particular properties of a pn junction to create a voltage-dependent conductivity. In a MOSFET the conductivity is affected by capacitive induction. In both cases, the concentration of free charge carriers (electrons or holes) is controlled by a voltage.

11.1.1 *Junction field-effect transistors*

In Chapter 9 we saw that the initial width of the depletion layer depends on the doping concentration of the materials. Further, the width can be modulated by the voltage across the junction. The JFET is based on this latter property. Figure 11.1 shows, schematically, the structure of a JFET. It consists of a p-doped (or n-doped) silicon substrate, with a thin layer of the complementary type. The layer is provided with two contacts, called source and drain. The path between these contacts is the channel. If the channel is n-type, the FET is an n-channel FET, otherwise it is a p-channel FET. In normal operation, the substrate and the channel are electrically separated from each other by a reverse voltage across the junction. The conductance of the channel between source and drain (the lateral conductance) depends on the channel dimensions, in particular the effective thickness, i.e. that part of the layer that

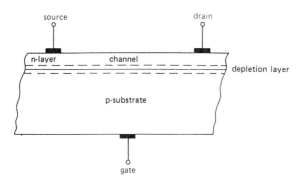

Figure 11.1 Schematic structure of an n-channel JFET.

contains free charge carriers. The depleted part of the channel does not contribute to the conductance. Upon increasing the reverse voltage across the pn junction, the depletion layer width increases as well, reducing the effective thickness of the channel and hence its conductance. The connection to the substrate is called the gate electrode or simply the gate.

To obtain a high sensitivity, that is a large change in the channel conductance when varying the gate voltage, the channel must be very thin, preferably around the width of a depletion layer (<1 μm). Modern technology offers the possibility of making such thin channel layers that can be completely depleted. At zero channel conductance, the FET is said to be pinched-off; the gate voltage at which this occurs is the pinch-off voltage V_p, an important parameter of the JFET.

Figure 11.2 shows the voltage–current characteristic of the channel, for low currents, and at various values of the gate–source voltage. Apparently, the JFET behaves in this region as a voltage-controlled resistance, the channel resistance. At pinch-off, the resistance is infinite; at $V_{GS} = 0$, the channel resistance has a relatively low value, typically in the order of several hundred ohms. The depletion layer width is determined by the voltage across the junction, and so by both the source and drain voltage (relative to the gate or the substrate). To explain the other curves in Figure 11.2, we should start with $V_{GS} = 0$, and see what happens when only the drain voltage is increased. A slight increase of V_{DS} will not affect the depletion layer width; its resistance remains constant. The drain current (from source to drain) increases linearly with the drain voltage, as with a normal resistor. This is the curve in Figure 11.2a at $V_{GS} = 0$ and in region 1 of Figure 11.2c. When further increasing the drain voltage, the voltage across the junction near the drain contact gradually increases as does the width of the depletion layer. The conductance of the channel therefore decreases. As the source–gate voltage is still zero, the width of the depletion layer in the vicinity of the source contact does not change. This explains the triangular shape of the depletion region in Figure 11.3a.

Due to the decreasing conductance, the drain current will increase less than

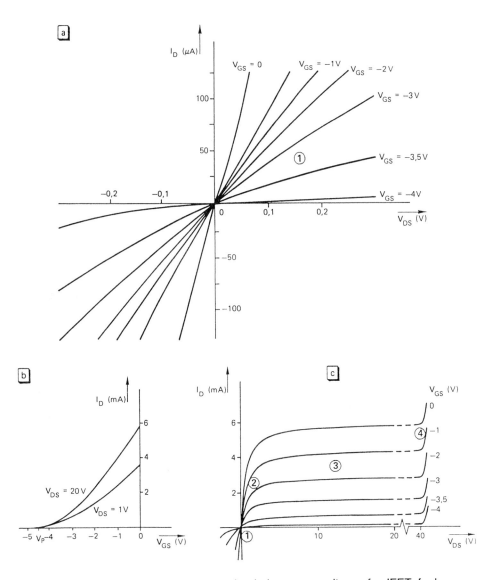

Figure 11.2 (a) The drain current versus the drain–source voltage of a JFET, for low current values, and for various values of the gate–source voltage; (b) drain current versus gate–source voltage for two values of V_{DS}; (c) drain current versus drain–source voltage for various values of V_{GS}. When the drain–source voltage becomes too high, breakdown occurs (region 4). Note that the voltages of an FET are usually given relative to the source voltage: so V_{GS} and V_{DS}.

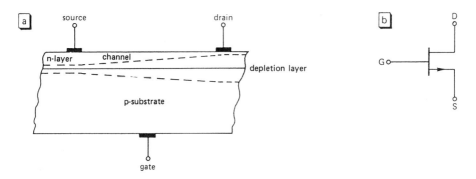

Fig. 11.3 (a) The shape of the depletion region for zero gate–source voltage and positive drain–source voltage; (b) circuit symbol for a JFET.

proportionally compared to the drain voltage (region 2 in Figure 11.2c). Finally, when the drain voltage is further decreased and thus the channel conductance is decreased, the current will reach a saturation level (region 3 in Figure 11.2c, upper curve). In this situation, the channel is almost pinched-off; the current flows through a narrow path near the drain, and cannot grow further, not even at higher drain voltage.

If we start the aforementioned process with a negative gate–source voltage, the initial channel width is smaller, corresponding to a higher value of the resistance. Consequently, the current–voltage curves in Figures 11.2a and c become less steep at a more negative gate–source voltage and the drain current is saturated at a much lower value (region 3, other curves).

In region 3, the drain current is independent of the drain voltage and determined only by the gate–source voltage. In this region, the JFET behaves as a voltage-controlled current source, similar to the bipolar transistor. This similarity is reflected in the circuit symbol of the JFET (Figure 11.3b).

Due to the different physical mechanism compared to that of the bipolar transistor, the JFET also requires a different way of biasing. In the bipolar transistor, one junction is forward biased, the other reverse biased. In the JFET, the channel must always be reverse biased to prevent conduction to the substrate. So, both source and drain must be reverse biased. This means that, for an n-channel JFET, both source and drain should be positive with respect to the gate (for a p-channel simply negative). An important consequence is the very low gate current, that consists of only the leakage current of the reverse-biased pn junction.

In the saturation region or pinch-off region, the theoretical relationship between the gate–source voltage and the drain current is:

$$I_D = I_{DS}\left(1 - \frac{V_{GS}}{V_p}\right)^2$$

with I_{DS} the current for $V_{GS} = 0$ (see also Figure 11.2b). Just as for the bipolar transistor, the transconductance g of a JFET is defined as the ratio between changes in I_D and V_{GS}:

$$g = \frac{dI_D}{dV_{GS}} = \frac{i_d}{v_{gs}} = \frac{-2}{V_p} \sqrt{I_{DS} I_D}$$

So, g is proportional to $\sqrt{I_D}$. A typical value for the pinch-off voltage is a few volts; I_{DS} ranges from several milliamps up to 100 mA. The transconductance of a JFET is, therefore, in the order of 1 mA/V at $I_D = 1$ mA. This is much less than the transconductance of the bipolar transistor at the same bias current. On the other hand, the gate current of the JFET is much lower than the base current; so the current gain of a JFET, I_D/I_G, is very high.

In most cases, the gate current of the JFET can be neglected; the small-signal behaviour of the JFET is described by only one equation:

$$i_d = g \cdot v_{gs}$$

The model of a JFET is very simple too (see Figure 11.4a). This model can be extended to account for all kinds of deviations from the ideal behaviour. As an example, the model of Figure 11.4b accounts for the influence of the drain voltage on the drain current (compare the Early effect in a bipolar transistor).

11.1.2 MOS field-effect transistors

The operation mechanism of a MOSFET differs in at least two aspects from that of the JFET. First, the channel conductivity is not controlled by the substrate voltage but by an isolated electrode, connected on top of the channel. The isolation consists of a very thin layer of silicon dioxide (Figure 11.5a). Secondly, the conducting channel is not a deposited layer but an induced one. At the surface of the silicon crystal, the

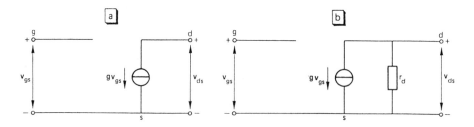

Figure 11.4 (a) A model of a JFET; (b) extended model, taking into account the drain differential resistance.

Properties of field-effect transistors

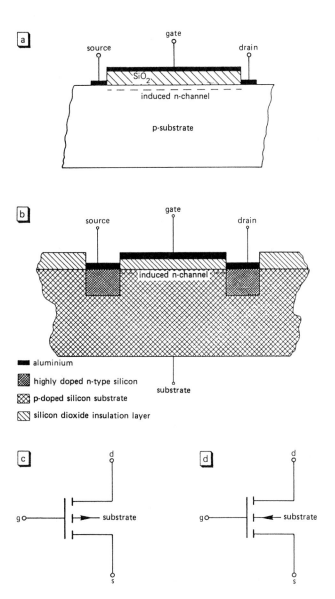

Figure 11.5 (a) Schematic structure of a MOSFET; (b) realization of an n-channel MOSFET; source and drain consist of highly doped n-type silicon with a very low resistance; (c) and (d) circuit symbols for an n- and p-channel MOSFET.

regular structure is broken; the atoms at the surface do not find neighbouring atoms to share the valence electrons. A large concentration of holes exists at the surface (which is still electrically neutral); the absent material acts as a donor. Electrons from the adjacent oxide easily recombine with those holes. When that happens, the top layer of the silicon becomes negatively charged, resulting in a more or less conductive channel near the surface. The concentration of electrons trapped at the surface of the p-silicon may become so high that the channel behaves as n-type silicon: this is called an inversion layer.

This phenomenon is employed for making a MOSFET. Two contacts, drain and source, form the end points of a channel that consists now of such an inversion layer. The structure is completed by a metal gate contact on top of the isolating oxide layer. The concentration of electrons in the channel can be decreased simply by a negative gate voltage; the negative voltage pulls holes from the p-region towards the surface where they recombine with the electrons, reducing the channel conductivity. At a sufficiently high gate voltage, the channel conductance becomes zero and the MOSFET is pinched-off, as in a JFET.

Apparently, this MOSFET type conducts at zero gate voltage; it is a normally on-type transistor, like a JFET (a bipolar transistor is normally off). It is possible to construct MOSFETs with an initially low charge carrier concentration in the induced channel. Only at a sufficiently positive gate voltage are enough electrons driven to the channel in order to make it conductive. This type is called a normally off MOSFET. Obviously, there are four types of MOSFETs: p- and n-channel, both normally on and normally off.

As the gate electrode is fully isolated from the drain and source electrodes by an oxide layer, the gate current is extremely small, in the order of 10 fA. Consequently, a MOSFET has an almost infinite current gain. A disadvantage of the MOSFET is the relatively low breakthrough voltage of the thin oxide layer (10–100 V); therefore, the gate contact may not be touched, to avoid electrostatic charge on the gate.

Another disadvantage is the poor noise behaviour. The MOSFET mechanism relies totally on the surface properties of the crystal; the slightest impurity will affect its operation. Nevertheless, MOSFETs are extensively used in digital integrated circuits, where noise is less important and where most interconnections are made internally, within the package. Furthermore, because of the simple layout, MOSFETs can be made very small, resulting in a high component density on the chip.

A MOSFET is realized as indicated in Figure 11.5b. The source and drain contacts are made on small areas of highly doped n-silicon in a lightly doped p-substrate. This results in a low contact resistance between the external electrodes and the channel material. However, two new pn junctions are introduced from both contacts to the substrate. To avoid any influence of these junctions, they must be kept reverse biased. MOSFETs have a special connection for that purpose, a substrate connection (different from the gate). By connecting the substrate to the most negative voltage in a circuit (preferably the negative power voltage source), the junctions are always reverse biased.

11.2 Circuits with field-effect transistors

We confine the discussions in this chapter to the junction field-effect transistor. All circuits explained in Section 10.2 can be realized with JFETs as well. Only the biasing differs; both the gate–source and the gate–drain junctions must be reverse biased; the gate of an n-channel FET should always be negative with respect to the source and drain. For a numerical determination of the bias, the characteristics of the JFET are required, for instance those given in Figure 11.2b and c.

11.2.1 Voltage-to-current converter

In the voltage-to-current converter of Figure 11.4, we replace the bipolar transistor by a JFET (Figure 11.6).

Suppose the characteristics of Figure 11.2 apply to this type of JFET. A bias current of (for instance) 2 mA occurs for $V_{GS} = -2$ V (the JFET operates in region 3). At $V_i = 0$ (bias condition), the source voltage is 2 V (V_{GS} is negative). The value of R_S must be such that $I_D R_S = V_S - V^-$, or $R_S = (2 - V^-)/2 \cdot 10^{-3}$. It is possible to choose $V^- = 0$ (no negative power source required), hence $R_S = 1$ kΩ. Evidently, V^+ must be sufficiently positive.

The transfer of the converter is found in a similar way to the circuit with the bipolar transistor. Using the model of Figure 11.4a, we find for the transfer:

$$\frac{i_o}{v_i} = \frac{1}{R_S + 1/g}$$

The rather low value of g compared to that of the bipolar transistor does not allow us to neglect it.

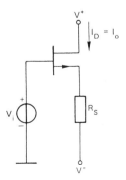

Figure 11.6 A voltage-to-current converter with a JFET.

11.2.2 *Voltage amplifier stage*

As both junctions of a JFET must be reverse biased, the biasing is rather easy (Figure 11.7).

The couple capacitors make the bias independent of the source and load circuits. Resistor R_G is inserted to assure that the (very small) gate current flows to ground. When this resistor is omitted, the gate current can flow nowhere and hampers a proper bias. R_G can be very large, for instance 1 MΩ; the gate voltage, $I_G R_G$, remains almost zero.

Example 11.1

Suppose we want a bias current of 4 mA. From the I_D–V_{GS} characteristic (Figure 11.2b), it follows a gate–source bias voltage of −1 V, hence R_S = 250 Ω. The drain–source voltage should be biased in the region of constant drain current, so at least 5 V (Figure 11.2c). The bias point is fixed halfway along the available range, that is between 6 V and V^+. For a power supply voltage of V^+ = 20 V, a proper bias value of V_D is 13 V, which is achieved by a drain resistance of $(20 - 13)/4 \cdot 10^{-3}$ = 1.75 kΩ. At a specified maximum gate current of, for instance, 200 pA, the gate resistance can be as high as 10 MΩ; the gate voltage is only $I_G R_G$ = 1 mV, a value that hardly influences the bias point.

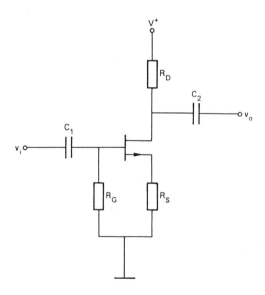

Figure 11.7 A voltage amplifier stage with JFET.

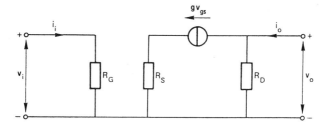

Figure 11.8 A model of the amplifier circuit from Figure 11.7; the couple capacitors are considered as short-circuits.

The voltage transfer, the input resistance and the output resistance can be calculated with the model of Figure 11.8, which uses the transistor model of Figure 11.4.

Voltage transfer ($i_o = 0$): $\quad v_o/v_i = \dfrac{-gR_D}{1+gR_S}$

Input resistance: $\quad r_i = R_G$

Output resistance: $\quad r_o = R_D$

Example 11.2
From the characteristic in Figure 11.2b we obtain a transconductance of about 2 mA/V. Using the bias conditions from Example 11.1, we find for the voltage transfer the value $v_o/v_i = -3.5/1.5 = -2.3$, the input resistance $r_i = 10$ MΩ and the output resistance $r_o = 1.75$ kΩ. The voltage gain can be increased somewhat by decoupling resistor R_S, resulting in a voltage transfer of $-g \cdot R_D = -3.5$. The rather high output resistance requires a sufficiently high input resistance at the next stage.

11.2.3 *Source follower*

The circuit of Figure 11.6 acts as a source follower when the source voltage rather than the drain current is taken as the output (Figure 11.9a).

The source follower has the same function as the emitter follower: a voltage transfer of 1 and a high input resistance and low output resistance.

We can analyse the properties of the source follower using the model of Figure 11.9b. This model also takes into account the effect of a finite value of r_d.

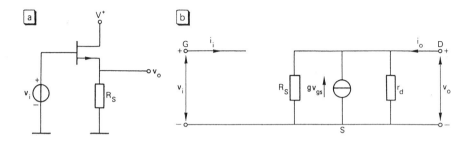

Figure 11.9 (a) Source follower, with voltage gain 1; (b) a model of the source follower.

- Voltage transfer (at $i_o = 0$): elimination of v_g and v_s from the equations:

$$v_o = g \cdot v_{gs}(R_s // r_d)$$
$$v_g = v_i$$
$$v_s = v_o$$

results in:

$$v_o = g(v_i - v_o)(R_s // r_d)$$

so:
$$\frac{v_o}{v_i} = \frac{g(R_s // r_d)}{1 + g(R_s // r_d)}$$

- Input resistance is infinite, because $i_i = 0$.
- The output resistance at $v_i = 0$ follows from the equation:

$$v_o = (i_o + g v_{gs})(R_s // r_d) = (i_o - g v_o)(R_s // r_d)$$

hence:
$$r_o = v_o / i_o = \frac{R_s // r_d}{1 + g(R_s // r_d)}$$

Example 11.3
Suppose the transconductance of the JFET applied in the circuit is 2 mA/V and $r_d = 100$ kΩ. To obtain a voltage transfer of 1, the term $g(R_S // r_d)$ must be as high as possible. Take, for instance, $R_S = 10$ kΩ. This results in $g(R_S // r_d) \approx 18.2$, so $v_o/v_i = 0.95$ and $r_o = 474$ Ω.

SUMMARY

Properties of field-effect transistors

- A junction field-effect transistor (JFET) consists of an n- or p-channel whose thickness (and thus the conductance) depends on the width of the depletion layer, which, in turn, is affected by the reverse voltage.
- The three terminals of a JFET are drain, source and gate; the channel is between the source and drain.
- The gate–source voltage for which the channel conductance is exactly zero is the pinch-off voltage V_p.
- The JFET behaves as a voltage-controlled current source; the relation between changes in the drain current and the gate voltage is $i_d = g \cdot v_{gs}$, with g the transconductance of the FET.
- For operation in analog signal-processing circuits, the FET is biased with DC voltages and currents; signals are fluctuations around the bias point.
- The gate current of a JFET is the very small leakage current of the reverse-biased channel–substrate junction. This gate current is much smaller than the base current of a bipolar transistor.
- The gate of a MOSFET is isolated from the channel by a thin oxide layer. The gate current of a MOSFET is extremely small.
- A MOSFET has a separate substrate terminal which (for n-channel types) must be biased on the most negative voltage in the circuit.

Circuits with field-effect transistors

- A junction field-effect transistor (JFET) is biased in such a way that the source–gate and the drain–gate junctions are reverse biased.
- The analysis of an electronic circuit with JFETs can be split in two: biasing and the small-signal behaviour.
- To analyse the small-signal behaviour of a transistor circuit, a transistor model is used where all fixed voltages (for instance from the power source) are zero.
- To avoid the source and load affecting the bias, parts of the circuit can be coupled by couple capacitors. Consequently, this sets a lower limit to the signal frequency.
- With JFETs, similar circuits can be made as with bipolar transistors. The main differences are:
  - a very small gate current, so a very high input resistance can be achieved;
  - smaller transconductance, so reduced voltage gain;
  - the influence of the drain voltage on the drain current is not negligible; this influence is represented by the internal drain resistance r_d.

Exercises

Properties of field-effect transistors

11.1 What is the pinch-off voltage of a JFET?

11.2 The region around a pn junction is called the depletion layer. Why?

11.3 Compare the JFET and the MOSFET with respect to the gate currents.

11.4 Find the drain voltage V_D and the drain current I_D in the circuit given below, for $V_{GS} = -4$ V.

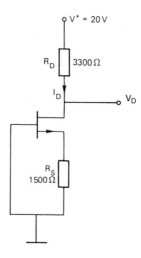

11.5 The channel resistance of a certain type of JFET varies from 800 Ω (at $V_{GS} = 0$) up to ∞ ($V_{GS} < V_p$). With the network shown below, design a voltage-controlled resistance, that varies from 1 kΩ to 10 kΩ, for the same range of V_{GS}. Find the values for R_1 and R_2.

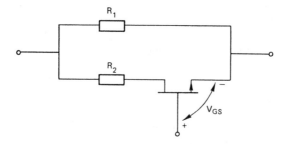

11.6 Explain the necessity of a substrate terminal of a MOSFET.

11.7 Field-effect transistors are also called unipolar transistors, contrasted with the bipolar transistors of Chapter 10. Explain these names.

Circuits with field-effect transistors

11.8 The I_D–V_{GS} characteristic of the JFET is drawn in the following figure. Find the expression for $I_D = f(V_{GS})$.

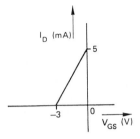

11.9 Calculate the drain current of the circuit in the following figure; the characteristic of the preceding exercise applies also to this JFET.

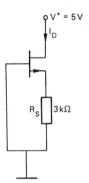

11.10 Calculate V_D in the following circuit in the figure below. The FET has the characteristic of Exercise 11.8.

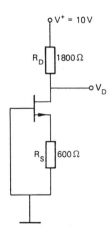

184   Field-effect transistors

11.11 Make a model of the source follower given in the figure below and find the voltage transfer, the input resistance and the output resistance under the conditions:
$g = 2$ mA/V and $i_g = 0$.

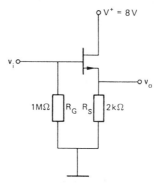

11.12 Given an AC voltage with DC component of 7 V (see figure below). Shift this signal down to an average level of 4 V, without affecting the AC part. This is achieved by the following circuit, a level shifter.
(a) Explain the operation of the circuit.
(b) For both JFETs: $I_D(V_{GS} = 0) = 2$ mA. Calculate R.
(c) What is the maximum allowable peak value of v_i?

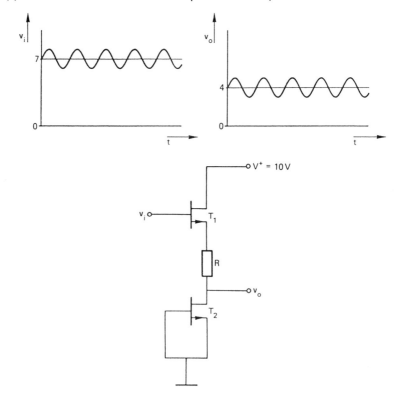

12 Operational amplifiers

Amplifiers with only one transistor have a number of shortcomings, particularly with respect to the gain, the input impedance, the output impedance and the offset. Better amplifiers are obtained by an appropriate combination of various circuits. Such configurations easily run into high complexity; their design has become a special discipline. Fortunately, the users of electronic systems need no longer make their own design, thanks to the advanced technology of integrated circuits. In particular, the operational amplifier allows simple configuring of various processing circuits of high quality. The basic concepts for such designs are given in the first part of this chapter. As a matter of fact, even an operational amplifier has its shortcomings, reflected in the designed circuits. The second part of this chapter deals with such errors and gives some remedies.

12.1 Amplifier circuits with ideal operational amplifiers

An operational amplifier is in essence a differential amplifier (Section 1.2) with very high voltage gain A (10^4–10^6), a high CMRR ($>10^4$) and a very low input current. The amplifier consists of a large number of circuit components, such as transistors, resistors and probably some capacitors. There are no inductors. All components are integrated on to a single silicon crystal, the chip, mounted in a metal or plastic encapsulation. Figure 12.1a shows the circuit symbol and Figures 12.1b and c show two common types in various encapsulations, together with the pin layout and pin functions.

The most important imperfections of an operational amplifier are the offset voltage (Section 1.2) and the input bias currents (small but noticeable DC currents flowing through the input terminals; Sections 10.1 and 11.1).

The offset voltage $V_{i,\mathrm{off}}$ can be taken into account by putting a voltage source in series with one of the two input terminals. The two bias currents $I_{\mathrm{bias},1}$ and $I_{\mathrm{bias},2}$ can be modelled by two current sources; an easier way to account for them is by arrows next to the input leads (Figure 12.1d).

The operational amplifier is nearly always used in combination with a feedback network. An open amplifier will run into saturation because of the high voltage gain and the non-zero input offset voltage; its output is either maximally positive or maximally negative, limited by the power supply voltage.

186 Operational amplifiers

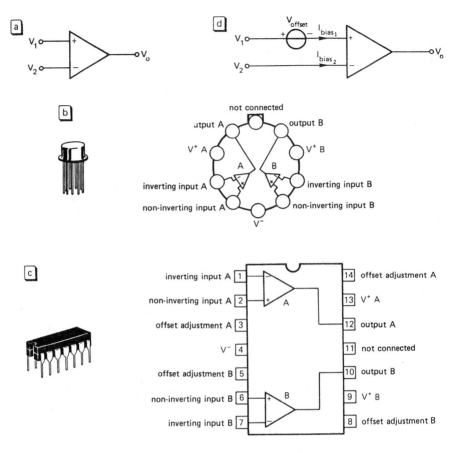

Figure 12.1 (a) The circuit symbol of an operational amplifier; $V_o = A(V_1 - V_2)$; (b) and (c) two types of dual operational amplifiers with corresponding pin layout and pin function indication; (d) modelling of the input offset voltage and the two bias currents.

In this section, we will assume an ideal behaviour of the operational amplifier. This means infinite gain, bandwidth, CMRR and input impedance, and zero input offset voltage, bias currents and output impedance. Manufacturers specify all these and a lot of other parameters. Appendix A.2.1 gives an example of the complete specification of a widely used type of operational amplifier.

The design of circuits with ideal operational amplifiers will be illustrated by a number of frequently applied circuits.

12.1.1 *Current-to-voltage converter*

Figure 12.2a shows the circuit diagram of a current-to-voltage converter with operational amplifier. Let V be the voltage across the input terminals of the operational amplifier, which here is also the voltage at the negative input, because the positive input is grounded. The output voltage is $V_o = A \cdot V$. The input current of an ideal amplifier is zero, hence, according to Kirchhoff's rule for currents, the input current I_i must flow through resistance R. According to Kirchhoff's rule for voltages, $-V + I_i R + V_o = 0$, hence $V = I_i R - V/A$ or $V = I_i R/(1 + A)$. The gain A is very high (∞ in the ideal case), so V is (almost) zero. This is a key property of an ideal operational amplifier; the voltage difference between the two input terminals is zero. The output voltage of the converter is simply $V_o = -I_i R$.

At first sight it seems strange to have an infinite gain and a finite output voltage. This is caused by the feedback; the output is partly fed back to the input. If V_o for some reason grows, the input voltage V increases as well; the current through R remains constant, hence its voltage; this tends to reduce the output change. The result is an equilibrium, where V is exactly zero. In practice, A is finite, but very large; so the voltage between the two input terminals is not zero, but very small and therefore negligible compared to other voltages in the circuit.

In the circuit of Figure 12.2, the plus terminal is connected to ground, so the voltage of the minus terminal also has zero potential; it is said to be virtually grounded. The voltage is zero, but it is not a real ground; no current can flow.

The two properties of the operational amplifier, zero input voltage and zero input currents, greatly simplify the analysis of such circuits. They offer the possibility of directly calculating the transfer of the circuit in Figure 12.2a. The procedure is as follows: I_i must flow entirely through R and, as $V = 0$, the output voltage is determined via $0 = I_i R + V_o$.

If a second current source is connected to the input of the converter, that current would flow entirely through R. This also holds for a third source, etc. (Figure 12.2b). This circuit acts as an adder for currents. The output voltage satisfies $V_o = -R(I_1 + I_2 + I_3)$.

12.1.2 Inverting voltage amplifier

The inverting voltage amplifier in Figure 12.3a can be considered as a combination of a voltage-to-current converter and a current-to-voltage converter.

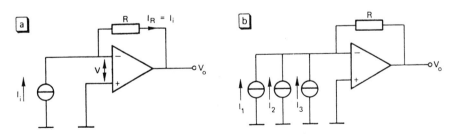

Figure 12.2 (a) Converter for the conversion of a current I_i into a voltage V_o; (b) extended circuit for the summation of currents $V_o = -R(I_1 + I_2 + I_3)$.

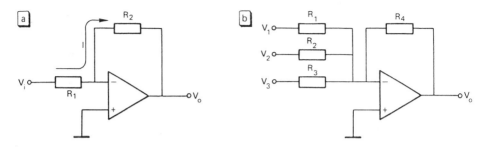

Figure 12.3 Inverting voltage amplifier ((a) with single input); (b) as adder, with multiple input.

The input voltage V_i is converted into a current with resistor R_1; the minus terminal of the operational amplifier is (virtually) grounded; its potential is zero, hence $I = V_i/R_1$. This current flows entirely through R_2, so $V_o = -R_2 I = -(R_2/R_1)V_i$. The gain of the amplifier circuit, $-R_2/R_1$, is merely set by the ratio of two external resistances and is independent of the parameters of the operational amplifier. The accuracy and stability are also determined only by the quality of the applied resistors, not that of the amplifier. For the special case of two equal resistances $R_1 = R_2$, the output is $V_o = -V_i$; the gain is -1, the voltage is inverted.

This circuit can be extended for the summation of voltages (Figure 12.3b). The output voltage of this circuit is $V_o = -(V_1/R_1 + V_2/R_2 + V_3/R_3)$, a weighted summation of the three input voltages.

The input resistance of the inverting amplifier of Figure 12.3a is exactly equal to R_1. The resistance values are determined by, on the one hand, the specified transfer and on the other by the lowest input resistance.

Example 12.1
Suppose the required gain of an amplifier is -50 and the input resistance must be at least 13 kΩ. Take $R_1 = 15$ kΩ, hence $R_2 = 750$ kΩ.

12.1.3 *Non-inverting voltage amplifier*

The output terminal of the operational amplifier in Figure 12.4a is directly connected to the negative input terminal (unity feedback). The input voltage is connected to the positive input terminal.

The voltage transfer of this circuit is found as follows: the voltages at the positive and the negative input terminals are equal (their difference is zero), hence $V_o = V_i$, the output follows the input voltage; the gain is 1. The circuit has a very high input impedance (that of the operational amplifier) and a very low output impedance (also that of the operational amplifier). The circuit acts as a buffer amplifier.

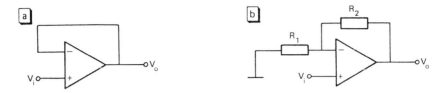

Figure 12.4 Non-inverting amplifiers; (a) with voltage transfer 1 (buffer amplifier); (b) with arbitrary voltage transfer >1, set by R_1 and R_2.

In the circuit of Figure 12.4b, only a part of the output is fed to the input. Again, the voltages on both input terminals of the operational amplifier are equal, so the voltage on the negative input terminal is V_i. This voltage is also equal to the fraction $R_1/(R_1+R_2)$ of the output (voltage divider), hence:

$$\frac{V_o}{V_i} = \frac{R_1+R_2}{R_1}$$

12.1.4 *Differential amplifier*

By combining an inverting and a non-inverting amplifier, it is possible to subtract voltages. Figure 12.5 shows the most simple configuration.

To derive the transfer, we use the principle of superposition (the circuit is linear). First we determine the output voltage due to V_1 only; the other input voltage is zero (grounded). Thus, the plus terminal is also zero (no current flows through R_3 and R_4). This situation is identical to the configuration of Figure 12.3a, for which $V_o = -(R_2/R_1)V_i$. Next we derive the output voltage due to V_2 only, where V_1 is considered

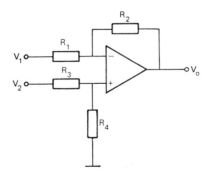

Figure 12.5 A differential amplifier circuit.

zero. Now, the voltage at the plus terminal of the operational amplifier is the output of the voltage divider with R_3 and R_4, hence $V_2 \cdot R_4/(R_3+R_4)$. With V_1 grounded, the circuit is identical to the configuration of Figure 12.4b, with transfer $(R_1+R_2)/R_1$ and the aforementioned input voltage. Hence:

$$V_o = \frac{R_1+R_2}{R_1} \frac{R_4}{R_3+R_4} V_2$$

The total output voltage is found by adding the contributions of V_1 and V_2:

$$V_o = \frac{R_2}{R_1} V_1 + \frac{R_4}{R_3+R_4} \frac{R_1+R_2}{R_1} V_2 \qquad (12.1)$$

We consider now a special case: $R_1 = R_3$ and $R_2 = R_4$. For this condition of the four resistors, the output is $V_o = (R_2/R_1)(V_2 - V_1)$, the amplified difference of both input voltages. The transfer of this differential amplifier is set by the four resistances R_1 to R_4, independent of the properties of the operational amplifier.

When we compare the properties of this new differential amplifier with the operational amplifier itself, we see that its input resistance is much lower, resistances R_1 and $R_3 + R_4$, respectively, that its gain is much lower, but more stable, because of the resistances and the CMRR is much lower, due to the tolerances of the resistors.

The CMRR is defined as the ratio of the differential mode gain and the common mode gain (Section 1.2). It will be derived for the new configuration.

Suppose the resistances have a relative inaccuracy ε_i ($i = 1$ to 4), so $R_i = R_i^*(1+\varepsilon_i)$, with R_i^* the nominal value of R_i. Suppose further that $\varepsilon \ll 1$, and $R_1^* = R_3^*$ and $R_2^* = R_4^*$ (the condition for a differential amplifier). A perfect differential mode voltage can be written as $V_d = V_1 - V_2$ or $V_1 = \tfrac{1}{2}V_d$ and $V_2 = -\tfrac{1}{2}V_d$. The transfer for this input voltage is $A_d = V_o/V_d \approx -R_2/R_1$.

A perfect common mode voltage is written as V_c or $V_1 = V_2 = V_c$; the transfer for such signals is:

$$A_c = V_o/V_c = \frac{R_1 R_4 - R_2 R_3}{R_1(R_3+R_4)} \approx \frac{R_2^*}{R_1^* + R_2^*}(\varepsilon_1 + \varepsilon_4 - \varepsilon_2 - \varepsilon_3)$$

As the sign of the relative errors ε is not known, we take the modulus $|\varepsilon|$ for each of them to find the worst case CMRR:

$$\text{CMMR} = A_d/A_c = \frac{1+R_2/R_1}{|\varepsilon_1|+|\varepsilon_2|+|\varepsilon_3|+|\varepsilon_4|}$$

Example 12.2
A differential amplifier is built according to Figure 12.5, with resistors having a specified inaccuracy of ±0.5%. The gain of this amplifier is −100. To find the CMRR, we conclude that $A_d = R_2/R_1 = 100$, so the guaranteed rejection ratio is $101/(4 \cdot 0.005) = 5050$.

12.1.5 Instrumentation amplifier

A major disadvantage of the differential amplifier in Figure 12.5 is the rather low input resistance. This can be solved by connecting buffer amplifiers (Figure 12.4a) to the inputs; the transfer is not changed, the input resistance is that of the buffers, so very high. Another shortcoming of the differential amplifier is its low CMRR, determined by the tolerances of the resistors. By inserting two buffers, the CMRR becomes even lower: suppose the buffers have slightly different gains; a common mode signal at their input gives a (small) differential signal at the output of the two buffers, which in turn is amplified by the differential amplifier. To get around this difficulty, we must use the arrangement of Figure 12.6.

The transfer and the CMRR are derived as follows; first, we consider a pure common mode input signal: $V_1 = V_2 = V_c$, connected to the plus terminals of operational amplifiers 1 and 2. The voltages at the minus terminals of these amplifiers are also V_c (assuming ideal amplifiers). So, the voltage across resistor R_a is zero, which means zero current through R_a and therefore no current through resistors R_b. This means that $V'_1 = V_1$ and $V'_2 = V_2$; the configuration behaves exactly like the original differential amplifier of Figure 12.5, with respect to common mode signals.

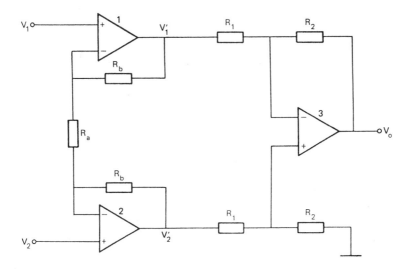

Figure 12.6 An instrumentation amplifier with high input resistance and high CMRR.

Now we consider a pure differential mode signal: $V_1 = -V_2 = \frac{1}{2}V_d$ or $V_1 - V_2 = V_d$. The voltage across resistor R_a is also V_d, which means a current through R_a of V_d/R_a. This current flows through the resistors R_b, producing a voltage across these resistors equal to $V_d(R_b/R_a)$. The inputs of the original differential amplifier are now $V_1' = V_1 + (R_b/R_a)V_d$ and $V_2' = V_2 - (R_b/R_a)V_d$, respectively. Its differential input is therefore $V_1' - V_2' = V_d(1 + 2R_b/R_a)$.

The differential gain of the configuration is increased by the factor $1 + 2R_b/R_a$, whereas the common mode gain is unaltered. Hence, the CMRR of the total configuration is increased by $1 + 2R_b/R_a$ with respect to the original differential amplifier.

The differential mode gain can only be adjusted by a single resistor, R_a, unlike the circuit of Figure 12.5, where two resistors must be equally varied to change the gain, in order to maintain a reasonable CMRR.

12.2 Non-ideal operational amplifiers

An actual operational amplifier has a number of deviations from the ideal behaviour. These deviations limit the applicability and require a careful evaluation of the specifications for a proper choice. It is of great importance to have good insight in the limitations of operational amplifiers and to know how to estimate their influence on the configuration designed. First we will discuss the main specifications of operational amplifiers. Next, we will look at their effect on the properties of the circuits from the preceding section and, where possible, give solutions for unwanted effects.

12.2.1 *Specifications of operational amplifiers*

The specifications of the various available types of operational amplifiers show strong divergence. Table 12.1 gives an overview of the main properties of three different types.

The specifications are typical, average values and are valid for 25°C. Minimum and maximum values are often specified (see Appendix A.2.1).

Type I in Table 12.1 is characterized by a low price; type II is designed for excellent input characteristics (low offset voltage and low bias current); type III is designed for high frequency applications.

The input components of type I are bipolar transistors, those of types II and III are JFETs. This can also be deduced from the values of the input bias currents, which correspond to the base current and gate currents of the applied transistors. The favourable high frequency characteristics of type III are reflected in the high value of f_t and the slew rate.

Table 12.2 shows an example of maximum allowable environmental conditions of type I.

Table 12.1 Overview of the main specifications of three types of operational amplifiers for 25°C

Parameter	Type I	Type II	Type III	Unit
V_{off}	1	0.5	2	mV
t.c. of V_{off}	20	2	50	µV/K
I_{bias}	80	0.01	0.2	nA
I_{off}	20	0.002	0.02	nA
t.c. of I_{off}	0.5	(doubles per 10 K)		nA/K
A_o	$2 \cdot 10^5$	$2.5 \cdot 10^5$	$2.5 \cdot 10^4$	–
R_i	$2 \cdot 10^6$	10^{12}	10^{10}	Ω
CMRR	90	80	60	dB
SVRR	96	80	60	dB
f_t	1	1	40	MHz
slew rate	0.5	3	330	V/µs

Explanation of the terms:
V_{off}: input offset voltage
I_{bias}: input bias current
I_{off}: difference between input bias currents
t.c.: temperature coefficient
A_o: low frequency voltage gain
R_i: input resistance between input terminals
SVRR: supply voltage rejection ratio (change of input offset voltage for 1 V change of power supply voltage)
f_t: bandwidth at unity feedback
slew rate: maximum of dV_o/dt

Table 12.2 Absolute maximal ratings for some of the parameters of type I from Table 12.1

Power supply	±18 V
Power dissipation	500 mW
Input differential voltage	±30 V
Output short-circuit duration	Indefinite
Operating temperature range	0 ... 75°C
Storage temperature range	−65 ... 150°C
Soldering temperature (60 s)	300°C

12.2.2 Input offset voltage

One of the most serious limitations of an operational amplifier is the offset voltage, in particular when processing small DC signals. Both the input offset voltage V_{off} and the input bias current I_{bias} contribute to the total offset of the circuit. This will be explained with Figure 12.7.

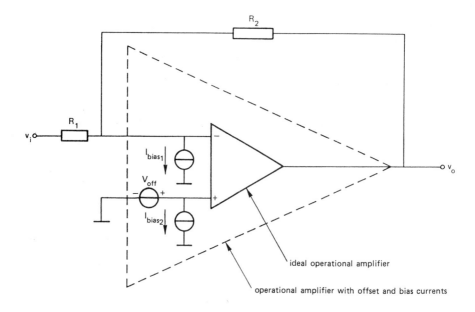

Figure 12.7 Inverting amplifier from Figure 12.4b, extended with an offset voltage source and two bias current sources.

We model the input offset voltage and the bias currents as external sources: V_{off} in series with one of the input terminals (it does not matter which) and I_{bias} parallel to the input terminals. These sources account for all the errors considered, so the operational amplifier itself is made error free: zero input currents and zero input voltage. The input is usually connected to a voltage source with low impedance. So, for the calculation of the errors, the input terminal is grounded.

The calculation is split in two: the contribution of the offset voltage only and the contribution of the bias currents. The resulting outputs are added according to the principle of superposition for linear systems and this output voltage is divided by the voltage gain, in order to find the equivalent total input error voltage.

First, $I_{bias} = 0$; the output voltage due to V_{off} is $V_{off}(1 + R_2/R_1)$. Next, $V_{off} = 0$; the voltage on the minus terminal is also zero. This means that there is no current through resistor R_1 (both terminals are zero). The bias current I_{bias1} must go somewhere, and the only path is through R_2. This results in an output voltage equal to $I_{bias1}R_2$. The other bias current, I_{bias2}, flows directly to ground and does not contribute to the output voltage. The total output error voltage is $V_{off}(1 + R_2/R_1) + I_{bias2}R_2$. We do not know the sign of V_{off} and I_{bias}, so we take the modulus to find the maximum error signal (worst case). After division by the voltage gain $-R_2/R_1$, we find for the maximum equivalent input error voltage:

$$\frac{R_2+R_1}{R_2}|V_{\text{off}}|+R_1|I_{\text{bias1}}|$$

Likewise, the error voltage of arbitrary other circuits can be derived.

The bias current of the operational amplifiers in Figure 12.4 will flow entirely through the input voltage source which is connected to the circuit. This causes an extra error voltage of $I_{\text{bias}} \cdot R_g$, where R_g is the source impedance. When measuring the voltage of a source with relatively high source resistance, an amplifier with very low bias current is required so as not to introduce a large voltage error.

The effect of both the offset and bias current can be significantly reduced. Reduction or compensation of the offset voltage is obtained by an additional voltage added to the input.

This compensation voltage can be taken from the power supply voltage (Figure 12.8a) using a proper voltage divider circuit. Some types have special connections that may be used to realize the compensation circuit more easily (Figure 12.8b). The

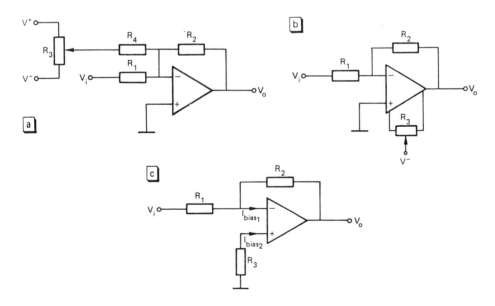

Figure 12.8 Various compensation methods: (a) with additional, adjustable voltage derived from the power supply; (b) with special connections at the operational amplifier; (c) bias current compensation.

compensation of the bias currents is based on the fact that both bias currents are (almost) equal and have the same sign (they flow either from or to the amplifier). A resistor R_3 is connected in series with the plus terminal (Figure 12.8c). This causes an extra offset voltage, $-I_{bias2}R_3$, at the plus terminal, resulting in an extra output error voltage of $-I_{bias}R_3(1+R_2/R_1)$. The contribution of I_{bias1} to the output is, as we have seen before, $I_{bias1}R_2$. Now if both bias currents are equal, the total output error voltage will be zero if R_3 satisfies the condition $R_2 = R_3(1+R_2/R_1)$ or $R_3 = R_1//R_2$. Evidently, the bias currents are not exactly equal; their difference is the bias offset current I_{off}. The output error voltage is, under the same condition for R_3, equal to $(I_{bias1} - I_{bias2})R_2 = I_{off}R_2$, and the equivalent input error voltage is $I_{off}R_1$. As the offset current is usually small compared to the bias currents, the effect of the bias currents is largely eliminated by simply adding resistor R_3.

These compensation methods are also applicable in most other circuits with operational amplifiers. After careful compensation or adjustment of the errors, the effect of the temperature coefficients of V_{off} and I_{off} remains, which is the reason for having these parameters specified as well.

Example 12.3

An inverting voltage amplifier in the configuration of Figure 12.8c uses a type 741 operational amplifier (type I in Table 12.1). $R_1 = 10\,k\Omega$, $R_2 = 1\,M\Omega$ and $R_3 = R_1//R_2$. At 20°C the output voltage is adjusted to zero (at zero input voltage) using the compensation circuit of Figure 12.8b. We derive the maximum possible input error voltage over a temperature range from 0 to 70°C.

The error input voltage is $V_{off,i} = 1.01\,V_{off} + 10^4\,I_{off}$. At 20°C these contributions cancel (due to adjustment). The maximum offset occurs at 70°C:
$|V_{off}| = \Delta T \cdot \text{t.c.}\,(V_{off}) = 50 \cdot 20\,\mu V$; $|I_{off}| = \Delta T \cdot \text{t.c.}\,(I_{off}) = 50 \cdot 0.5\,nA$, hence $V_{off,i}$ at that temperature can be as high as 1.26 mV.

12.2.3 *Finite voltage gain*

So far, we have assumed an infinite voltage gain of the operational amplifier. At low frequencies this is an acceptable assumption, however, not for higher frequencies. Figure 12.9 shows the amplitude transfer characteristic of a typical low cost operational amplifier (type 741).

The characteristic has a remarkably low value of the $-3\,dB$ frequency. Is it still justified to assume an infinite gain? We will investigate this for the case of the non-inverting voltage amplifier of Figure 12.4b. For this configuration, the following equations apply: $V_i = V_+$ (the voltage at the plus terminal); $V_- = V_o R_1 / (R_1 + R_2) = \beta V_o$ (β is the fraction of the output that is fed back to the input); $V_o = A(V_+ - V_-)$. Elimination of V_+ and V_- finally gives $V_o/V_i = A/(1+A\beta)$. Only if $A\beta$ is large compared to 1, may we approximate the transfer by $V_o/V_i = 1/\beta = 1 + R_2/R_1$, as found earlier.

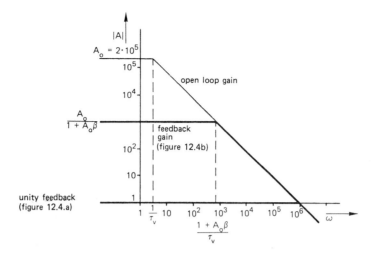

Figure 12.9 Amplitude transfer characteristic of an operational amplifier without and with feedback.

A less crude approximation is:

$$\frac{V_o}{V_i} = \frac{A}{1+A\beta} = \frac{1}{\beta}\frac{1}{1+1/A\beta} \approx \frac{1}{\beta}\left(1-\frac{1}{A\beta}\right)$$

The relative deviation from the ideal value is about $1/A\beta$. The term $A\beta$ is often encountered in calculations on control systems; it is called the open loop gain of the system: the transfer of the amplifier and the feedback network together. The higher the loop gain, the less is the transfer affected by the amplifier itself. This is the main reason for striving to obtain the highest possible voltage gain of an operational amplifier. For the buffer amplifier in Figure 12.4a, $\beta = 1$, so the actual transfer deviates only a fraction $1/A$ from 1.

As appears from Figure 12.9, A decreases with increasing frequency. Apparently, the amplifier behaves as a first-order low-pass filter with complex transfer function $A(\omega) = A_o/(1+j\omega\tau_v)$. Substitution of this expression in the transfer function of the non-inverting amplifier configuration gives:

$$\frac{V_o}{V_i} = \frac{A(\omega)}{1+A(\omega)\beta} = \frac{A_o}{1+j\omega\tau_v + A_o\beta} = \frac{A_o}{1+A_o\beta}\frac{1}{1+j\omega\tau_v/(1+A_o\beta)}$$

This transfer has a first-order low-pass characteristic as well: the -3 dB frequency is $\omega = (1+A_o\beta)/\tau_v$. The bandwidth is a factor $1+A_o\beta$ more than that of the open amplifier. The low frequency transfer is $A_o/(1+A_o\beta)$ and is a factor $1+A_o\beta$ less than that of the open amplifier. Obviously, the product of bandwidth and gain remained the same, irrespective of the feedback. This can also be seen in Figure 12.9. At unity feedback, $\beta = 1$, so $V_o/V_i = A_o/(1+A_o) \approx 1$. The bandwidth is $(1+A_o)/\tau_v \approx A_o\tau_v$. This is the unity gain bandwidth, denoted as f_t (in Hz). From the unity gain bandwidth, the bandwidth of a circuit with arbitrary gain follows directly.

Example 12.4
An operational amplifier has a unity gain bandwidth f_t = 2 MHz. The bandwidth of an amplifier with gain 100 is 20 kHz; at a gain of 1000, the bandwidth is only 2 kHz.

Only amplifiers that behave as a first-order low-pass filter have a constant gain–bandwidth product (GB product). Many amplifiers, in particular those for high frequency applications, have a second-order or even higher-order transfer function. These amplifiers are not always stable at arbitrary feedback. However, they have some extra connections for adding compensating networks, for instance a small capacitor to guarantee stability at the chosen gain factor. This is called external frequency compensation. The manufacturer gives directives in the specification sheets.

An integrated operational amplifier is built up of at least two amplifier stages, so they behave as a second-order system. With an internal frequency compensation network, a first-order characteristic is obtained, resulting in a guaranteed stability for an arbitrary gain.

SUMMARY

Amplifier circuits with ideal operational amplifiers

- An ideal operational amplifier has an infinite voltage gain, CMRR and input impedance; the voltage offset, bias currents and output impedance are zero.
- At proper feedback, the voltage between the two input terminals of an operational amplifier is zero.
- Operational amplifiers are used for the realization of various signal operations. These operations are fixed by external components; they are almost independent of the properties of the operational amplifier itself.
- The voltage transfer of the inverting voltage amplifier of Figure 12.3a is $-R_2/R_1$; that of the non-inverting amplifier of Figure 12.4b is $1 + R_2/R_1$.
- The inverting voltage amplifier has a rather low input resistance: R_1; the input resistance of the non-inverting amplifier is very high.
- The differential gain of the differential amplifier in Figure 12.5 is $-R_2/R_1$, under the condition $R_1 = R_3$ and $R_2 = R_4$.
- The CMRR and the input resistance of the differential amplifier in Figure 12.5 can be substantially increased by adding two operational amplifiers, in the arrangement in Figure 12.6.

Non-ideal operational amplifiers

- The output offset voltage of the non-inverting voltage amplifier is composed of the terms $(1 + R_2/R_1)V_{off}$ and $I_{bias}R_2$. The contribution of V_{off} can be reduced by the compensation methods of Figures 12.8a and b; the contribution of I_{bias} is reduced to $I_{off}R_2$ by adding the compensation resistor R_3 according to Figure 12.8c. These compensation methods are also applicable to other amplifier configurations.
- The equivalent input offset voltage or error voltage is the output error voltage divided by the voltage transfer.
- An operational amplifier with a first-order amplitude transfer characteristic has a constant gain–bandwidth product, specified as the unity gain bandwidth f_t.

Exercises

Amplifier circuits with ideal operational amplifiers

12.1 What is meant by a virtual ground? The minus terminal of an inverting amplifier is virtually grounded. Why? The minus terminal of a non-inverted amplifier is not virtually grounded. Explain this.

12.2 All operational amplifiers in the following figures a–f can be considered as ideal. Find their output voltage V_o.

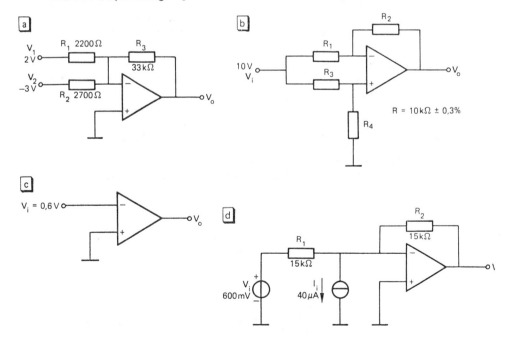

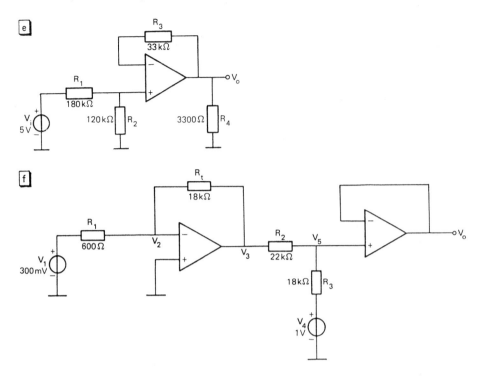

12.3 Design an amplifier according to Figure 12.3a, with voltage gain −50 and a minimum input resistance of 5 kΩ. Take only resistance values from the E12 series.

12.4 Design a circuit with only one operational amplifier that is able to add three voltages such that $V_o = -10 V_1 - 5 V_2 + 2 V_3$.

12.5 Derive the transfer for common mode and differential mode signals of the circuit in Figure 12.6, for the case when the two resistors R_b are not equal. Is it necessary for a high CMRR that they are equal?

Non-ideal operational amplifiers

12.6 The specifications of the operational amplifier in the circuit below are $V_{off} = 0.4$ mV; $I_{bias} = 10$ nA; $I_{off} = 1$ nA; $f_t = 1.5$ MHz. Calculate V_o due to V_{off} only.

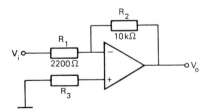

12.7 Calculate V_o in the circuit of Exercise 12.6, due to I_{bias} only. Find the proper value of R_3 for optimal compensation.

12.8 Find the input resistance in the circuit of Exercise 12.6. Find also the output resistance.

12.9 Find the bandwidth of the same circuit.

12.10 Design a circuit according to Figure 12.8a such that the voltage gain is -30 and the input resistance is at least 10 kΩ. The output offset voltage should be adjustable over a range from -1.2 to 1.2 V; the power supply voltages are 15 and -15 V.

13 Frequency-selective transfer functions with operational amplifiers

In Chapter 8, several circuits for frequency-selective signal processing with passive components were discussed. Some disadvantages of such circuits, like the unfavourable input and output impedances or the obligatory use of inductors, can be overcome by using operational amplifiers.

The first part of this chapter describes circuits for processing in the time domain, such as the integrator and the differentiator. The second part of this chapter deals with circuits with a high frequency selectivity.

13.1 Circuits for time-domain operations

The basic configurations for the circuits in this section are the inverting and the non-inverting amplifiers from Chapter 12 (Figure 13.1). The complex transfer functions of these circuits are expressed as $H = -Z_2/Z_1$ and $H = (Z_1+Z_2)/Z_1$, respectively.

13.1.1 *The integrator*

The circuit of Figure 13.1a acts as an electronic integrator when Z_1 is a resistance and Z_2 a capacitance (Figure 13.2).

The transfer function of this circuit is $H = -1/j\omega RC = -1/j\omega\tau$. Its modulus is $1/\omega\tau$, which is inversely proportional to the frequency; the modulus decreases by 6 dB/octave over the full frequency range. Disregarding the minus sign in the transfer, its argument has the constant value of $-\pi/2$. Any sinusoidal input signal $v_i = \hat{v}\sin\omega t$ gives an output signal equal to $v_o = -(1/\omega\tau)\hat{v}\sin(\omega t - \pi/2) = -(1/\omega\tau)\hat{v}\cos\omega t$, which is proportional to the integral of v_i.

This circuit acts as an integrator for other signals as well, which can be shown as follows. Because of the virtual ground of the inverting input terminal, the input signal v_i is converted into a current v_i/R, which flows through the capacitance C. The output of the amplifier is $v_o = -v_c$. The current through a capacitor is $i = C(dv_c/dt)$. As $i = v_i/R$, it follows that $C(dv_c/dt) = v_i/R$, so the output voltage satisfies the equation:

$$v_o = -\int \frac{v_i}{RC} dt = \frac{-1}{\tau} \int v_i \, dt$$

Circuits for time-domain operations 203

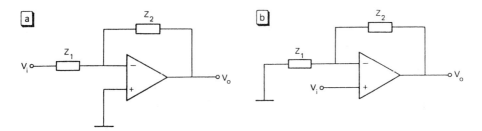

Figure 13.1 The basic configurations of the circuits in this chapter, (a) with inverting transfer and (b) with non-inverting transfer.

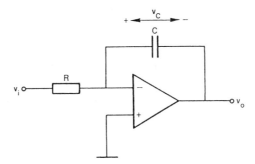

Figure 13.2 Basic electronic integrator.

The foregoing only holds for an ideal operational amplifier. Imperfections of the amplifier cause deviations from its ideal behaviour as an integrator. The influences of the bias current I_{bias} and the offset voltage V_{off} can be derived from Figure 13.3. First, suppose $I_{bias} = 0$. The current through R equals V_{off}/R, as the non-inverting input is virtually grounded. This current flows through the capacitance C. Next, to find the effect of I_{bias}, the offset is supposed to be zero. In that case, I_{bias} flows directly into C, because the voltage across R is zero. So, at zero input, the total current through C is $I_{bias} + V_{off}/R$, resulting in an output voltage equal to:

$$V_o = \frac{1}{C} \int \left(I_{bias} + \frac{V_{off}}{R} \right) dt + V_{off}$$

The polarities of I_{bias} and V_{off} are mostly unknown, so the worst case output offset is:

$$V_o = \frac{1}{C} \int \left(|I_{bias}| + \frac{|V_{off}|}{R} \right) dt + |V_{off}|$$

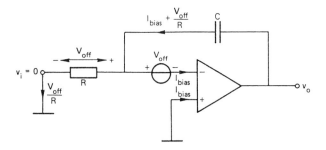

Figure 13.3 To determine the effect of the offset voltage V_{off} and the bias current I_{bias} the input signal is set to zero.

Example 13.1
An integrator is designed according to Figure 13.2, with $R = 10\ k\Omega$ and $C = 0.01\ \mu F$. The operational amplifier used in this circuit has an offset voltage of less than 1 mV and a bias current less than 100 nA. At zero input voltage, the output will increase 20 V during each second; so within 1 s after switching on the amplifier will be in saturation.

To prevent the circuit from running into saturation due to the offset voltage and the bias current, it is necessary to choose an amplifier with low values of V_{off} and I_{bias}. Besides that, the following measures can be taken (see Figure 13.4):

- An additional resistor R_2 in series with the non-inverting input of the operational amplifier; this reduces the effect of the bias current (Section 12.2.2).
- Offset compensation, in a similar way as to that with the inverting amplifier (see, for instance, Figure 12.8a and b).
- An additional resistor R_1 in parallel to C. The transfer becomes $-(R_1/R)/(1+j\omega R_1 C)$. Figure 13.4b shows the transfer characteristic. At low frequencies, the transfer is limited to R_1/R. A disadvantage of this method is the confined integrating range: only signals with frequencies much higher than $1/(2\pi R_1 C)$ Hz are integrated.

An ideal integrator has a very high gain at low frequencies (infinite at DC). However, the finite gain of the operational amplifier limits the usable range of the integrator at low frequencies. The frequency-dependent gain of the amplifier can be written as:

$$A(\omega) = A_0/(1+j\omega\tau_v)$$

with A_0 the DC gain or open loop gain and τ_v the first-order time constant of the

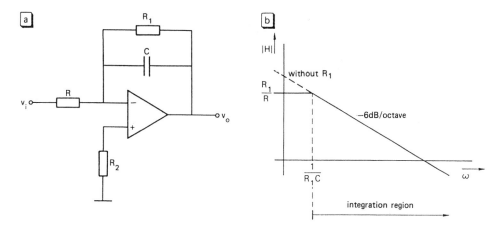

Figure 13.4 (a) The resistor R_1 in this circuit reduces the effect of the offset voltage and R_2 compensates for the bias current; (b) the transfer characteristic shows that the circuit acts as an integrator only for frequencies much higher than $1/(2\pi R_1 C)$ Hz.

amplifier. The transfer function of the integrator from Figure 13.4a can be derived from the following equations:

$$V_o = \frac{A_0}{1+j\omega\tau_v}(V_+ - V_-)$$

$$V_i - \frac{V_i - V_-}{R} = \frac{V_- - V_o}{Z_1}$$

In these formulas, $Z_1 = R_1/(1+j\omega R_1 C)$, and V_- and V_+ are the voltages at the inverting and non-inverting inputs of the amplifier, respectively. V_- is zero because the input current is supposed to be zero. Eliminating V_- from these two equations and with the approximations $A_0 \gg 1$, $A_0 R \gg R_1$, $R_1 \gg R$ and $R_1 C \gg \tau_v/A_0$, the transfer of the integrator circuit appears to be:

$$V_o/V_i = \frac{R_1}{R} \cdot \frac{1}{1+j\omega R_1 C} \cdot \frac{1}{1+j\omega\tau_v/A_0}$$

The frequency characteristic is depicted in Figure 13.5. Apparently, the integrator has a limited range, for which $1/R_1 C \ll \omega \ll A_0/\tau_v$.

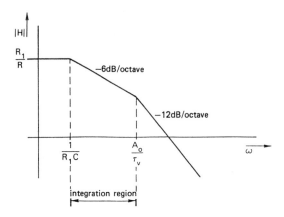

Figure 13.5 The frequency characteristic of the integrator in Figure 13.4a, in the case of an operational amplifier with a low frequency cut-off. The working range of the integrator lies within the two −3 dB frequencies.

Example 13.2

For the circuit in Figure 13.4, the components have the following values: $R_1 = 1\ \text{M}\Omega$, $C = 0.1\ \mu\text{F}$, $R = 1\ \text{k}\Omega$, $A_0 = 10^5$ and $\tau_v = 0.1$ s. In this design, the −3 dB frequencies lie at 10 and 10^6 rad/s. This means an integration range from roughly 15 Hz to 150 kHz.

13.1.2 Differentiator

By exchanging the positions of the resistance and the capacitance in Figure 13.2, the circuit turns into the differentiator of Figure 13.6.

The input voltage v_i is converted into a current $i = C(dv_i/dt)$, flowing through capacitor C. This current flows into the feedback resistor R, resulting in an output voltage v_o equal to $-Ri = -RC(dv_i/dt)$. The transfer in the frequency domain is $H = -Z_2/Z_1 = -j\omega\tau$. The modulus of the transfer is directly proportional to the frequency (a raise of 6 dB per octave) and the argument (disregarding the minus sign) is $\pi/2$ over the whole frequency range.

As in the case of the integrator, the differentiator also suffers from the non-ideal characteristics of the operational amplifier, setting a limit to its useful range of operation. The bias current causes a constant output voltage equal to $-I_{\text{bias}}R$. The offset voltage appears unchanged at the output, because there is no current through R (at zero bias current) (see Figure 13.6b). Not knowing the polarity of V_{off} and the direction of I_{bias}, the output offset voltage is less than $|V_{\text{off}}| + |I_{\text{bias}}|R$.

A serious disadvantage of this circuit is the high gain for signals with high frequency components. Thermal noise (which has a wide band) and fast changing interference signals are strongly amplified by the differentiator. Furthermore, steep input signals (as from square- or pulse-shaped voltages) can bring the differentiator into saturation.

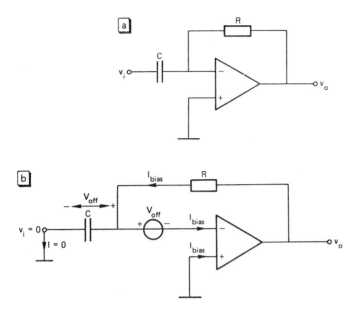

Figure 13.6 (a) Basic electronic differentiator; (b) to determine the effect of the offset voltage V_{off} and the bias current I_{bias}, the input signal is set to zero.

Finally, the frequency-dependent transfer of the operational amplifier may give rise to a pronounced peak in the frequency characteristic of the circuit or even to instability.

For all these reasons, the differentiator of Figure 13.6 is not recommended. The circuit of Figure 13.7 behaves somewhat better. Here, the range of the differentiator is narrowed to low frequencies only. By adding R_1, the high frequency gain is limited to a certain maximum. If the frequency dependency of the operational amplifier is not taken into account, the transfer function of the circuit in Figure 13.7 satisfies:

$$H(\omega) = -(Z_2/Z_1) = -R/(R_1 + 1/j\omega C) = -j\omega\tau/(1 + j\omega\tau)$$

with $\tau = R_1 C$. The frequency characteristic of this transfer is shown in Figure 13.7b.

Example 13.3

Our goal is the design of a differentiator for frequencies up to 100 Hz (about 600 rad/s); the gain may not be more than tenfold. These requirements are met if, for instance, $\tau = R_1 C = 10^{-3}$ s. The ratio R/R_1 should be 10. For a stable transfer, it is preferable to have a low gain at high frequencies. A suitable design is $R_1 = 10\,k\Omega$, $R = 100\,k\Omega$ and $C = 0.1\,\mu F$.

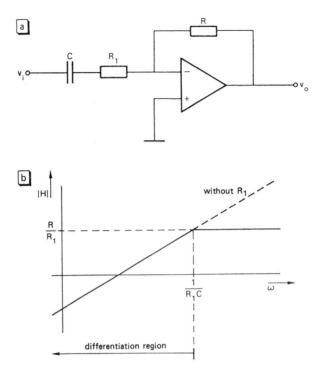

Figure 13.7 (a) To reduce noise and high frequency interference, resistor R_1 is added to the circuit of Figure 13.6; (b) the frequency characteristic of the differentiator shows that the range of the differentiator is restricted to frequencies below $1/(2\pi R_1 C)$.

13.1.3 Circuits with PD, PI and PID characteristics

Control systems often require a specific transfer characteristic. Such a control characteristic may contain a proportional region (P, a frequency-independent transfer), a differentiating region (D) and an integrating region (I), or a combination of these three possibilities (denoted as PI, PD and PID). These names reflect the type of transfer in the time domain of the controller, which generates the control signals. All these characteristics can be realized with the configuration of Figure 13.1a and are discussed below (Figure 13.8):

- PI characteristic (Figure 13.8a). $Z_1 = R_1$ and $Z_2 = R_2 + 1/j\omega C_2$. The transfer becomes $H = -(R_2/R_1 + 1/j\omega R_1 C_2)$. To prevent the integrator from saturating, the gain for DC signals must be restricted (see the dotted line in Figure 13.8a).
- PD characteristic (Figure 13.8b). Z_1 consists of a resistor R_1 in parallel with capacitor C_1; $Z_2 = R_2$. The transfer equals $H = -(R_2/R_1 + j\omega R_2 C_1)$. Measures

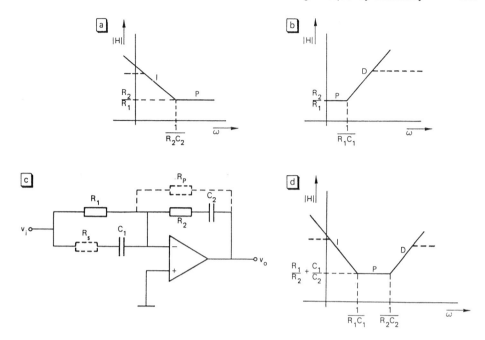

Figure 13.8 Transfer characteristics of (a) PI circuit; (b) PD circuit; (c) PID circuit; (d) circuit for a PID characteristic.

discussed in the paragraph on the differentiator have to be taken to guarantee stability of the system.
- PID characteristic (Figure 13.8c). $Z_1 = R_1/(1 + j\omega R_1 C_1)$; $Z_2 = R_2 + 1/j\omega C_2$. The transfer function satisfies $H = -(C_1/C_2 + R_2/R_1 + 1/j\omega R_1 C_2 + j\omega R_2 C_1)$.

To restrict the working range of the integrator to high frequencies only, a resistor R_p is set in parallel to Z_2; the working range of the differentiator is restricted to low frequencies by a resistor R_s in series with C_1. The transfer characteristic of this circuit is depicted in Figure 13.8d.

The low output impedance of the circuits discussed above permits cascading several circuits without significant loading effects. The complex transfer function of such a serial system is simply the product of the individual transfer functions. For other and special filter characteristics, as well as for the respective pulse and step responses, the reader is referred to the numerous textbooks on this subject.

13.2 Circuits with high frequency selectivity

This section shows how to achieve a transfer function with a high selectivity without using inductors, by applying active components (operational amplifiers). Second-order band-pass filters with a very small bandwidth and low-pass filters of higher order, respectively, are discussed.

13.2.1 *Resonance filters*

Passive filters consisting of inductors and capacitors can have a high selectivity in a very small frequency band, due to the effect of resonance. Band-pass filters with high selectivity, without coils, require the use of active components such as operational amplifiers.

The general transfer function of a second-order network can be written as:

$$H(j\omega) = \frac{a_0 + a_1(j\omega) + a_2(j\omega)^2}{b_0 + b_1(j\omega) + b_2(j\omega)^2}$$

The coefficients a_0, a_1 and a_2 in this expression are not all equal to zero. If the numerator contains only the factor $a_1 j\omega$, the filter is of the band-pass type, because the modulus of the transfer approaches zero for $\omega \to 0$ as well as for $\omega \to \infty$. If $a_2 = 0$ (and possibly also $a_1 = 0$), $|H| = a_0/b_0$ for $\omega = 0$, whereas the transfer approaches zero for $\omega \to \infty$. Such a filter has, therefore, a low-pass characteristic.

Similarly, it can be shown that the transfer has a high-pass characteristic if $a_0 = 0$. First, band-pass-type filters will be discussed, so $a_0 = a_2 = 0$. The denominator of $H(j\omega)$ can be rewritten as $1 + 2j\omega z/\omega_0 - \omega^2/\omega_0^2$ (see Section 6.1.2) or as $1 + j\omega/Q\omega_0 - \omega^2/\omega_0^2$; in this section the latter expression is used. If the numerator is written as $H_0 j\omega/Q\omega_0$, then the transfer function of the second-order band-pass filter becomes:

$$H(j\omega) = H_0 \frac{j\omega/Q\omega_0}{1 + j\omega/Q\omega_0 - \omega^2/\omega_0^2}$$

The transfer is fixed with three parameters: H_0, ω_0 and Q. H_0 is the transfer at $\omega = \omega_0$. This is equal to the maximum transfer. Therefore, ω_0 is called the resonance frequency. In Section 6.1.2 this maximum transfer has been derived for a second-order function with $a_0 = 1$ and $a_1 = a_2 = 0$; in that case the maximum transfer occurs at a slightly lower frequency. The bandwidth of the filter (the frequency span between both -3 dB points) can be derived by calculating the frequencies $\omega \pm \Delta\omega$ for which $|H| = \frac{1}{2}H_0\sqrt{2}$. This bandwidth appears to be equal to $B = 2\Delta\omega = \omega_0/Q$, if $\Delta\omega \ll \omega_0$ (see Figure 13.9). The parameter Q increases with decreasing bandwidth; therefore, Q is called the quality factor of the filter (see also Section 6.1.2). With only passive components, the maximally attainable Q is 0.5.

The number of possible inductorless band-pass filters is considerable. We will restrict to only two classes, namely filters built up with one operational amplifier with both positive and negative feedback.

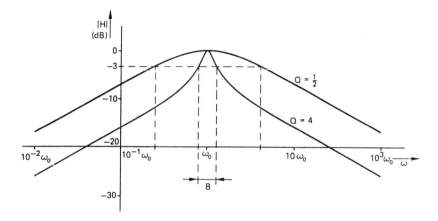

Figure 13.9 The frequency characteristic of a second-order band-pass filter with transfer $H(j\omega) = j\omega/\omega_0\, Q/(1 + j\omega/\omega_0\, Q - (\omega/\omega_0)^2)$, for different values of Q.

Band-pass filters with frequency-selective positive feedback

Figure 13.10 shows the control diagram of an amplifier with positive feedback. The feedback network consists of passive components and has a transfer equal to $\beta(\omega)$. The voltage transfer V_o/V_i can be calculated with the equations $V_o = K(V_i + V_t)$ and $V_t = \beta V_o$, yielding:

$$V_o/V_i = \frac{K}{1 - K\beta(\omega)}$$

If the feedback network has a band-pass characteristic (for instance, one of the networks from Section 8.1.3), then $H(\omega)$ also has one.

The properties of such a band-pass filter are clarified with the following example, shown in Figure 13.11.

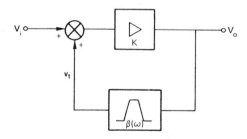

Figure 13.10 Basic operation of a system with positive feedback. The transfer function has a band-pass characteristic because the feedback network is a band-pass-type filter.

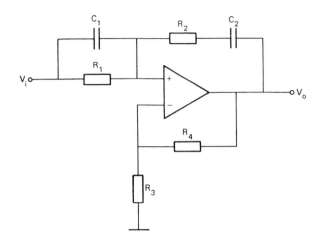

Figure 13.11 An example of a band-pass filter, made up with an operational amplifier with positive feedback.

The gain K and the summation of the input signal and the feedback signal are realized in a single operational amplifier. The passive band-pass filter consists of the components R_1, R_2, C_1 and C_2. Assuming ideal properties of the operational amplifier, the following equations hold:

$$\frac{V_i - V_+}{Z_1} = \frac{V_+ - V_o}{Z_2}$$

$$V_- = \frac{R_3}{R_4 + R_3} V_o$$

$$V_+ = V_-$$

with $Z_1 = R_1 // j\omega C_1$ and $Z_2 = R_2 + 1/j\omega C_2$; V_- and V_+ are the voltages at the inverting and non-inverting inputs of the operational amplifier, respectively. Eliminating V_- and V_+ from these equations, the transfer function is found to be equal to:

$$H = V_o/V_i = \frac{R_4 + R_3}{R_3} \cdot \frac{(1 + j\omega\tau)^2}{1 + j\omega\tau(2 - R_4/R_3) - \omega^2\tau^2}$$

where it is assumed that $R_1 = R_2 = R$ and $C_1 = C_2 = C$. From the denominator, it appears that $\omega_0 = 1/\tau$ and $Q = R_3/(2R_3 - R_4)$. The transfer at resonance is $H_0 = 2(R_4 + R_3)/(2R_3 - R_4)$. The sensitivity in Q to varying resistance values is high, in particular when the difference between $2R_3$ and R_4 is small. The same holds for H_0, the transfer at resonance. The system is unstable if $R_4 \geqslant 2R_3$.

The resonance frequency and the quality factor of the filter can be varied independently: Q only depends on R_3 and R_4, whereas with R and C the resonance frequency can be tuned. A condition for this independency is the equality of both resistances R and capacitances C.

The filter from Figure 13.11 is the Wien filter, because it is derived from a Wien measuring bridge. Many other types of filters can be designed according to the general principle of Figure 13.10. The common property of these filters is the high sensitivity for varying parameters at high Q. Therefore, it is difficult to realize stable filters with a high quality factor. An advantage of this type of filter is the relatively low value of K; for the Wien filter this value is somewhat below 3 ($K = 3$ corresponds to an infinite Q factor), allowing the design of such filters for relatively high frequencies; the gain–bandwidth product of an operational amplifier is constant, and so a low gain corresponds with a high bandwidth.

Band-pass filters with frequency-selective negative feedback

A high quality factor can also be achieved with negative feedback. To that end, the passive feedback network should have a notch characteristic. Figure 13.12 shows the control diagram of such a system. Again, the operation will be explained by an example (see Figure 13.13).

The notch filter in this circuit is hardly recognizable as such, because of the combination of the amplifier and filter functions with a small number of passive components. The negative feedback is easily recognized; there is just one feedback path from the output to the inverting input of the operational amplifier. The transfer function of this filter appears to be:

$$H = V_o/V_i = -\frac{1}{R_3} \cdot \frac{R_1 + R_2 + j\omega(R_1 + R_3)R_2C_1}{1 + j\omega(R_1 + R_2)C_2 - \omega^2 R_1 R_2 C_1 C_2}$$

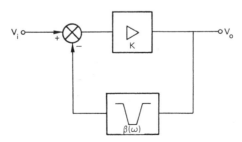

Figure 13.12 Basic principle of a filter configuration with negative feedback. The transfer function has a band-pass characteristic because the feedback network has a notch characteristic.

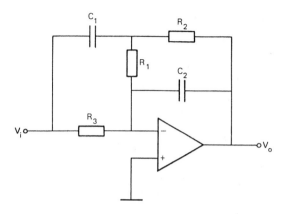

Figure 13.13 An example of a band-pass filter with operational amplifier and frequency-selective negative feedback.

The resonance frequency is $\omega_0 = 1/\sqrt{R_1 R_2 C_1 C_2}$, the quality factor is $Q = \sqrt{R_1 R_2 C_1/C_2}/(R_1 + R_2)$. The maximum value of Q is $\frac{1}{2}\sqrt{C_1/C_2}$, at $R_1 = R_2$. The quality factor is determined by the ratio of two capacitance values and is, therefore, very stable. A disadvantage of this filter is the mutual dependency of ω_0 and Q. These parameters cannot be varied independently, as can be seen from the expressions for ω_0 and Q.

An important property of the band-pass filter with negative feedback is its unconditional stability, also at high Q. However, a high Q requires a high gain K. When an operational amplifier is used, this means a restriction to the height of ω_0, because the bandwidth of the amplifier with feedback decreases with increasing gain.

13.2.2 Active Butterworth filters

In Section 8.2, filter types with different approximations to the ideal behaviour have been discussed. These filters can be designed with resistors and capacitors only, when using operational amplifiers.

The discussion in this section is restricted to Butterworth filters of the second and third order.

A second-order low-pass filter is depicted in Figure 13.14. This is the so-called Sallen and Key filter. From the transfer function, the conditions for a Butterworth characteristic can be derived. The transfer is described by:

$$H(j\omega) = V_o/V_i = \frac{1}{1 + j\omega(R_1 + R_2)C_2 - \omega^2 R_1 R_2 C_1 C_2}$$

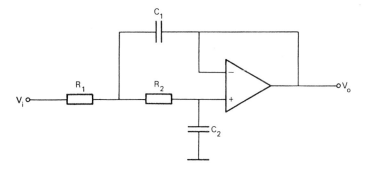

Figure 13.14 An example of an active second-order low-pass Butterworth filter.

The modulus of $H(j\omega)$ is:

$$|H(j\omega)| = [(1 - \omega^2 R_1 R_2 C_1 C_2)^2 + \omega^2 (R_1 + R_2)^2 C_2^2]^{-1/2}$$

This is a Butterworth characteristic (see Section 8.2.2) if the factor with ω^2 equals zero: $(R_1 + R_2)^2 C_2^2 = 2R_1 R_2 C_1 C_2$, or $C_1/C_2 = (R_1 + R_2)^2/2R_1 R_2$. To simplify the design, $R_1 = R_2 = R$, so $C_1 = 2C_2$.

The circuit can be extended to a third-order filter by adding a first-order low-pass section (Figure 13.15). At the output, an additional amplifier stage is connected, to achieve a low output impedance, and to offer the possibility of choosing an arbitrary gain. To derive the Butterworth condition of this filter, the transfer function is calculated. To simplify the calculation, we put directly $R_1 = R_2 = R_3 = R$. The modulus of the transfer function then becomes:

$$|H|^2 = \frac{(1 + R_5/R_4)^2}{[(1 - \omega^2 R^2 C_1 C_2)^2 + 4\omega^2 R^2 C_2^2](1 + \omega^2 R^2 C_3^2)}$$

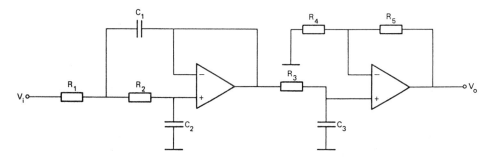

Figure 13.15 An example of an active third-order low-pass filter with Butterworth characteristic, consisting of a second-order and a first-order filter in cascade.

This expression has a Butterworth characteristic if the factors with ω^2 and ω^4 are zero. This results in:

$$-2C_1C_2 + 4C_2^2 + C_3^2 = 0$$

and

$$C_1^2 C_2^2 - 2C_1 C_2 C_3^2 + 4C_2^2 C_3^2 = 0$$

Both equations are satisfied for $C_1 = 4C_2$ and $C_3 = 2C_2$.

It will be clear from the foregoing that the derivation of the Butterworth conditions for filters of even higher order becomes very time-consuming. Furthermore, the equations can no longer be solved analytically, so numerical methods are required. In the literature much information can be found on the design of filters of higher order and other types, such as Bessel, Butterworth and Chebyshev filters.

SUMMARY

Circuits for time-domain operations

- With the basic inverting configurations of Figure 13.1, simple frequency-selective transfer functions can be realized, such as integrators, differentiators, band-pass filters and circuits with PD, PI and PID characteristics.
- The complex transfer function of the integrator of Figure 13.2a is $-1/j\omega RC$; in the time-domain: $v_o = -(1/RC) \int v_i \, dt$.
- Reduction of the effects of bias currents and offset voltages in the integrator circuit are achieved by:
  - compensation techniques similar to those with amplifiers;
  - restriction of the integrating range by adding a resistor in parallel with the capacitor.
- The complex transfer function of the differentiator of Figure 13.5 is given by $-j\omega RC$; in the time-domain: $v_o = -RC(dv_i/dt)$.
- For a proper operation of the differentiator, its range must be limited, for instance by adding a resistor in series with the capacitor.

Circuits with high frequency selectivity

- Second-order resonance filters are described by three parameters: the resonance frequency ω_0, the quality factor Q and the transfer at resonance H_0. The quality factor is the ratio between ω_0 and the bandwidth B; $Q = \omega_0/B$.

- Inductorless filters can be realized by applying active elements (amplifiers). Two basic configurations are positive feedback through a passive band-pass network and negative feedback through a notch network. In the first type, the filter parameters are sensitive to variations in component values, and the filter may even become unstable; the second type has a guaranteed stability and is less sensitive to component variations.
- Inductorless filters with Bessel, Butterworth and Chebyshev characteristics can be realized using operational amplifiers. The design of such filters is done with the aid of tables and plots that give the ratio of the component values at a given order and configuration.

Exercises

Circuits for time-domain operations

13.1 In the integrator circuit given in the figure below, the component values are $C = 1\,\mu F$ and $R = 10\,k\Omega$. The specifications of the operational amplifier are $|V_{off}| < 0.1\,mV$ and $|I_{bias}| < 10\,nA$. The input is supposed to be zero. At $t = 0$, the output voltage $v_o = 0$. What is the value of v_o after 10 s?

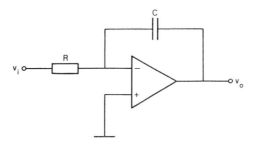

13.2 In the circuit shown in Exercise 13.1, a resistor R is connected between ground and the non-inverting input of the operational amplifier. What is v_o at $t = 10$ s, under the same conditions as in the preceding problem?

13.3 The circuit shown in Exercise 13.1 is now extended by a resistor $R_0 = 1\,M\Omega$ in parallel with C. What is v_o, under the same conditions as before?

13.4 (a) Give the Bode plot (amplitude characteristic only) of the following circuit shown in the figure below (asymptotic approximation).
(b) The circuit is used as a differentiator, so the phase shift should be 90°. At which frequency is this phase shift error more than 10°?

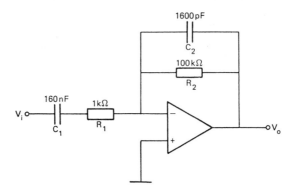

13.5 In the following PI circuit (see figure below), the component values of R_1, R_2 and C have to be chosen so that the following requirements are satisfied:
- input resistance $R_i > 10\ \text{k}\Omega$;
- proportional gain 2;
- integration range at least up to 100 Hz.

The operational amplifier may be considered as ideal.

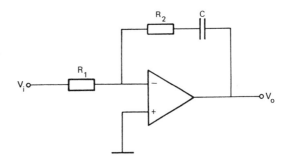

Circuits with high frequency selectivity

13.6 The transfer function of a certain second-order band-pass filter is given by:

$$H = V_o/V_i = \frac{(1 + j\omega\tau)^2}{3 + j\omega\tau/4 - 3\omega^2\tau^2} \quad \text{with } \tau = 10^{-3}\ \text{s}.$$

Find the resonance frequency ω_0, the transfer at very low and at very high frequencies and the quality factor Q.

13.7 Find the transfer function of the circuit in the figure below, and estimate the resonance frequency and the quality factor. $\tau = RC = 10$ ms, $R_1 = R_2 = R$ and $C_1 = C_2 = C$.

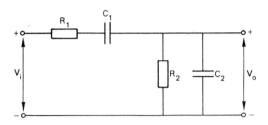

13.8 Convert the circuit of Figure 13.14 into a second-order high-pass filter with a Butterworth characteristic.

13.9 The circuit shown in the figure below is a filter consisting of two integrators and a differential amplifier. This filter is called a dual-integrator loop or state-variable filter. Find the transfer function. Which components should be made adjustable to vary Q independently of ω_0, and which used for tuning ω_0 independently of Q?

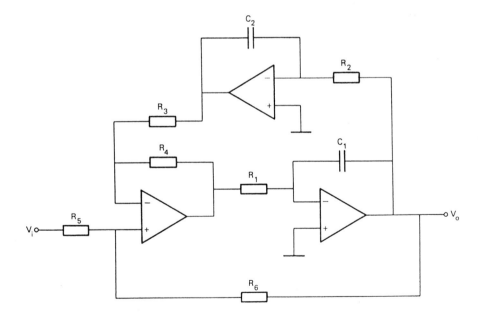

14 Non-linear signal processing with operational amplifiers

This chapter explains how to make circuits for non-linear transfer functions. We will first discuss some circuits for specific non-linear operations, such as the comparator, the Schmitt trigger and special non-linear circuits where pn diodes are used as electronic switches. Some particular functions, like logarithmic and exponential functions, employ the exponential relation between the current and the voltage of a pn diode or bipolar transistor (Chapters 9 and 10). Such electronic systems are available as complete integrated circuits, in common with those for multiplication and division. These circuits will be discussed in the second part of this chapter, together with some circuits for arbitrary non-linear functions.

14.1 Non-linear transfer functions

This section deals with circuits for the realization of some widely used non-linear functions. We will then discuss the comparator, the Schmitt trigger, active voltage limiters and rectifiers.

14.1.1 *Voltage comparators*

A voltage (or short) comparator is an electronic circuit that responds to a change in the polarity of a voltage difference. The circuit has two inputs and one output (Figure 14.1a). The output has only two levels: high or low, depending on the polarity of the voltage difference between the input terminals (Figure 14.1b). The comparator is frequently used for the determination of the polarity with respect to a reference voltage (Figure 14.1c).

Sometimes, an operational amplifier without feedback is used as a comparator. The high gain makes the output either maximally positive or maximally negative, depending on the input signal. However, an operational amplifier is rather slow; it takes too much time to return from the saturation state.

The specially designed comparators have a much faster recovery time, with response times as low as 10 ns. They have an output level compatible with the levels used in digital electronics (0 and 5 V). Other properties correspond to a normal operational amplifier; the circuit symbol is similar to that of the operational amplifier. Table 14.1 shows some specifications of two different types of comparators, a fast type and an accurate type.

Non-linear transfer functions 221

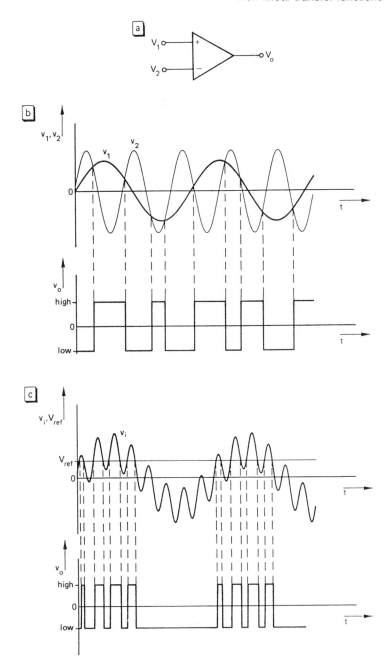

Figure 14.1 (a) The circuit symbol of a comparator; (b) comparator output voltage v_o as a function of two sinusoidal input voltages v_1 and v_2; (c) the output for an input voltage v_i that is compared to a reference voltage V_{ref}.

Table 14.1 Specifications of two types of comparators: a fast one (type I) and an accurate one (type II)

		Type I	Type II
Voltage gain	A	–	$2 \cdot 10^5$
Voltage input offset	V_{off}	± 2 mV ± 8 μV/K	± 1 mV, max. ± 4 mV
Input bias current	I_{bias}	5 μA	25 nA, max. 300 nA
Input offset current	I_{off}	0.5 μA $\pm$ 7 nA/K	3 nA, max. 100 nA
Input resistance	R_i	17 kΩ	–
Output resistance	R_o	100 Ω	80 Ω
Response time	t_r	2 ns	1.3 μs
Output voltage, high	$V_{o,h}$	3 V	–
Output voltage, low	$V_{o,l}$	0.25 V	–

Integrated circuits exist with two to four comparators on one chip. As the pins for the power supply are combined, the total number of pin connections can be kept reasonably low. There are also comparators with two complementary outputs; when one output is high, the other is low and vice versa. Other types have an additional control input terminal to switch off the whole circuit. If this strobe input is high (or low, depending on the type), the output of the comparator is high too, irrespective of the input polarity. At low (or high) strobe input, the circuit operates as a normal comparator.

The comparator gives information not only about the polarity of the input voltage, but also about the moment of polarity change. This property makes the comparator a useful device in various kinds of counter circuits. Imperfections of the comparator (offset, time delay) cause uncertainty in the output; the same holds for input signal noise (Figure 14.2).

The time error increases with decreasing slope of the input signal and increasing noise level.

14.1.2 Schmitt trigger

Noise in the input signal causes fast, irregular changes of the comparator output (Figure 14.3a). To reduce or eliminate this output jitter, hysteresis is intentionally added to the comparator function (Figure 14.3b). The output switches from low to high as soon as v_i exceeds the upper reference level V_{ref1}, and from high to low as soon as v_i drops below the lower level V_{ref2}. For a proper operation, the hysteresis $V_{ref1} - V_{ref2}$ must be more than the noise amplitude. However, a large hysteresis causes large timing errors in the output signal.

The required hysteresis is achieved by an operational amplifier with positive feedback (Figure 14.4). A fraction β from the output voltage is fed back to the positive input terminal: $\beta = R_1/(R_1 + R_2)$. The circuit is called a Schmitt trigger; its operation is as follows: suppose the most positive output voltage is V^+ (about the positive

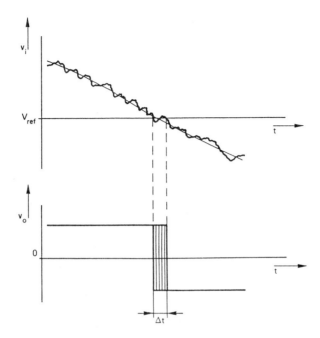

Figure 14.2 Uncertainty in the timing of the comparator, due to noise in the input signal.

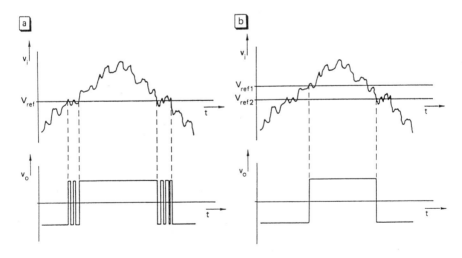

Figure 14.3 (a) Jitter in the output of a comparator due to input noise; (b) reduction of jitter by introducing hysteresis.

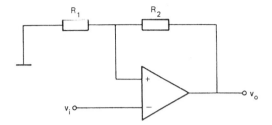

Figure 14.4 A Schmitt trigger, realized with an operational amplifier with positive feedback (comparator with hysteresis).

power supply voltage), and the most negative output is V^-. The voltage at the plus terminal is either βV^+ or βV^-. When v_i is below the voltage on the plus terminal, then v_o equals V^+ (because of the high gain); this is a stable situation as long as $v_i < \beta V^+$. If v_i increases up to βV^+, the output decreases sharply and so does the voltage at the plus terminal. The voltage difference between both input terminals decreases much faster than v_i increases, so, within a very short time, the output becomes maximally negative (V^-). As long as $v_i > \beta V^-$, the output remains $v_o = V^-$, a new stable state.

The comparator levels of the Schmitt trigger are, apparently, βV^+ and βV^-; the hysteresis gap is $\beta(V^+ - V^-)$. Thanks to the positive feedback, even a rather slow operational amplifier can have a fast response time as the Schmitt trigger. Evidently, a comparator circuit is even better.

The switching levels can be varied by connecting R_1 to a reference voltage source V_{ref}. Both levels shift by a factor $V_{ref}R_1/(R_1 + R_2)$. In most cases, the hysteresis is small compared to the input or output signal, so R_1 is small compared to R_2. The switching levels are approximated by $V_{ref} + (R_1/R_2)v_{o,max}$ and $V_{ref} + (R_1/R_2)v_{o,min}$, respectively.

14.1.3 *Voltage limiters*

In Section 9.2.1, we studied voltage limiters comprising only resistors and diodes. Compared to the simplicity of those circuits, their limiting level is not accurately fixed. After all, this level is determined by the diode threshold voltage, which shows dispersion and is temperature dependent (about $-2.5\,\text{mV/K}$). These disadvantages can be overcome using operational amplifiers. Figure 14.5 depicts the basic arrangement for an active limiter, with adjustable limiting levels and a transfer in the linear range.

The operation is as follows: suppose diodes D_1 and D_2 are either both reverse biased (infinite resistance) or both conducting (zero resistance, voltage V_D). The circuit has two stable states. At a positive input voltage, current flows through R_1 and D_1 to the output of the amplifier. The amplifier has a proper feedback via D_1. The minus terminal of the operational amplifier is virtually grounded and the output voltage

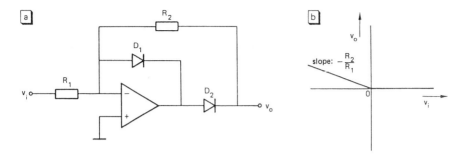

Figure 14.5 (a) Active voltage limiter; (b) corresponding transfer characteristic.

of the amplifier equals $-V_D$, the threshold voltage of D_1. This voltage is sufficiently negative to maintain D_2 as reverse biased; there is no current flowing through R_2, hence the output voltage of the circuit is zero.

A negative input voltage causes a current to flow from the output of the amplifier via D_2, R_2 and R_1 to the input. Now, the operational amplifier is fed back via the conducting diode D_2, whereas diode D_1 is reverse biased because v_o is positive. The output voltage in this state equals $v_o = -(R_2/R_1)v_i$. Figure 14.5b shows the transfer characteristic.

The circuit in Figure 14.5a limits the output voltage to a minimum value, which is zero in this case. Limitation up to a maximum value (zero) is achieved by reversing the connections of both diodes. The resulting transfer function is depicted in Figure 14.6b, curve 1.

Figure 14.6a shows a configuration that allows shifting of the characteristic in horizontal and vertical directions. The voltage V_{ref1} shifts the curve horizontally. As explained before, the current flows either through D_1 or through D_2 and R_2.

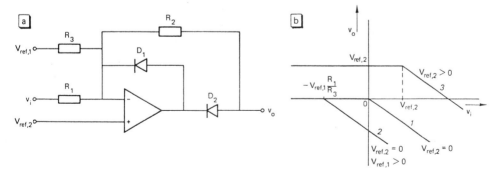

Figure 14.6 (a) Voltage limiter with adjustable limiting levels for the input and output voltage; (b) corresponding transfer characteristics. Curve 3 holds for non-connected input voltage V_{ref1} (floating terminal).

According to Kirchhoff's rule for currents, this must be equal to $V_{ref1}/R_3 + v_i/R_1$. The break point in the characteristic occurs at zero current, hence for $v_i = -V_{ref1}R_1/R_3$ (Figure 14.6b, curve 2). A vertical shift is obtained with a voltage V_{ref2} connected to the plus terminal of the operational amplifier (Figure 14.6b, curve 3).

Apparently, the transfer characteristic does not depend on the diode threshold voltages, due to the high gain of the amplifier; a slight change of the input voltage is enough to switch the diodes. The inaccuracy of the transfer characteristic is only determined by the offset of the operational amplifier.

14.1.4 Rectifiers

The goal of a double-sided rectifier is the generation of a transfer function that satisfies $y = |x|$. For a single-sided rectifier, $y = x$ ($x > 0$) and $y = 0$ ($x < 0$); x and y are voltages or currents.

A double-sided rectifier with passive components is the Graetz diode bridge, discussed in Section 9.2. It operates only for signals that exceed the threshold voltage of the diodes. A simple rectifier circuit without that disadvantage is depicted in Figure 14.7. At negative input currents, D_1 is forward biased and D_2 is reverse biased. The voltage at the plus terminal of the operational amplifier is zero, and so the circuit behaves as the current-to-voltage converter of Figure 12.2: $v_o = -Ri_i$. At positive input currents, D_2 is forward biased and D_1 reverse biased. In this situation, the circuit behaves as the buffer amplifier of Figure 12.4, with input signal i_iR and the same output signal.

The circuit of Figure 14.5 can serve as an inverting single-sided rectifier. When the inversion is not wanted, another inverter may be connected to its output. The inverting rectifier can be extended to a double-sided voltage rectifier by adding twice the input signal to the inverted single-sided rectified signal. This is illustrated in Figure 14.8 with a triangular input voltage.

When applying the circuits described above, the imperfections of the operational amplifier should be kept in mind. In particular, the limited bandwidth and slew rate may cause trouble when processing signals with relatively high frequencies.

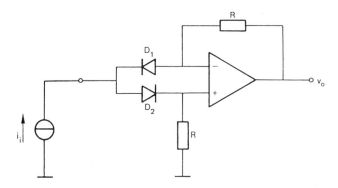

Figure 14.7 A current–voltage converter that operates as a double-sided rectifier.

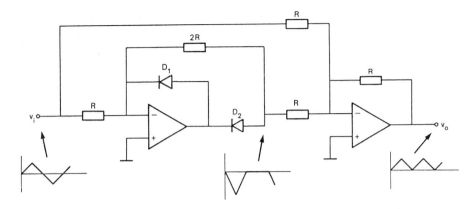

Figure 14.8 A double-sided voltage rectifier, consisting of a single-sided rectifier and a summing circuit.

Beside the circuits discussed in this section, there is quite a number of other configurations for rectifiers and limiters described in many electronics textbooks. Some of these circuits have severe disadvantages. For instance, the operational amplifier has no feedback in one of the stable states, slowing its speed down substantially; others work only at a particular load resistance. So, despite the charming simplicity of such circuits, care should be taken when applying them.

14.2 Non-linear arithmetic operations

14.2.1 Logarithmic converters

Figure 14.9 displays the principle of a logarithmic voltage converter. As the plus terminal of the ideal operational amplifier is connected to ground, the minus terminal is virtually grounded. So, the current through R is exactly v_i/R, and flows entirely through the diode. For this diode, the relation between the voltage and the current is given as $I_D = I_0 e^{qV_D/kT}$ or $V_D = (kT/q)\ln(I_D/I_0)$. This results in an output voltage equal to:

$$v_o = -V_D = -\frac{kT}{q}\ln\frac{I_D}{I_0} = -\frac{kT}{q}\ln\frac{v_i}{RI_0}$$

$$= \left(-\frac{kT}{q}\log\frac{v_i}{RI_0}\right)(\log e)^{-1}$$

At room temperature, $kT/q \approx 25$ mV; further, $1/\log e \approx 2.3$, hence,

$$v_o = -0.06\log(v_i/RI_0)$$

The output voltage decreases by about 60 mV at ten times the input voltage. The validity of this exponential relation extends over a current range from about 10 nA to several milliamps. At elevated currents, the series resistance of the diode disturbs the exponential relation; at very low currents the relation is not valid either.

The relation between the collector current I_C and the base–emitter current V_{BE} of a bipolar transistor is also exponential, valid over a much wider range: from several picoamps to several milliamps. Figure 14.9b shows how this property of a bipolar transistor is employed in a logarithmic converter. The collector voltage is virtually grounded, so the base–collector junction is not forward biased, a condition for the proper operation of the transistor. Similar to the circuit with the diode, the output of this circuit is:

$$v_o = -\frac{kT}{q} \ln \frac{v_i}{RI_0} = -0.06 \log(v_i/RI_0)$$

Both circuits operate for positive input voltages only. To allow negative input voltages, the polarity of the diode in Figure 14.9a must be reversed and in Figure 14.9b the npn transistor must be replaced by a pnp type; however, the circuits maintain their unipolar operation.

The major shortcoming of the logarithmic converters of Figure 14.9 is their strong temperature sensitivity. Two terms are responsible for this: kT/q and the leakage current I_0. The latter can easily be compensated for by a second converter with identical structure, operating at the same temperature, for instance as depicted in Figure 14.10.

Subtracting one output from the other results in a voltage proportional to the ratio of the two input signals, and independent of I_0, provided both transistors have the same leakage current. The remaining temperature coefficient, due to the factor kT/q, is only 1/300 or 0.33% per K (at room temperature, 300 K).

Other shortcomings are caused by imperfections of the operational amplifier, such as the bias currents and the offset voltage. Both contribute to an additional current through the diode or transistor, which amounts in total to:

$$I_D = \frac{v_i}{R} + \left|\frac{V_{off}}{R}\right| + |I_{bias}|$$

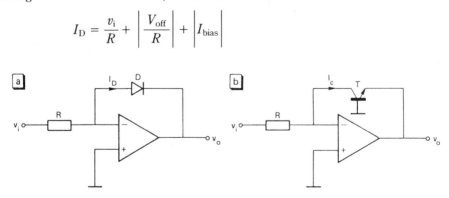

Figure 14.9 Basic circuit of a logarithmic converter, (a) with a pn diode; (b) with a bipolar transistor.

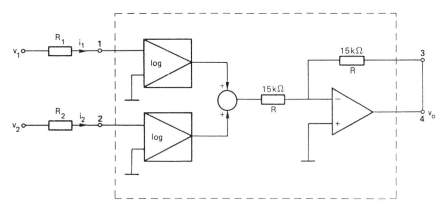

Figure 14.10 Basic scheme of an integrated logarithmic converter, internally adjusted to base 10.

We put the modulus of I_{bias} and V_{off} because we do not know the polarity of these quantities. The output of the diode circuit becomes $v_o = V_{off} - V_D$ (V_D includes the additional current due to V_{off} and I_{bias}); the output of the transistor circuit amounts $v_o = -V_{BE}$: although the collector voltage is $|V_{off}|$, it does not affect the base–emitter voltage. The influence of both V_{off} and I_{bias} can be compensated in a similar way as in linear amplifier circuits (Section 12.2.2).

Figure 14.10 gives the internal structure of a commercial-type logarithmic converter. The blocks with 'log' are circuits according to Figure 14.10b, but without resistor R, which is connected externally. These converters exist in two types: 'P' and 'N', for positive and negative input voltages, respectively. Table 14.2 lists the main specifications of such a converter. This converter is composed of two logarithmic converters, a subtractor (differential amplifier) and an output amplifier. The output voltage is proportional to the ratio between the two input currents I_1 and I_2, that is why it is called a log ratio converter. The transfer (or scaling factor) can be determined (between certain boundaries) by the user of the system, simply by putting an external resistor between pins 3 and 4. The low input impedance of the integrated circuit also allows the conversion of input voltages, by connecting external resistors in series with the inputs.

Example 14.1
A voltage v_i should be converted into a voltage $v_o = -2\log(v_i/V_{ref})$, using the converter of Figure 14.10 and corresponding specifications of Table 14.2. The voltage v_i ranges from 10 mV to 10 V; there is a reference voltage available of 10 V.

To minimize errors due to bias currents and offset voltages, we make the currents as large as is allowed, for instance $I_1 = I_2 = 1$ mA. This is achieved with $R_1 = R_2 = 10$ kΩ. The scaling factor is set at 2 (twice the nominal value) by connecting a resistor of 15 kΩ between terminals 3 and 4. At the lowest input voltage,

Table 14.2 Specifications of a log ratio converter

Transfer function	$V_o = -K \log I_1/I_2$
Input current range	$I_1, I_2; (+/-)$ 1 nA ... 1 mA
Scaling factor K	1 V/decade ± 1% ± 0.04%/K
Bias current	I_{bias1}, I_{bias2}: 10 pA (doubles per 10 K)
Input offset voltage	$V_{off,i} = \pm 1$ mV(max.) ± 25 µV/K
Output offset voltage	$V_{off,o} = \pm 15$ mV(max., adjustable to zero) ± 0.3 mV/K

the output is $-2 \log 10^{-3} = 6$ V. The output error at that input voltage consists of two parts: one due to $V_{off,o}$: $2 \cdot (\pm 15)$ mV (the gain is twice the nominal value), and the other due to $V_{off,i}$: $v_o = -2 \log(10 \text{ mV} \pm 1 \text{ mV})/10$ V which lies between the values 5.98 and 6.09 V, hence a maximum error of about 90 mV. Other errors are negligible compared to these offset errors.

14.2.2 Exponential converters

The pn diode and the bipolar transistor can also be applied for the realization of an exponential converter (sometimes called an antilog converter) as is shown in Figure 14.11.

Assuming an ideal operational amplifier, the output voltage equals $v_o = IR = I_0 R \, e^{-qv_i/kT}$. The minus sign comes from the fact that $v_o = V_{EB} = -V_{BE}$. The circuit operates for negative input voltages only; a circuit with a pnp transistor instead of an npn transistor operates for positive input voltages only.

The strong temperature sensitivity of I_0 can be compensated for using the same method as applied for the logarithmic converter, by adding a second, identical circuit. The remaining temperature coefficient due to the factor kT/q is sometimes reduced by a built-in temperature compensation circuit. The transfer of such an exponential converter can be written as $v_o = -V_{ref} e^{-v_i/K}$, with V_{ref} an internal or external reference voltage and K a scaling factor. Some of the available types can also be used as a logarithmic converter, depending on the external connections. Figure 14.12 shows

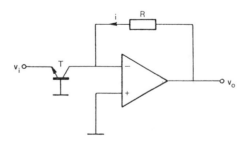

Figure 14.11 Basic circuit of an exponential voltage converter.

Non-linear arithmetic operations

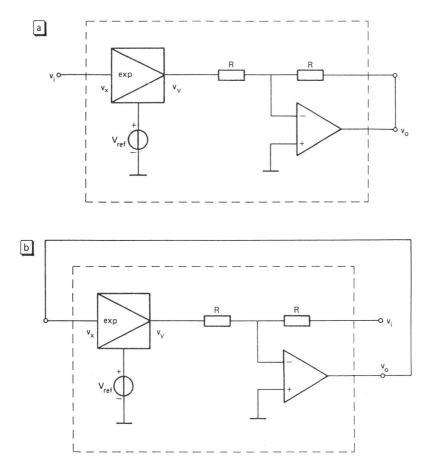

Figure 14.12 An integrated exponential converter: (a) connected as an exponential converter; (b) connected as a logarithmic converter.

that possibility. The system consists of an exponential converter with the transfer as mentioned before, a reference voltage V_{ref} and an operational amplifier. In Figure 14.12a, the device is connected as an exponential converter and in Figure 14.12b as a logarithmic converter. The transfer in the exponential mode is $v_o = -v_y = V_{ref} e^{-v_i/K}$. For the log mode, $v_o = v_x$ and $v_i = -v_y$, so $v_o = -K \ln(v_i/V_{ref})$. The operational amplifiers are assumed to be ideal.

14.2.3 *Multipliers*

Most commercial analog multipliers are based on a combination of logarithmic and exponential transfer functions. Figure 14.13 depicts the functional diagram of a

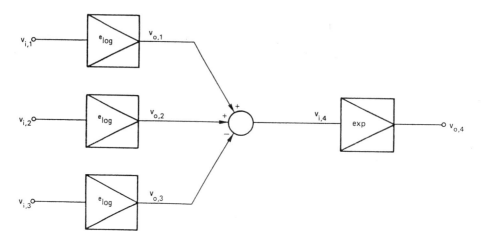

Figure 14.13 Functional diagram of an analog multiplier based on exponential and logarithmic converters and a summing circuit.

multiplier composed of the exponential and logarithmic converters of Figures 14.9 and 14.11. The transfer of the logarithmic converters is $v_{o,j} = -(kT/q)\ln(v_{i,j}/I_0 R_j)$, $j = 1, 2, 3$, whereas the transfer of the exponential converter is $v_{o,4} = I_0 R_4 e^{-qv_{i,4}/kT}$. The voltage after the summation point is $-(kT/q)\ln(v_{i,1}v_{i,2}R_3/v_{i,3}I_0 R_1 R_2)$, so, the output voltage is:

$$v_{o,4} = \frac{v_{i,1}v_{i,2}}{v_{i,3}} \frac{R_3 R_4}{R_1 R_2}$$

Despite the strong temperature dependence of the individual converters, the output does not depend on the leakage current I_0 and the temperature T (given identical converters and the same temperature). An additional advantage is the possibility of the circuit to act as a divider. The main disadvantage is the unipolarity; the multiplication is performed only in one quadrant.

To eliminate this latter shortcoming, multipliers are constructed that can handle both polarities. Such multipliers are also based on logarithmic and exponential relations of a bipolar transistor, and are available as integrated circuits.

Table 14.3 is part of the specification list of a low-cost analog multiplier integrated circuit.

This integrated device has three inputs: the X, Y and Z input. The meaning of the Z input will be discussed in Section 14.2.4. The offset voltages of the three inputs can be individually adjusted to zero, by connecting compensation voltages to three extra terminals. The manufacturer specifies in the data sheet how to do this. The cross-over in the specification list is the output voltage due to only one input voltage

Table 14.3 Specifications of an analog multiplier

Transfer function	$V_o = K v_x v_y$; $K = 0.1$ V^{-1}
Scale error	$\pm 2\% \pm 0.04\%/$K
Max. input voltages	± 10 V
Non-linearity	$v_x = v_o = 20$ V (peak–peak); $\pm 0.8\%$
	$v_y = v_o = 20$ V (peak–peak); $\pm 0.3\%$
Cross-over (peak–peak value 50 Hz)	$v_x = 20$ V, $v_y = 0$; $v_o < 150$ mV
	$v_y = 20$ V, $v_x = 0$; $v_o < 200$ mV
Input resistance X input	10 MΩ
Y input	6 MΩ
Z input	36 kΩ
Output resistance	100 Ω
Output offset voltage	adjustable to zero
t.c. of the offset voltage	± 0.7 mV/K
Bias current X, Y input	3 µA
Z input	± 25 µA
Bandwidth	750 kHz

where all other input voltages are zero. The non-linearity of the device is specified for the individual X and Y channel, at maximum input voltage of both channels.

14.2.4 *Other arithmetic operations*

The multiplier discussed in the preceding section can also be applied for division and for the realization of square power and square root transfer functions. A quadratic transfer function is achieved by simply connecting the input signal to both inputs of the multiplier: $v_o = K v_i^2$. Division and extraction of the square root is obtained by a feedback configuration using an operational amplifier. Most commercial devices have a built-in amplifier for that purpose.

Figure 14.14 shows the functional structure of a divider circuit. Here, the Z input from the section before is shown as well. At proper feedback, the voltage at the minus terminal of the operational amplifier is zero, so, $v_z R_2/(R_1 + R_2) + K v_x v_y R_1/(R_1 + R_2) = 0$. Furthermore, $v_y = v_o$, so $v_o = -v_z R_2/K v_x R_1$; the output is proportional to the ratio of v_z and v_x. The proportionality factor can freely be chosen with R_1 and R_2. The divider operates only for positive values of v_x; for negative values the feedback changes into feedforward, resulting in instability of the system. The error of the division operation depends on the errors of the multiplier and the operational amplifier. Assume an output offset voltage of the multiplier $V_{\text{off, a}}$ and an input offset voltage of the operational amplifier $V_{\text{off, b}}$. Then:

$$\frac{(K v_x v_y + V_{\text{off, a}}) R_1}{R_1 + R_2} + \frac{v_z R_2}{R_1 + R_2} = V_{\text{off, b}}$$

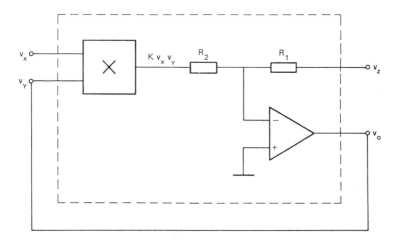

Figure 14.14 An example of a divider circuit: $v_o = -(R_2/KR_1) \cdot (v_z/v_x)$. The multiplier is in the feedback loop of the operational amplifier. The IC can also be connected as multiplier and as square root.

and thus, with $v_o = v_y$:

$$v_o = -\frac{v_z R_2}{K v_x R_1} + \frac{V_{\text{off,b}}(R_1 + R_2)}{K v_x R_1} - \frac{V_{\text{off,a}}}{K v_x}$$

It appears that the error in v_o can be very large at low values of v_x.

The circuit of Figure 14.14 can be used for making a square root transfer function. The X and Y inputs are connected together. As now $v_x = v_y = v_o$, the output voltage is $v_o = -R_2 v_z/K v_o R_1$, or $v_o = \sqrt{-R_2 v_z/KR_1}$. Evidently, v_z can only be negative.

Transfer functions with other powers than 2 or ½ can be realized with combinations of logarithmic and exponential converters as well. The general principle is depicted in Figure 14.15. Here, m is an amplifier ($m > 1$) or an attenuator ($m < 1$). For $m = 2$ and $m = ½$, the transfer corresponds to a square power and a square root, respectively. These circuits have a better accuracy compared to those with analog multipliers. A disadvantage is their relative complexity and unipolarity.

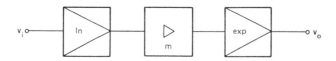

Figure 14.15 An arbitrary power function is realized with a logarithmic converter, an amplifier/attenuator and an exponential converter: $v_o = K v_i^m$.

Example 14.2
The mass flow Φ_m of a gas satisfies the equation $\Phi_m = F\sqrt{P\Delta p/T}$, with P the total pressure, Δp the pressure difference across the flow meter and T the absolute temperature; F is a constant. The quantities P, Δp and T can be measured with electronic sensors. Assume the output signals of those sensors are x_P, $x_{\Delta p}$ and x_T, respectively. An electronic circuit for the determination of Φ_m is shown in Figure 14.16. The operation follows from the figure. The output signal is
$x_o = (x_P x_{\Delta p} / x_T x_{ref})^{1/2}$.

14.2.5 Piecewise linear approximation of arbitrary transfer functions

An arbitrary transfer function $y = f(x)$ can be approximated with segments of straight lines (Figure 14.17). The smaller the segments (or the more segments within the interval $[x_{min}, x_{max}]$), the better is the approximation.

The approximation is achieved by a combination of equally shaped elementary functions (Figure 14.17b), in which the transfer function of the limiter from Section 4.2.3 can be recognized. To simplify the explanation, we will take that circuit as the starting point of the segmentized approximation.

From Figure 14.17b, it is clear that the segmented function is realized by weighted addition of the elementary functions. For each of these subfunctions, we take the circuit as in Figure 14.14a, with fixed values of R_1, R_2 and R_3, and adjustable reference voltages. The addition is performed with an operational amplifier according to the method from Figure 12.3b. If needed, a normal linear transfer can be added as well. Figure 14.18 shows an example of such a configuration with corresponding transfer characteristic, containing two break points (or three segments). The reference voltages $V_{ref,j}$ determine the position of the break points; with resistances $R_{i,j}$ the slope of the characteristic can be adjusted from the jth break point. The slope $-R_t/R_{i,j}$ of the linear function is reduced at each next break point with an amount

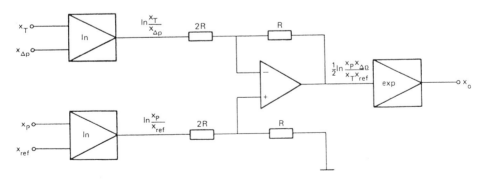

Figure 14.16 An analog electronic circuit for the determination of $\phi_m = F\sqrt{P\Delta p/T}$; the logarithmic converters are of the type as given in Figure 14.10.

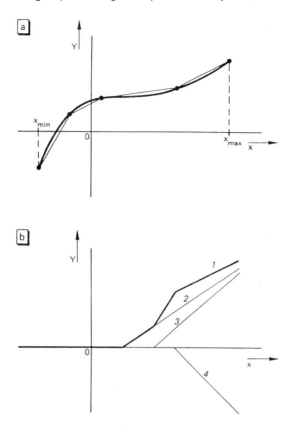

Figure 14.17 (a) Piecewise linear approximation by a number of segments; (b) an example of function 1, composed of elementary functions 2, 3 and 4.

$-R_t/R_{i,j}$ times the (negative) slope of the jth limiter. The whole characteristic can be rotated around the origin by varying R_t.

SUMMARY

Non-linear transfer functions

- A comparator is a differential amplifier with very high gain and fast response. Its output is either high or low, depending on the polarity of the input voltage.
- A Schmitt trigger is a comparator with premeditated hysteresis. Like the comparator, it is used for the determination of the sign (polarity) of a voltage difference.

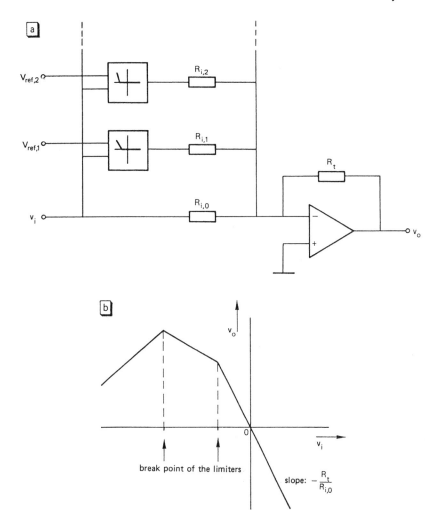

Figure 14.18 (a) A circuit configuration for the weighted summation of elementary transfer functions from Figure 14.6b, using the circuits in Figure 14.5a; (b) piecewise linear approximation of a non-linear transfer function.

- The Schmitt trigger has a better noise immunity due to its hysteresis; this hysteresis introduces an additional error in time or in amplitude, when comparing two signals.
- A Schmitt trigger is realized with an operational amplifier with positive feedback.
- In active limiters and rectifiers made from diodes and operational amplifiers, the switching levels are independent of the diode forward voltage.

Non-linear arithmetic operations

- Logarithmic and exponential signal converters are based on the exponential relation between the current and the voltage of a pn diode or the even more accurate exponential relation between the collector current and the base–emitter voltage of a bipolar transistor.
- The inherently strong temperature sensitivity of a logarithmic and an exponential converter, due to the leakage current, is eliminated by compensation with a second, identical diode or transistor.
- Most analog signal multipliers are based on a combination of logarithmic and exponential converters; they are available as single-quadrant or as four-quadrant multipliers.
- Analog multipliers can also be used for division and square rooting. Many integrated multipliers have built-in circuits for that purpose and can be employed by the user by making the appropriate external connections.
- When applying analog multipliers, special care should be taken with the imperfections in the integrated circuits, such as voltage offset and non-linearity.
- Arbitrary non-linear transfer functions can be realized by a piecewise linear approximation, built up with a combination of limiters and a summing circuit.

Exercises

Non-linear transfer functions

14.1 A voltage v_g from a source with source resistance $R_g = 1$ kΩ is compared with a reference voltage V_{ref} whose inaccuracy is ± 2 mV. A comparator is used with specifications as listed in Table 14.1, type I. Calculate the total inaccuracy of this comparison expressed in mV, over a temperature range from 0 to 70°C.

14.2 Find the switching levels and the hysteresis of the Schmitt trigger depicted in the figure below. The operational amplifier has limited output voltages of -15 and 15 V.

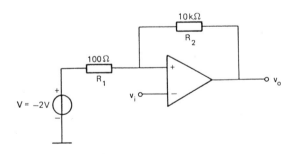

14.3 Draw in one figure the output voltage, the voltage at the plus terminal and the input voltage of the circuit from Exercise 14.2. Assume a triangular input voltage with amplitude 5 V and zero mean value.

14.4 For the circuit in Figure 14.6, $R_1 = R_2 = \tfrac{1}{2}R_3 = 10\,\text{k}\Omega$; $V_{ref1} = -3\,\text{V}$, $V_{ref2} = 2\,\text{V}$. Make a sketch of the transfer v_o/v_i.

14.5 Draw the voltages at the points indicated by the arrows in Figure 14.8, for the case when all diodes are reversely connected, but for the same input signal.

14.6 Find the transfer function of the circuit given in the figure below. Make a sketch of its transfer function v_o/v_i. The forward voltage of the diodes is 0.5 V.

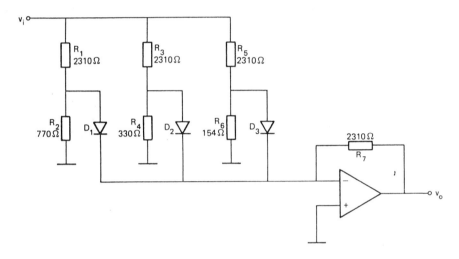

Non-linear arithmetic operations

14.7 A logarithmic converter with transfer $v_o = K_L \ln(v_i/V_L)$ and an exponential converter with transfer $v_o = K_E\, e^{v_i/V_E}$ are called complementary if the total transfer of the converters in series equals 1. Find the condition for this with respect to K_L, K_E, V_L and V_E.

14.8 Find the transfer functions v_o/v_i of the two circuits shown below. The scaling factor is $K = 1$; the operational amplifier has ideal properties.

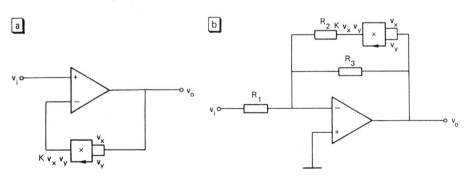

240 Non-linear signal processing with operational amplifiers

14.9 Find the transfer function of the following system shown below. The transfer of the log converter is $v_p = -\log(v_i/10)$ and that of the exponential converter is $v_o = -10^{-v_q/10}$; $A = 5$.

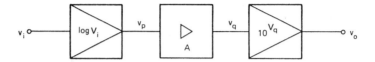

14.10 Design a circuit with logarithmic and exponential converters and with only one operational amplifier, that realizes the function:

$$V_o = \frac{V_3}{V_1^{1/3} \cdot V_2^{2/3}}$$

14.11 The bias currents of a log ratio converter with transfer $K\log(I_1/I_2)$ are 1 nA at maximum. $K = 1$ V. Calculate the output voltage for $I_1 = I_2 = 1\ \mu A$. What is the error in the output for $I_1 = 10\ \mu A$ and $I_2 = 100\ \mu A$?

15 Electronic switching circuits

This chapter deals with electronic switches and circuits built up with such switches. In the first part we will discuss some general properties of electronic switches and introduce various components operating as electronically controllable switches. In the second part of the chapter, two circuits are described that make use of electronic switches: time multiplexers and sample-hold circuits. Both these circuits are widely used in electronic measurement systems.

15.1 Electronic switches

Many components can serve as switches. We have already met most of them in preceding chapters. In this section, we will look in particular at their properties as an electronic switch, reviewing the reed switch, photoresistor, pn diode, bipolar transistor, junction field-effect transistor, MOSFET and thyristor. We start with some general properties of electronic switches.

15.1.1 *Properties of electronic switches*

When on, an ideal switch is a perfect short-circuit between the two terminals 1 and 2 (Figure 15.1a) and when off a perfect isolation. Furthermore, the ideal switch has zero response time (it switches directly upon a command). Finally, the control terminal 3 is isolated from the circuit terminals which, in turn, are isolated from ground (floating switch). Obviously, an actual switch only partially approaches this ideal behaviour. The major aberrations are modelled in Figure 15.1b. This model may be quite complicated, but still does not take into account all dynamic properties of the switch. Usually, the dynamic properties of a switch are specified in a time diagram. Figure 15.2 explains the commonly used specifications with respect to the dynamic behaviour. The quantity x_i is connected to the one terminal (input); x_o is the quantity at the other terminal (output).

The quantity to be switched on and off may be a voltage or a current. In practice, two types of switches are distinguished: voltage switches and current switches. This distinction is based on the particular imperfections rather than on the construction. For instance, a switch with a large offset voltage V_{off} can be used better as a current switch than as a voltage switch.

242 Electronic switching circuits

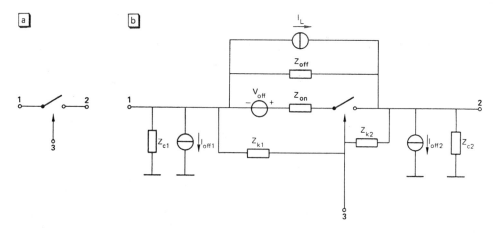

Figure 15.1 (a) Circuit symbol of an electronically controllable switch; (b) model of some imperfections of an actual switch. Z_{on} = on-impedance; Z_{off} = off-impedance; Z_c = leakage impedance to ground; Z_k = cross-over impedance between circuit terminals and control terminal; V_{off} = offset voltage; I_{off} = leakage current from circuit terminals to ground; I_L = leakage current through the switch.

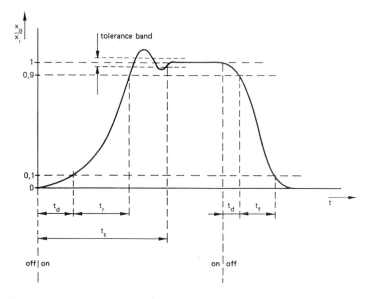

Figure 15.2 Dynamic properties of a switch, in terms of time delays. $t_d(on)$ = turn-on delay time; t_r = rise time; t_s = settling time (time period between the on-command and the output being within a specified error band around the steady state); $t_d(off)$ = turn-off delay time; t_f = fall time.

Electronic switches 243

The major imperfections of an electronic switch are its on- and off-resistance. In combination with the connected circuits, they may cause serious transfer errors. Figure 15.3 shows three different configurations, where a voltage from a signal source with source resistance R_g is switched to a load R_L (input resistance of a system that is connected to the output terminals of the switch).

For each of these configurations we can derive the transfer in both the on- and off-state, using the formula of the voltage divider.

Series switch (Figure 15.3a)

On: $\quad v_o/v_g = \dfrac{R_L}{R_g + r_{on} + R_L}$

Off: $\quad v_o/v_g = \dfrac{R_L}{R_g + r_{off} + R_L}$

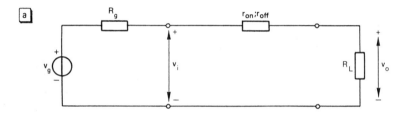

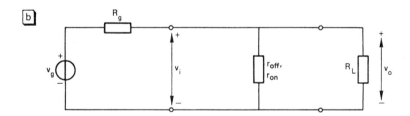

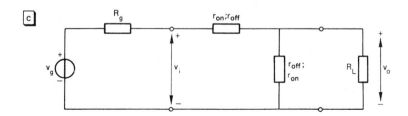

Figure 15.3 Voltage switching with (a) a series switch; (b) a shunt switch; (c) a series shunt switch.

With a perfect switch, the transfer in the on-state should be equal to $R_L/(R_g+R_L)$. So, to estimate this ideal transfer, the on-resistance of the switch must satisfy $r_{on} \ll R_g + R_L$. As usually $R_g \ll R_L$, this reduces to $r_{on} \ll R_L$. The ideal transfer in the off-state is zero; hence, the requirement for the switch is $r_{off} \gg R_L$.

Shunt switch (Figure 15.3b)

On: $$v_o/v_g = \frac{R_L // r_{off}}{R_g + R_L // r_{off}} = \frac{R_L}{R_g R_L / r_{off} + R_g + R_L}$$

Off: $$v_o/v_g = \frac{R_L // r_{on}}{R_g + R_L // r_{on}} = \frac{r_{on}}{r_{on}(1 + R_g/R_L) + R_g}$$

To estimate the ideal transfer, the on-resistance must satisfy $R_g R_L / r_{off} \ll R_g + R_L$ or, as usually $R_g \ll R_L$, the requirement is $r_{off} \gg R_g$. For zero transfer in the off-state: $r_{on}(1 + R_g/R_L) \ll R_g$ or, as $R_g \ll R_L$: $r_{on} \ll R_g$.

Series shunt switch (Figure 15.3c)

On: $$v_o/v_g = \frac{r_{off} // R_L}{R_g + r_{on} + r_{off} // R_L} = \frac{R_L}{R_g R_L/r_{off} + r_{on} R_L/r_{off} + r_{on} + R_g + R_L}$$

Off: $$v_o/v_g = \frac{r_{on} // R_L}{R_g + r_{off} + r_{on} // R_L} = \frac{r_{on}}{(1 + r_{on}/R_L)(r_{off} + R_g) + r_{on}}$$

For the same reasons, the requirements for the on- and off-resistances are $r_{off} \gg R_g$; $r_{off} \gg r_{on}$ and $r_{on} \ll R_L$. The transfer in the off-state can be approximated by r_{on}/r_{off}, using the conditions $r_{on} \ll R_L$ and $r_{off} \gg R_g$.

The requirements are summarized in Table 15.1. In most cases, R_L is large and R_g is small. This interferes with the requirement $r_{off} \gg R_L$ for the series switch and $r_{on} \ll R_g$ for the shunt switch. The series shunt switch does not have this problem: the only requirement is $r_{off} \gg r_{on}$, independent of the source and load, and easily met with most switch types. A disadvantage of the series shunt switch is the need for two complementary switches. In particular applications, however, this is even an advantage, as is shown in Section 15.2.

15.1.2 *Components as electronic switches*

Reed switch

The reed switch is a mechanical switch, consisting of two tongues (reeds) of nickel–iron, encapsulated in a glass tube filled with nitrogen or another inert gas

Table 15.1 Requirements for the on- and off-resistances for the three configurations of Figure 15.3

	r_{on}	r_{off}
Series	$\ll R_L$	$\gg R_L$
Shunt	$\ll R_g$	$\gg R_g$
Series shunt	$\begin{cases} \ll R_L \\ \ll r_{off} \end{cases}$	$\begin{cases} \gg R_g \\ \gg r_{on} \end{cases}$

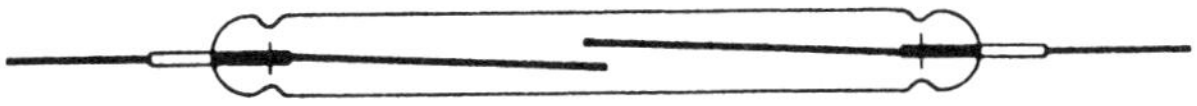

Figure 15.4 The reeds or tongues of a reed switch close when a magnetic field is applied.

(Figure 15.4). The reeds are normally a set distance apart, but when magnetized by an externally applied magnetic field they attract each other and make contact. The magnetic field is usually produced by a current flowing through a little coil surrounding the glass tube.

The major properties of a reed switch are:

- very low on-resistance (0.1 Ω);
- very high off-resistance (>10^9 Ω);
- very low offset voltage (<1 μV, mainly due to thermoelectric voltages)
- high reliability (over 10^7 switching operations).

A major disadvantage of this otherwise ideal switch is the low switching speed; a switching frequency of 100 Hz can be achieved, but not much more. The reed switch is not suitable for high speed switching operations, but is an excellent device for switching functions in, for instance, automatic measurement systems (occasionally self-calibration, automatic range switching) and telephone switching boards. The reed switch is an inexpensive component, available in various encapsulations. There are types with several switches in a single, IC-like encapsulation, including the driving coils.

Photoresistor

The photoresistor, introduced in Section 7.2, can be used as a switch, when combined with a switchable light source such as an LED. The major properties of this component as a switch are:

- rather high on-resistance (up to 10^4 Ω);
- moderate off-resistance (in the order of 10^6 Ω);
- very low offset voltage (<1 μV).

Due to the inherent slowness of the photoelectric effect (especially from light to dark), the switching rate is limited to about 100 Hz.

PN diode

The high resistance of a pn diode when reverse biased and the rather low differential resistance when forward biased make the diode suitable for switching operations (see also Section 9.2). The main properties as a switch are:

- on-resistance: equal to the differential resistance r_d, inversely proportional to the forward current, and 25 Ω at 1 mA;
- off-resistance: high, in the order of 10^8 Ω;
- offset voltage: large, equal to the threshold voltage V_k of the forward-biased junction, about 0.6 V.

Another disadvantage is the absence of a separate control terminal; the diode is self-switching, switching as a result of the voltage across it.

Figure 15.5 shows a switch circuit of four diodes, that permit independent switching (with control currents i_{s1} and i_{s2}), and in which the offset voltage V_k is compensated. The properties of this switching bridge are:

- on-resistance: $2r_d//2r_d$, hence r_d;
- offset voltage: difference between the values V_k; typically 1 mV;
- offset current: $i_{s1} - i_{s2}$.

Diodes have small delay times, so high switching rates can be achieved (up to several GHz).

Bipolar transistor

To understand how a bipolar transistor can act as a switch, we will refer to the I_C–V_{CE} characteristic (Figure 15.6a). There are three states: the saturation region, the pinch-off region and the active (or linear) region. When used as a switch, the transistor is either saturated (on) or pinched-off (off).

We will consider only its use as a shunt switch (Figure 15.6b). In this case, the collector–emitter voltage is $V_{CE} = v_i - I_C R_C$. This is the equation of the so-called load line in Figure 15.6a. The base current controls the switch. The transistor is off for $I_B = 0$; the collector current is also zero except for the small leakage current and the transistor is biased at point A. When changing v_i, point A shifts along the pinch-off line. The current does not change, so the device behaves as a high resistance: $v_o = v_i$.

At a large base current, the collector current is even larger; the transistor is biased at point B on the saturation line. The slope of the load line is fixed with R_C, so point B moves along the saturation line when varying v_i; the resistance is low. V_{CE} hardly changes at changing v_i, and as V_{CE} is almost zero, v_o is zero too.

The bipolar transistor has a high switching speed. Special types have switching rates of several GHz. Leakage currents and offset voltage make this switch less attractive for high precision applications.

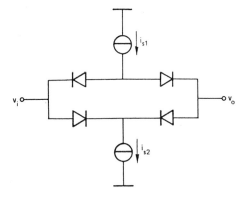

Figure 15.5 The diode bridge is switched off when i_{s1} and i_{s2} are zero. At a rather arbitrary positive current, the switch is on; its on-resistance is r_d.

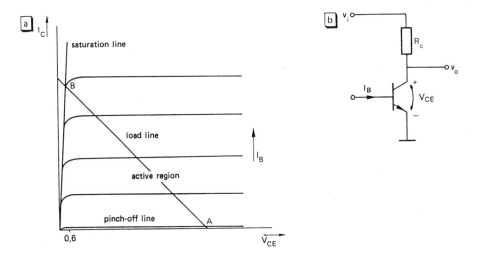

Figure 15.6 The bipolar transistor as switch: (a) current–voltage characteristic; (b) as a shunt switch.

Junction FET

Chapter 11 described the junction FET as a voltage-controlled resistance; the channel resistance between drain and source depends on the gate voltage (Figure 11.2a). For $V_{GS} = 0$, this resistance is rather low (50–500 Ω). For V_{GS} below the pinch-off voltage, the channel resistance is almost infinite ($>10^8$ Ω). Hence, the ratio between the off- and on-resistance is high. The advantages of the JFET over the bipolar transistor are the very low control current (the gate current) and the low offset voltage (typically 1 μV).

Figure 15.7 illustrates the use of the JFET as a series switch and a shunt switch. In Figure 15.7a, the source voltage is zero, so the gate can simply be controlled by a voltage relative to ground. In Figure 15.7b, the source voltage varies with the input voltage v_i. To keep the JFET on, V_{GS} must remain zero, irrespective of v_i. This is accomplished with a resistor between gate and source. The switch is on when $V_{GS} = 0$, so for $i_s = 0$. The switch is off as soon as $i_s R$ exceeds the pinch-off voltage. In this configuration, not only the control source but also the signal source must be able to supply the required control current i_s. Here, the advantage of powerless control is cancelled.

MOSFET

Like the JFET, the MOSFET can be employed as a switch. The MOSFET requires a rather high control voltage to switch the device and also an additional substrate voltage (Section 11.1.2). Further, the on-resistance is usually higher than that of the JFET. Advantages of the MOSFET are the smaller dimensions and fewer fabrication steps, which make it attractive for integration, together with other components. The energy consumption is extremely small.

MOSFET switches are applied in (integrated) multiplexers (Section 15.2) and digital integrated circuits (Chapters 19 and 20). The rather high on-resistance demands additional buffer stages in analog applications, to minimize transfer errors due to loading.

Thyristors

A thyristor can be considered as a diode that, when forward biased, only conducts after the occurrence of a voltage pulse on a third terminal (the control gate or control input). After this pulse, the thyristor still conducts. If the forward current falls below a certain threshold value, the thyristor switches off; this state is maintained until the next control pulse.

A thyristor is made up of a four-layer structure of alternately p- and n-type silicon. The explanation of the physical operation of this device is not covered in this book.

Figure 15.8a shows the circuit symbol of the thyristor (like a diode, now with an additional connection), whereas Figure 15.8b illustrates the switching of a rectified sine wave.

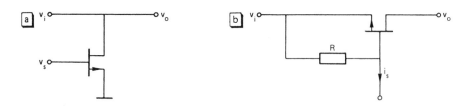

Figure 15.7 A JFET as (a) a shunt switch; (b) a series switch.

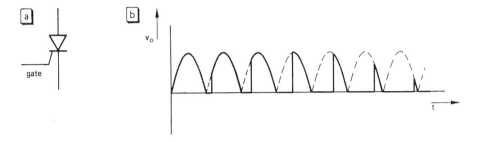

Figure 15.8 (a) Circuit symbol of the thyristor; (b) power control: the (mean) power is reduced in this example by gradually changing the phase of the control pulse relative to the sine wave.

The thyristor is used in particular for power control, from low power systems (like incandescent lamps) to very high power systems (for instance electric locomotive engines). The average output power depends on the surface area below the sine waves in Figure 15.8b, which is controllable by the phase between the control pulses and the sine wave.

There are similar devices that conduct in two directions (triac), like two thyristors connected in antiparallel.

15.2 Circuits with electronic switches

This section deals with two special circuits in which electronic switches are essential components: time multiplexers (introduced in Section 1.1) and sample-hold circuits. At the end of this section, we will analyse an important phenomenon occurring in most switching circuits: transients due to capacitive cross-over.

15.2.1 *Time multiplexers*

Time multiplexers are used to scan a number of measurement signals to connect them successively to a common information channel, in order to share the expensive parts of an electronic measurement system (Section 1.1). Multiplexing (the prefix 'time' will be omitted in the rest of this section) of digital signals is performed by logic circuits (Chapter 19). Analog multiplexers require an accurate signal transfer for the selected channel. They are available as integrated circuits.

An analog multiplexer consists of a set of electronic switches that are switched on successively. Figure 15.9a shows a configuration for the multiplexing of measurement signals from a differential amplifier. Figure 15.9b is suitable for immediate multiplexing of differential voltages (double-poled switches), also called a differential multiplexer. The switching pairs of a differential multiplexer must have equal on- and off-

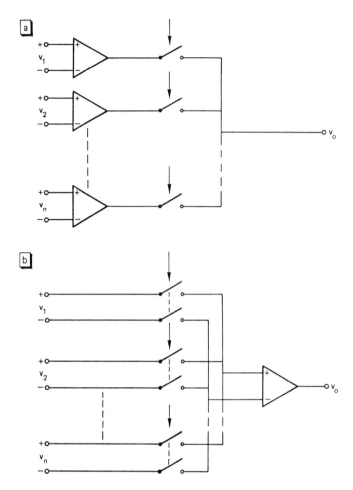

Figure 15.9 A multiplexer for n channels (a) with differential amplifiers at the input and single-poled switches; (b) with double-poled switches and a single differential amplifier.

resistances, in order not to degenerate the CMRR of the circuit. Some types of multiplexers permit the user to choose between single- or double-mode multiplexing.

An integrated multiplexer contains some additional circuits: a decoder to select the proper switch and switch drivers to supply the required voltages to the electronic switches (Figure 15.10). Usually, the channel selection is controlled from a computer that sends a binary-coded signal to the input of the decoder. With p binary lines, 2^p different codes can be transmitted. A multiplexer with n individual channels needs only $^2\log n$ control inputs for the selection of one channel out of n. Thus, the decoder reduces substantially the number of connections (pins) of the multiplexer IC.

Circuits with electronic switches 251

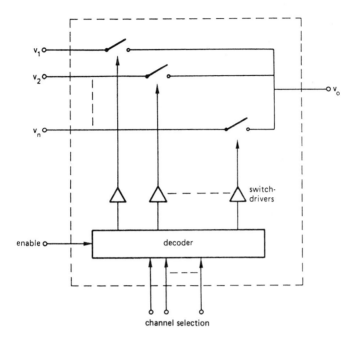

Figure 15.10 Simplified internal structure of an integrated multiplexer.

Many multiplexers have an additional control input, the enable input. With this binary control signal all channels can be switched off simultaneously, irrespective of the channel selection. This possibility allows the multiplexing of a number of multiplexer ICs and hence the extension of the number of channels.

Table 15.2 shows the properties of a multiplexer circuit in which each switch is composed of an n- and p-channel MOSFET in parallel. Most analog multiplexers can be used as demultiplexers as well (see, for instance, the circuit in Figure 15.10).

15.2.2 *Sample-hold circuits*

A sample-hold circuit is a signal processing device with two different transfer modes (two states). In the sample/track mode, the output follows the input; usually the transfer is 1. In the hold mode the output keeps the value at the moment of the hold command. Figure 15.11 illustrates the operation of a sample-hold circuit.

When looking at the possible errors of this circuit, we can distinguish four phases: track phase, track-hold transition, hold-phase and hold-track transition (Figure 15.12).

1. Track phase: in this state, errors occur similar to those of an operational amplifier, such as offset voltage, drift, bias currents, noise, gain errors and limited bandwidth.

Table 15.2 Specifications of an integrated 16-channel multiplexer

Number of channels	16
Contacts	break before make
Voltage range	−15 to +15 volts
r_{on}	700 Ω + 4%/K
Δr_{on}	<10 Ω
I_{off} (input and output)	0.5 nA
CMRR (DC)	125 dB
(60 Hz)	75 dB
C_i(off)	2.5 pF
C_o(off)	18 pF
$C_{transfer}$(off)	0.02 pF
Settling time (0.01%)	800 ns
(0.1 %)	250 ns
Power dissipation	525 mW

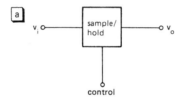

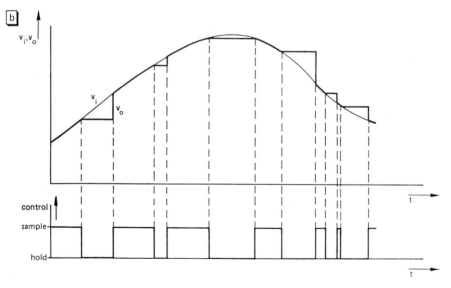

Figure 15.11 (a) The circuit symbol of a sample-hold circuit; (b) an example of the output voltage v_o at corresponding input voltage v_i and control signal.

2. Track-hold transition: the main errors are the delay time t_1 (also the aperture delay time) and the uncertainty in that delay time, the aperture jitter t_2.
3. Hold phase: during this phase the hold voltage can drift slowly away; this is denoted by the term droop.
4. Hold-track transition: the specified times are given in Figure 15.12.

Figure 15.13 shows a simple version of a sample-hold circuit, composed of a switch, a hold capacitor C_H and a buffer amplifier. The capacitor acts as the analog memory of the voltage to be held. It is charged by the input source via the switch. The time constant of the charging (Section 4.2) is $(R_g + r_{on})C_H$; the smaller C_H, the faster the capacitor is charged to the input voltage.

When the switch is off, the capacitor remains charged, as both the switch and the amplifier have high resistance. However, the bias current of the buffer amplifier tends to discharge the capacitor (or charges, depending on the direction of the bias current). During the hold period T_H, the voltage of the capacitor changes with an amount $\Delta v_o = T_H I_{bias}/C_H$. To have a small droop at given I_{bias}, the capacitance must be as large as possible.

Table 15.3 gives an overview of some specifications belonging to a particular type of sample-hold device with the structure as in Figure 15.14.

For switching to the hold mode, a voltage of more than 2 V is required at the control input; to change to the track mode, the circuit needs a control voltage below

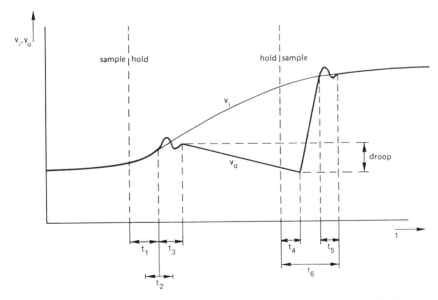

Figure 15.12 The various transition times and delay times of a sample-hold circuit: t_1 = turn-off delay time or aperture delay time; t_2 = aperture uncertainty or aperture jitter; t_3 = settling time; t_4 = turn-on delay time; t_5 = settling time; t_6 = acquisition time.

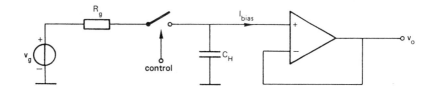

Figure 15.13 A simple sample-hold circuit with a capacitor as memory device.

Table 15.3 Some specifications of an integrated sample-hold circuit with the structure of Figure 15.14

analog input:	
V_{off}	6 mV
I_{bias}	3 µA
R_i	30 MΩ
hold → sample:	
t_{acq} (0.01%)	25 µs
(0.1 %)	6 µs
sample → hold:	
aperture delay	150 ns
aperture jitter	15 ns
settling time (0.01%)	0.5 µs
hold mode:	
I(droop)	100 pA
track mode:	
CMRR	60 dB
bandwidth	1.5 MHz
slew rate	3 V/µs
R_o	12 Ω

0.8 V. The input is provided with an integrated input amplifier. The user can set the transfer in the track mode to an arbitrary value, by connecting proper values of resistors R_1 and R_2. The capacitor C_H should also be connected externally.

15.2.3 *Transient errors*

Electronic switches often produce unwanted pulse-shaped voltages superimposed on the measurement signal, due to capacitive cross-over of the rectangularly shaped control signals of the switches. This cross-over is illustrated in Figure 15.15 for a series switch, controlled by a voltage v_s.

We can calculate the output signal due to only the control signal; the input voltage is made zero or the input terminal is connected to ground, which amounts to the same thing. Figures 15.15b and c show the situations directly after the off- and

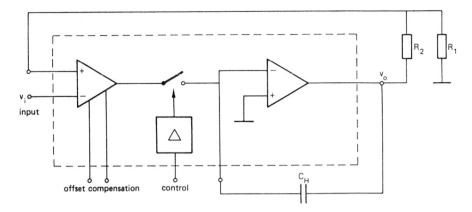

Figure 15.14 An integrated sample-hold circuit; all components are integrated, except for the hold capacitor and some resistors, that have to be connected externally.

on-command, respectively. It is supposed that the switch itself responds immediately. R_L and C_L represent the load of the switch (the input impedance of the connected circuit). In both situations, the circuit corresponds with a differentiating network from Figure 8.7. The output voltage due to a step in the control voltage is given as:

$$v_o = \alpha v_s e^{-t/\tau}, \quad \text{with} \quad \alpha = C/(C + C_L)$$

After switching off, $\tau = \tau_{off} \approx R_L(C + C_L)$; after switching on, $\tau = \tau_{on} \approx r_{on}(C + C_L)$. The step responses are depicted in Figure 15.15d. For $C_L = 0$ (no capacitive load), the height of the output pulse is equal to the height of the control voltage. A capacitive load reduces the cross-over pulses somewhat. The resulting sharp pulses can be capacitively coupled into other parts of the circuit. So it is important to keep them as low as possible.

In the series shunt switch of Figure 15.16 both switches produce cross-over pulses; however, they compensate each other. The circuit model in Figure 15.16a contains only the relevant components; Figure 15.16b shows the model used for the analysis. This model is valid for both on- and off-states: in either state one switch is off, the other on, and the switch resistances are in parallel. The output voltage is obtained by adding the contributions of the two signal voltages:

$$v_o = \alpha v_s e^{-t/\tau} - \alpha' v_s e^{-t/\tau}$$

with $\alpha = C/(C + C' + C_L)$, $\alpha' = C'/(C + C' + C_L)$ and $\tau = \tau' = r_{on}(C + C' + C_L)$. Here it is assumed that $r_{on} \ll R_L$. The compensation is perfect if C and C' are equal: in that case $\alpha = \alpha'$, hence $\Delta\alpha$ in Figure 15.16 is zero. This holds only for the condition that both switches respond exactly at the same time, and that they have equal on-resistances.

256   Electronic switching circuits

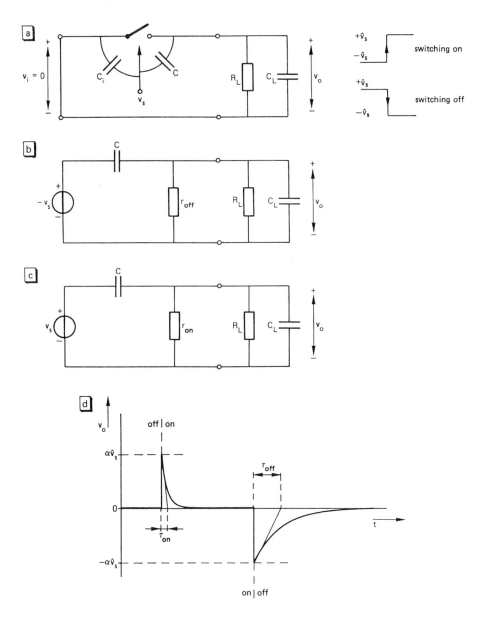

Figure 15.15 Transients due to switching: (a) to account for the cross-over of v_s to the output, the input is short-circuited; (b) circuit model directly after switching off; (c) circuit model directly after switching on; (d) output voltage due to capacitive cross-over.

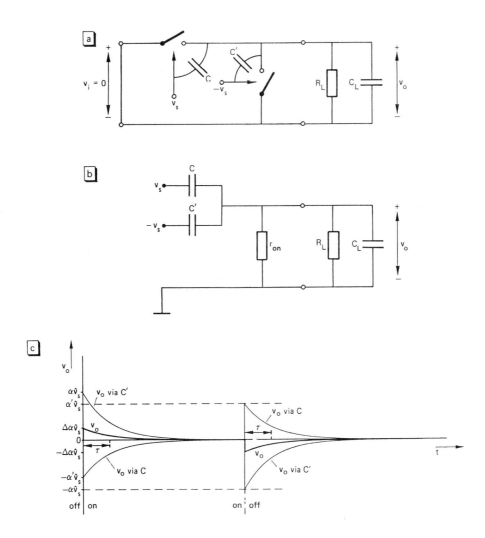

Figure 15.16 Series shunt switch, with compensation of transients; (a) switch with short-circuited input; (b) model for both on- and off-states; (c) output signal due to control voltages v_s.

SUMMARY

Electronic switches

- Important parameters of electronically controlled switches are the on- and off-resistances, the offset voltage and the leakage current; for dynamic behaviour, the delay times and the acquisition time are major characteristics.
- The on- and off-resistances produce transfer errors, when combined with a source resistance and load resistance.
- The following components can be employed as an electronic switch: reed switch, photoresistor, pn diode, bipolar transistor, JFET, MOSFET and thyristor.
- The reed switch and the photoresistor are slow switches; transistors are faster and diodes have the highest speed.
- When used as a switch, the gate–source voltage of the JFET acts as the control quantity: the JFET switches between a low channel resistance (50–500 Ω) and a very high value.
- The MOSFET is widely used as an electronic switch, particularly in integrated circuits with many active components, like microprocessors.
- The thyristor can be considered as a diode that only conducts after a pulse on the control gate. It switches itself off when the voltage or current falls below a certain value. They are used for power control, by controlling the moment of ignition with respect to the phase of the power signal.

Circuits with electronic switches

- A time multiplexer is a multiple switch with multiple inputs and a single output. The channel selection is performed via a decoder.
- A sample-hold circuit has two states (modes): track and hold mode. The output tracks the input during the track mode; this value is kept at the output when switched to the hold mode.
- Sample-hold circuits are available as complete integrated circuits. Important parameters are the properties as a follower (or voltage amplifier), delay and acquisition times and droop (the drift during the hold mode).

Exercises

Electronic switches

15.1 Given a voltage source with source resistance 10 Ω. This voltage must be connected to a circuit with input resistance of 50 kΩ, using an electronic switch in series with the source and load. Derive the conditions for the on- and off-resistance of the switch, to meet the following requirements: maximal transfer error in the on-state, 0.1%; maximal transfer of 0.1% in the off-state.

15.2 The same question as in Exercise 15.1, but now for a shunt switch.

15.3 The same question as in Exercise 15.1, now for a series shunt switch comprising two identical switches (Figure 15.3c).

15.4 Create a table similar to Table 15.1, but now for a current switch. The current source (parallel) resistance is R_g, and the load resistance is R_L ($R_L \ll R_g$).

15.5 What is the on-resistance of the diode bridge at a control current of 5 mA?

15.6 The following figures show two possible combinations of an electronic switch and an operational amplifier. Discuss the effect of the on-resistance and the offset voltage of the switch on the accuracy of the transfer in the on-state. Which configuration is preferred, and why?

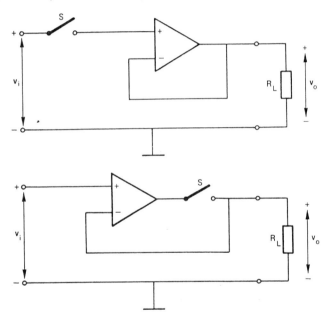

15.7 The pinch-off voltage of the type of JFET applied in the circuit below has a specified range between -2 and -6 V. The control current of this switch is $i_s = 2$ mA. Find the proper value of R, to guarantee correct switching.

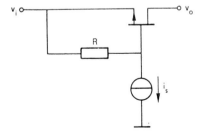

260   Electronic switching circuits

15.8 The control quantity for the JFET from Exercise 15.7 is a voltage, connected according to the figure below. The input voltage varies between −3 and 3 V. Find the conditions for the upper and lower levels of the control voltage, for a proper operation of the switch.

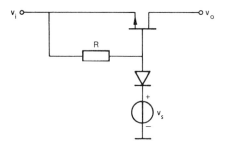

Circuits with electronic switches

15.9 The output voltage of a double-poled multiplexer is connected to the differential amplifier as given in the figure below. The on-resistance of the multiplexer is specified as $r_{on} = 500\ \Omega \pm 5\ \Omega$. The operational amplifier is ideal. Calculate the CMRR of the whole circuit.

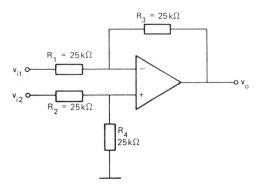

15.10 The specifications of the next sample-hold circuit are $V_{off} = 100\ \mu V$; $I_{bias} = 1\ \mu A$ (positive in the direction of the arrow); $C_H = 1\ \mu F$ and $r_{on} = 100\ \Omega$. The circuit is connected to a voltage source with source resistance 50 Ω. Calculate the absolute error in the output voltage when the circuit is in the track mode.

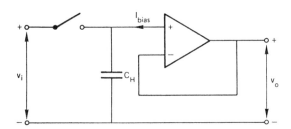

15.11 The circuit of Exercise 15.10 is now connected to a voltage source with sinusoidal output; the amplitude is 8 V, its frequency is 100 Hz. The offset voltage is adjusted to zero. Just at the moment that v_i reaches its peak value, a hold command is given. Find v_o after 1 and after 100 periods of the input signal.

16 Signal generation

In Chapter 2, we made a distinction between periodic and aperiodic signals. Measurement signals are almost invariably aperiodic. Sometimes they look periodic, for instance an ECG (electrocardiogram) or the signal from a vibration sensor mounted on a rotating machine. It is simply the deviation from periodicity that provides the relevant information in these examples.

Pure periodic signals also have an important function in instrumentation. Often they serve as an auxiliary signal (for instance, the carrier of a modulated signal, see Chapter 17), or as a test signal (for instance, when analysing the frequency transfer function of a system, Chapter 21). For this reason it is worth while to know how periodic signals are generated. An instrument that produces a periodic signal is called a signal generator. If only sinusoidal signals are produced, such an instrument is called an oscillator. The principles of oscillators are explained in the first part of this chapter. There are instruments for the generation of other periodic signals. For instance, a function generator produces various periodic signals such as square wave, triangle, ramp and also sine-shaped signals. These instruments are described in the second part of this chapter.

16.1 Sine wave oscillators

There are various ways to generate a sinusoidal signal. One of these is solving a second-order differential equation using analog electronic circuits, a principle that is outlined in this section. A second method starts with a symmetric, rectangular or triangular signal. The sine shape is obtained either by filtering the superharmonics (leaving the fundamental) or reshaping the signal using resistance-diode networks, as described in Chapter 14. Obviously, the accuracy of this last method depends strongly on the quality of such non-linear converters. A third way is the synthesis of arbitrary periodic signals by a computer. The processor generates a series of successive codes, that are converted into an analog signal by a DA converter (Chapter 18).

16.1.1 Principle of harmonic oscillators

The general solution of the linear differential equation:

$$a_0 \frac{d^2 x}{dt^2} + a_1 \frac{dx}{dt} + a_2 x = 0 \tag{16.1}$$

is

$$x = \hat{x} e^{-\alpha t} \sin(\omega t + \varphi) \tag{16.2}$$

with $\alpha = a_1/2a_0$, $\omega = \sqrt{a_2/a_0 - a_1^2/4a_0^2}$ and $\hat{x}$ and φ arbitrary constants. In accordance with the sign of α, x is a sinusoidal signal with an exponentially decreasing amplitude ($\alpha > 0$) or an exponentially increasing amplitude ($\alpha < 0$). Only when $\alpha = 0$, $x(t)$ is a pure sine wave with constant amplitude. The coefficient α is therefore called the damping factor. It is easy to design an electronic circuit whose voltages and currents satisfy Equation (16.1). Less easy is the design of such a circuit where $a_1 = 0$, and the voltages and currents are pure sine waves with a constant amplitude.

The derivative and the integral of a signal are obtained with inductances and capacitances. Active elements are required to nullify the damping factor. Furthermore, some kind of feedback is necessary. The principle is illustrated with an example of less practical significance, but useful for the explanation. Figure 16.1 shows the block diagram of an electronic system with two differentiators and one amplifier. The output is connected straight to the input. Hence:

$$v_o = K\tau^2 \frac{d^2 v_o}{dt^2} \tag{16.3}$$

The solution of this linear, homogeneous differential equation is $v_o = \hat{v} \sin(\omega t + \varphi)$, with $\omega^2 = -1/K\tau^2$. Evidently, K must be negative (an inverting amplifier). For $K = -1$, the frequency of the signal produced is $f = 1/2\pi\tau$. Instead of differentiators, we can also take integrators. We have seen in Chapter 13 that an integrator is more stable and produces less noise than a differentiator. The differential equation remains the same.

The amplitude can have any value between the signal limits of the system. The amplitude is not fixed by the circuit parameters. As long as a_1 in Equation (16.1) is zero, the amplitude, once present, remains constant. However, in an actual circuit the component values vary steadily due to, for instance, temperature fluctuations, so the condition $a_1 = 0$ will not be met for a long time. This means that the system must somehow be controlled in order to fix the amplitude at a predescribed value. There is another need of such a control circuit. When switching on the system, the amplitude is zero. As long as the damping factor α is zero (or positive), the amplitude remains zero. Therefore, α must be negative for a short time, to let the amplitude rise to the desired value. Having reached that value, α must be made zero again.

264 Signal generation

In the circuit of Figure 16.1, the term a_1 is obtained by simply adding a fraction γ of v_2 to the voltage v_3. The output voltage becomes:

$$v_o = K\left(\tau^2 \frac{d^2 v_o}{dt^2} + \gamma\tau \frac{dv_o}{dt}\right) \qquad (16.4)$$

Whether the output amplitude increases or decreases depends on the fraction γ; for $\gamma = 0$, a sine wave with steady output is generated.

Figure 16.2 shows the set-up of an oscillator with electronic amplitude control. The control system consists of the following parts:

- amplitude detector; its output $\hat{v}_o$ is a measure for the amplitude of the generated sine wave, for instance the peak detector of Section 9.2.2 or a rectifier with low-pass filter;
- a reference voltage V_{ref};

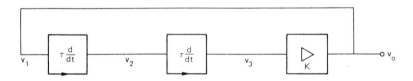

Figure 16.1 An example of an oscillator with two differentiators and an amplifier.

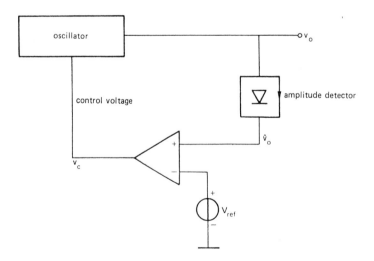

Figure 16.2 Principle of an oscillator with amplitude control.

- a control amplifier, which amplifies the difference between the reference voltage V_{ref} and the peak value $\hat{v}_o$;
- a control element in the oscillator; this can be an electronically controllable resistance (JFET, thermistor; photoresistor) or an analog multiplier. This control element affects the factor a_1 in Equation (16.1) or γ in Equation (16.4) and hence the damping factor α.

Since the signal power is determined by the square of the signal voltage (or current) amplitude, it is possible to utilize a control element based on heat dissipation. A widely used element for this purpose is the thermistor. The thermistor is part of the oscillator network and operates in such a way that at increasing amplitude (hence decreasing resistance) a further increase is stopped. The method is extremely simple (a single component) but the control is slow, due to the thermal nature. Further, the amplitude in the steady state depends on the parameters of the thermistor and the heat resistance to the environment (hence the environmental temperature). This method is not suitable for a high amplitude stability.

16.1.2 Harmonic oscillator circuits

The relations between voltages and currents in an electronic network are given as linear differential equations. Solving these equations is rather time consuming, and the main reason for the introduction of complex variables (Chapter 4). A harmonic oscillator generates (in the steady state) a pure sine wave, so we can analyse such oscillators by using complex variables. For example, Equation (16.3), which belongs to the circuit of Figure 16.1, can be written as $V_o = K\tau^2(j\omega)^2 V_o$, from which follows $\omega^2 = -1/K\tau^2$.

Any harmonic oscillator is composed of at least one amplifier and one (passive) network with a frequency-selective transfer. In most cases an oscillator can be modelled as in Figure 16.3. When mutual loading can be neglected (or when A or β discount for this effect), then $V_o = AV_i$ and $V_i = \beta(\omega)V_o$, hence $A\beta(\omega) = 1$. This complex equation is called the oscillation condition. The solution of this equation provides the conditions for oscillation and the oscillation frequency. This will be illustrated with some examples.

Wien oscillator

The feedback network of a Wien oscillator consists of two resistors and two capacitors, arranged as a voltage divider with a band-pass characteristic (Figure 16.4). The transfer of this Wien network is, for $R_1 = R_2 = R$ and $C_1 = C_2 = C$:

$$\beta(\omega) = V_o/V_i = \frac{1}{3 + j\omega\tau + 1/j\omega\tau}$$

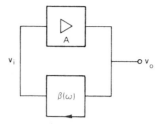

Figure 16.3 Basic diagram of a harmonic oscillator; the amplifier output is fed back to the input via a network with frequency selective transfer.

The oscillation condition is $A\beta(\omega) = 1$, hence:

$$3 + j\omega\tau + \frac{1}{j\omega\tau} = A$$

The real parts of the left- and right-hand side of this equation must be equal, just like the imaginary parts. This results in two equations:

$$A = 3 \tag{16.5}$$

$$\omega = 1/\tau \tag{16.6}$$

Equation (16.5) describes the condition for constant amplitude. If $A > 3$, the amplitude increases, but for $A < 3$ it decreases. Equation (16.6) gives the frequency of the generated signal. Figure 16.4b shows a possible configuration of a Wien oscillator without amplitude control. Derivation of the oscillation conditions directly from this circuit results in the equations $\omega = 1/\tau$ and $R_4 = 2R_3$. This is not a surprising result; the amplifier without a Wien network behaves as a non-inverting amplifier with gain $1 + R_4/R_3 = 3$ for the condition found earlier.

Phase-shift oscillator

The basic idea of a phase-shift oscillator is depicted in Figure 16.5. The feedback network consists of three cascaded low-pass RC filters. If the values of the resistors and capacitors are chosen so that the time constants are equal, but the sections do not load each other (compare Figure 8.13), the transfer is $\beta(\omega) = 1/(1 + j\omega\tau)^3$. The oscillation condition is simply $K = (1 + j\omega\tau)^3$. Splitting this equation into real and imaginary parts results in $K = -8$ and $\omega^2 \tau^2 = 3$.

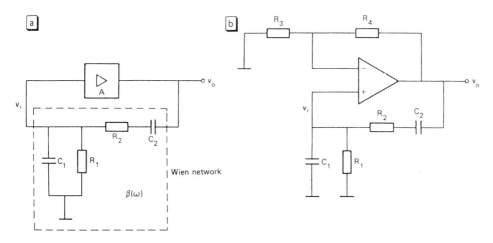

Figure 16.4 (a) An oscillator according to the principle of Figure 16.3, with Wien network; (b) a Wien oscillator with one operational amplifier.

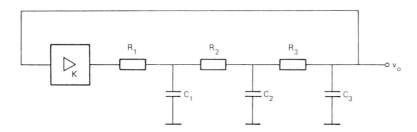

Figure 16.5 A phase-shift oscillator with three low-pass RC sections.

Two-integrator oscillator

The last example of a harmonic oscillator is the two-integrator oscillator or dual integrator loop (Figure 16.6). The oscillator condition can be found easily: $(-1/j\omega\tau)^2(-1) = 1$, hence $\omega = 1/\tau$. The oscillation frequency can be varied by simultaneous changes to the resistors R_1 and R_2 or the capacitors C_1 and C_2. The same effect is achieved with two adjustable voltage dividers (potentiometers) at the inputs of the integrators (Figure 16.7). Let k be the attenuation of the voltage dividers; the oscillation condition is $(1/j\omega\tau)^2 k^2(-1) = 1$, hence $\omega = k/\tau$. The oscillation frequency is proportional to the attenuation of the potentiometers. In Figure 16.7 an amplitude stabilization circuit is also given. A fraction γ of v_2 is added via R_5 to the inverting amplifier 1 (compare Equation (16.4)). The value of γ is multiplied by the output of the control amplifier, whose inputs are the rectified oscillator output voltage and a

268   Signal generation

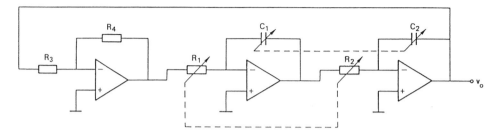

Figure 16.6 A two-integrator oscillator consisting of an amplifier and two integrators. The frequency is adjustable with both capacitors C or resistors R. Usually, C_1 and C_2 are switched in steps of a factor of 10 (coarse frequency adjustment), whereas R_1 and R_2 are potentiometers for fine tuning of the frequency.

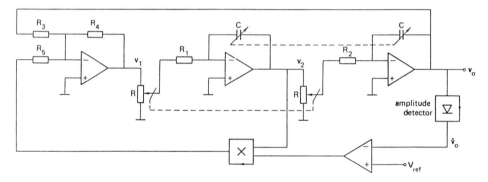

Figure 16.7 A possible configuration of the two-integrator oscillator with amplitude stabilization circuit.

reference voltage. The control circuit tends to make the difference between V_{ref} and $\hat{v}_o$ zero.

16.2 Voltage generators

Periodic non-sinusoidal signals are frequently used in instrumentation systems. In combination with an oscilloscope (Chapter 21), it is possible to visualize step and pulse responses of a system, or to measure its rise and delay times. This is done by connecting a periodic square wave or pulse signal to the input of the test system and observing the output on an oscilloscope. Triangular or ramp signals allow the determination of a system's non-linearity. They are also useful as a control signal for various actuators to test systems or products. Pulse and square wave signals are widely used in digital systems, for instance for synchronization.

We discuss in this section a number of generators for non-sinusoidal, periodic voltages. Most of these instruments are based on the periodic charging and discharging of a capacitor.

16.2.1 *Triangle voltage generators*

A triangle generator is based on the periodic charging and discharging of a capacitor with a constant current (Figure 16.8). The voltage across a capacitance is, at constant current, a linear time function. This voltage is connected to a Schmitt trigger (Section 14.2.2) with output levels V^+ and V^- (Figure 16.9). The control circuit controls the switches in such a way that for $v_s = V^+$, the capacitor is charged with current I_1 and for $v_s = V^-$ it is discharged with I_2. This results in a triangular voltage across C. Any time v_o passes the switching levels of the Schmitt trigger, the two states are automatically interchanged. The peak values of the triangle are equal to the switching levels of the Schmitt trigger. The slope of the triangle can be adjusted by varying the values of the currents I_1 and I_2.

Figure 16.10 shows a simple configuration with one integrator, a Schmitt trigger and an inverting amplifier.

At equal charging and discharging currents and $V^+ = -V^-$, a symmetrical triangle voltage is obtained. The circuit generates simultaneously a square wave voltage.

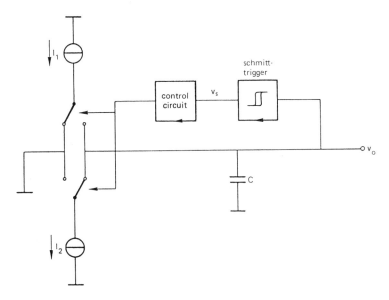

Figure 16.8 Functional diagram of a triangle generator. The triangular voltage is produced by periodically charging and discharging a capacitor by a constant current.

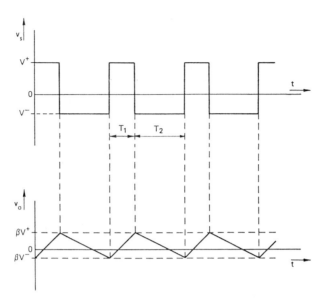

Figure 16.9 The voltages from Figure 16.8. V^+ and V^- are the positive and negative power supply voltages; βV^+ and βV^- are the switching levels of the Schmitt trigger. The rise and fall times of v_o are determined by I_1, I_2 and C.

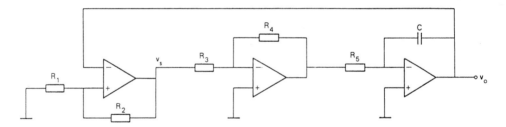

Figure 16.10 A simple configuration for a triangle generator, using a Schmitt trigger, an inverting amplifier and an integrator.

The ratio between the time T_1 and the total period time $T = T_1 + T_2$ of such periodic signals is the duty cycle of the signal. A symmetric signal has a duty cycle equal to 50%. By varying one of the currents in Figure 16.8, the duty cycle is changed.

16.2.2 *Ramp generator*

A ramp voltage can be considered as a triangular voltage with one vertical slope. Such a short fall time is obtained by discharging a charged capacitor over a switch. The

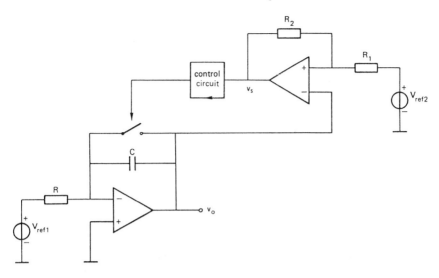

Figure 16.11 The functional diagram of a sawtooth generator. The output voltage is produced by charging a capacitor with a constant current and discharging it via a switch.

switch is controlled by a Schmitt trigger (Figure 16.11). For $v_s = V^+$, the switch is on and for $v_s = V^-$ it is off. The capacitor is part of an integrator, whose input is connected to a fixed reference voltage V_{ref1}. The output of the integrator rises linearly with time, until the switch turns on and the capacitor discharges, making the output return to zero. The process starts again upon releasing the switch.

Figure 16.12 shows various voltages for a negative reference voltage. The switch goes on as soon as v_o reaches the upper switching level of the Schmitt trigger. The capacitor discharges in a very short time, the output voltage drops to zero and remains zero as long as the switch is on. This means that the lower switching level of the Schmitt trigger (which is βV^- for $V_{ref2} = 0$, see Section 14.2.2) must be higher than zero; otherwise, the Schmitt trigger remains in the state $v_s = V^-$; the switch will never switch off and the output remains zero. This explains the need of the second reference voltage V_{ref2}. The switching levels of the Schmitt trigger are determined by R_1, R_2 and V_{ref2}.

Example 16.1

The power supply voltages of the operational amplifiers in Figure 16.11 are $V^+ = 15$ V and $V^- = -15$ V. The output levels of the Schmitt trigger are assumed to be equal to the supply voltages, hence 15 and -15 V. The switching levels are calculated as follows. Both levels are described by:

$$V_{ref2} \frac{R_2}{R_1 + R_2} + V_s \frac{R_1}{R_1 + R_2}$$

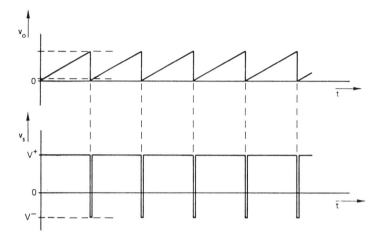

Figure 16.12 The output voltage v_o of the integrator and the output voltage v_s of the Schmitt trigger in the circuit of Figure 16.11. The dotted lines in v_o are the switching levels of the Schmitt trigger. For $v_s = V^+$, the switch is on; for $v_s = V^-$, it is off.

The upper level occurs for $v_s = 15$ V, so it amounts to:

$$V_{ref2} \frac{R_2}{R_1 + R_2} + 15 \frac{R_1}{R_1 + R_2}$$

which is also the upper peak of the ramp v_o. The lower switching level occurs for $v_s = -15$ V, hence:

$$V_{ref2} \frac{R_2}{R_1 + R_2} - 15 \frac{R_1}{R_1 + R_2}$$

The minimum value of v_o is zero (discharged capacitor). To let the ramp run from 1 to 10 V, the upper level must satisfy:

$$V_{ref2} \frac{R_2}{R_1 + R_2} + 15 \frac{R_1}{R_1 + R_2} = 10$$

and the lower level:

$$V_{ref2} \frac{R_2}{R_1 + R_2} - 15 \frac{R_1}{R_1 + R_2} = 1$$

These equations set the ratio between R_1 and R_2 as well as V_{ref2}: 3/7 and 7.9 V, respectively.

The width of the pulse-shaped voltage v_s is determined by the discharge time of the capacitor and the delay times of the switch and the Schmitt trigger. This pulse is very narrow, causing the very steep upgoing edge of v_o. The Schmitt trigger could be replaced by a comparator, a Schmitt trigger without hysteresis. However, due to the on-resistance of the switch (Section 15.1), the discharging period is not zero, but an exponentially decaying curve. Due to the hysteresis of the Schmitt trigger, the switch goes off again only if v_o is sufficiently close to zero, irrespective of the discharge time. The frequency of the generated ramp voltage is determined by the time constant RC, by V_{refl} and by the hysteresis of the Schmitt trigger. The control circuit shifts the output of the Schmitt trigger to the appropriate levels to activate the switches.

16.2.3 Square wave and pulse generators

Most circuits for the generation of square wave and pulse-shaped signals are composed of resistors, capacitors and some digital circuits; they can also be made up with an operational amplifier. The principle of operation is essentially the same. Figure 16.13a shows a square wave generator with a Schmitt trigger. The output of the Schmitt trigger (the operational amplifier together with R_1 and R_2) is connected to the input via an integrating RC network. Again, the capacitor is periodically charged and discharged, however, not with a constant current but via the resistor R. The voltage v_c across the capacitor approaches exponentially the levels V^+ and V^-, respectively. The switching levels are βV^+ and βV^-. The frequency of the generated signal can be derived from Figure 16.13b. It appears that the period time is:

$$T = \tau \ln\left(\frac{V^- - \beta V^+}{V^- - \beta V^-} \cdot \frac{V^+ - \beta V^-}{V^+ - \beta V^+}\right)$$

with $\tau = RC$.

16.2.4 Voltage-controlled oscillators

Many applications need a generator whose frequency can be controlled by a voltage. Such a generator is called a voltage-controlled oscillator (VCO) or sweep generator. The latter name refers to the possibility of changing linearly or logarithmically the frequency of the VCO. The control voltage is a triangular or ramp voltage that sweeps the frequency of the VCO between two adjustable values.

The oscillators from Section 16.1 are not suitable for a VCO: their frequency is determined by resistances and capacitances, which can hardly be electronically controlled (at least not over a very wide range). The generators with a periodically charging and discharging capacitor are better suited to a VCO. The charging time and hence the frequency are mainly determined by the charge current. For instance, the frequency of the ramp generator of Figure 16.11 is directly proportional to the reference voltage V_{refl}.

The principle of the VCO is based on that of the triangle generator of Figure 16.8.

274 Signal generation

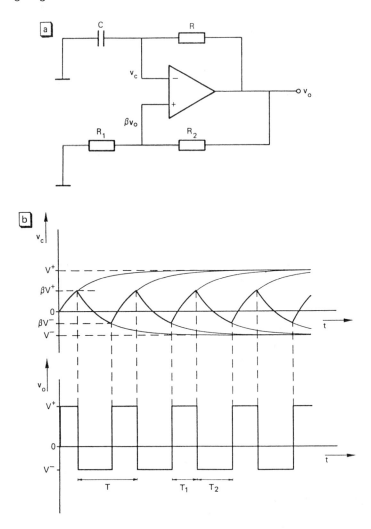

Figure 16.13 (a) The circuit and (b) the corresponding voltages of a square wave generator constructed with a Schmitt trigger.

The frequency of the triangular voltage is proportional to the charge current I_1 and the discharge current I_2. By employing a voltage-to-current converter to produce these currents, a VCO is obtained.

VCOs are available as integrated circuits. Figure 16.14 shows a block diagram with the connections of such a circuit. The circuit contains two buffered outputs, one for a triangular voltage and the other for a square wave signal. The frequency is controlled by the input voltage v_i. The sweep range is, depending on the type, a factor 3–10. The

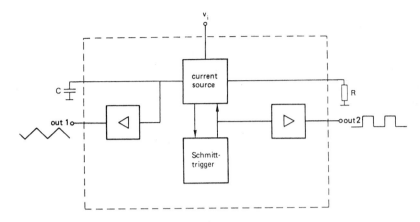

Figure 16.14 An integrated voltage-controlled oscillator (VCO) with triangular and square wave outputs.

sweep range can be shifted by external resistances or capacitances, over a range from 1 Hz to 1 MHz.

There are also VCO types with a much wider sweep range; their output is usually a pulse or square wave only. The frequency sweep may be more than a factor of 1000; the frequency range runs to over 10 MHz.

Such circuits are usually called voltage-to-frequency converters, in particular when they have an accurate relation between the control (or input) voltage and the output frequency. They are used for converting sensor signals into frequency-analog signals, in order to improve the noise immunity when transmitting sensor signals over a large distance.

SUMMARY

Sine wave oscillators

- One method to generate a sine wave is based on the electronic solution of a second-order differential equation with a sine wave as its solution. Other methods are filtering of the harmonics from a triangular or rectangular signal wave; sine shaping by non-linear transfer circuits and synthesizing by a computer.
- A harmonic oscillator comprises an amplifier, a frequency-selective network and an amplitude control circuit. The parameter to be controlled is the damping factor of the second-order system.
- The oscillation condition is $A\beta(\omega) = 1$, where A and β are the transfers of the amplifier and the feedback network, respectively. Solving this equation gives the condition for the gain and the oscillation frequency.

- Frequently used types of oscillators are the Wien oscillator (an amplifier with a Wien network), the phase-shift oscillator (an amplifier with a cascaded series of RC networks) and the two-integrator oscillator or dual integrator loop (a loop of two integrators and an inverting amplifier).

Voltage generators

- A triangular or ramp voltage can be produced by alternately charging and discharging a capacitor with a constant current. The switching moments are determined by the switching levels of a Schmitt trigger; the peak-to-peak value by its hysteresis interval.
- The duty cycle of a pulse-shaped signal is the ratio between the 'on' time and the period time.
- A voltage-controlled oscillator (VCO) is a generator whose frequency can be varied with a voltage. The sensitivity of a VCO is expressed in Hz/V or kHz/V.

Exercises

Sine wave oscillators

16.1 What is the reason for the need of an amplitude control circuit in a harmonic oscillator?

16.2 For $\alpha < 0$, the solution of Equation (16.2) is a sine wave with exponentially growing amplitude. However, when the system is switched on, the initial amplitude is zero. Why will the circuit start oscillating?

16.3 Find the oscillation conditions for circuits a–c shown in the figure below. Derive the oscillation frequency and the condition for the component values. The operational amplifiers can be considered as ideal. Further, to simplify the calculations, take $R_1 = R_2 = R$; and $C_1 = C_2 = C$.

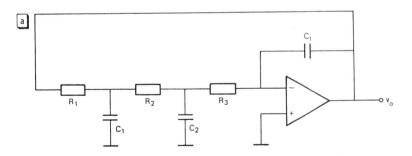

Exercises 277

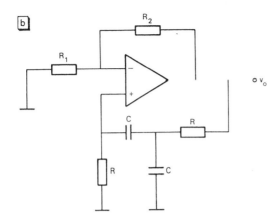

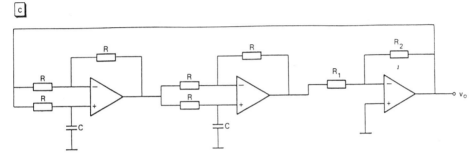

16.4 Refer to the sine wave oscillator of Figure 16.4b. The amplitude control is realized by replacing R_4 by a thermistor. The following properties of the applied components are given: $R_3 = 500\ \Omega$; $R_4 = R_{th} = 10^4\ e^{5200(1/T-1/T_0)}$. The heat resistance between the thermistor and the ambient is 100 K/W. $T_0 = 273$ K; the ambient temperature is 300 K. Find the amplitude of the output signal.

Voltage generators

16.5 Given the following triangle generator (shown in the figure below) with $R_3 = 5R_4$, $R_1/(R_1 + R_2) = \beta$ and $R_5C = \tau$, discuss the changes in v_o, at increasing value of R_1.

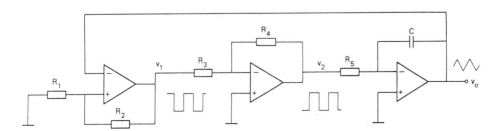

Signal generation

16.6 Calculate the duty cycle of the voltages v_1 and v_2 in the figure of Exercise 16.5, for the case $V^+ = 15$ V and $V^- = -5$ V.

16.7 Calculate the frequency, the average value and the amplitude of the triangular voltage v_o from the circuit in Exercise 16.5, when given $V^+ = 15$ V, $V^- = -5$ V, $R_5 = 10$ kΩ, $C = 100$ nF and $\beta = 0.5$.

16.8 Refer to the ramp generator depicted in the figure below. The output frequency of v_o is 1 kHz. Find the value of R. All components may be considered as ideal; the output levels of the Schmitt trigger are $v_s = V^+ = 18$ V or $v_s = V^- = -12$ V.

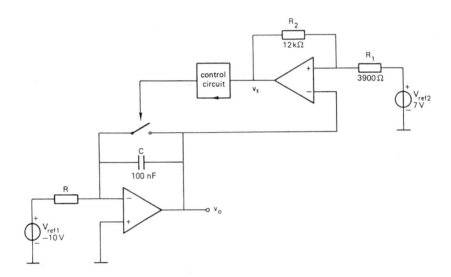

16.9 Refer to the generator in Exercise 16.8. By adding only one voltage source, the average output voltage can be adjusted to zero. How, and what value?

16.10 Refer to the ramp generator of Exercise 16.8. Find the amplitude and the frequency of the output signal v_o for $R = 100$ kΩ; the delay time of the switch is 2 ms (for the rest the switch is ideal).

16.11 Modify the circuit of Exercise 16.8 so that the output has a negative slope.

16.12 Derive the duty cycle of the output signal in Figure 16.13, as a function of V^+, V^-, β and $\tau = RC$. Find the condition for 50% duty cycle.

17 Modulation and demodulation

Modulation is a particular type of signal conversion, that makes use of an auxiliary signal, the carrier. One of the parameters of this carrier is varied analogously to the input (or measurement) signal. The result is a shift of the complete signal frequency band over a distance related to the carrier frequency. Due to this property, modulation is also referred to as frequency conversion.

The modulated waveform offers a number of advantages over the original waveform and is, for that reason, widely used in various electrical systems. One application of modulation is frequency multiplexing, a method for a more efficient way of signal transmission (Section 1.1). Telecommunication is inconceivable without modulation. A very important advantage of modulated signals is their better noise and interference immunity. Particularly in instrumentation systems, modulation offers the possibility of bypassing offset and drift. This is the main reason for discussing it in this book.

The carrier is a signal with a simple waveform, for instance a sine wave, square wave or pulse-shaped signal. Several parameters of the carrier can be modulated with the input signal, for instance the amplitude, the phase, the frequency, the pulse height or the pulse width. These types of modulation are referred to as amplitude modulation (AM), phase modulation, frequency modulation (FM), pulse height and pulse width modulation, respectively. Figure 17.1 shows some examples; in this figure, x_i is the input signal of the modulator, x_d the carrier signal and x_o the output or modulated signal.

In an FM signal (Figure 17.1d), the original information is embedded in the zero crossings of the modulated signal; in an AM signal the information is mainly included in the peak values. These peak values are easily affected by additive interference; the positions of zero crossings are much less sensitive to interference. That is why FM signals have a better noise immunity than AM signals. Nevertheless, even AM is a powerful technique in instrumentation to suppress interference; therefore, this chapter is confined to amplitude modulation only. The first part of the chapter deals with the basic concepts of amplitude modulation and methods for modulation and the reverse operation (demodulation). The second part of the chapter describes some applications of modulation and demodulation in particular measurement instruments.

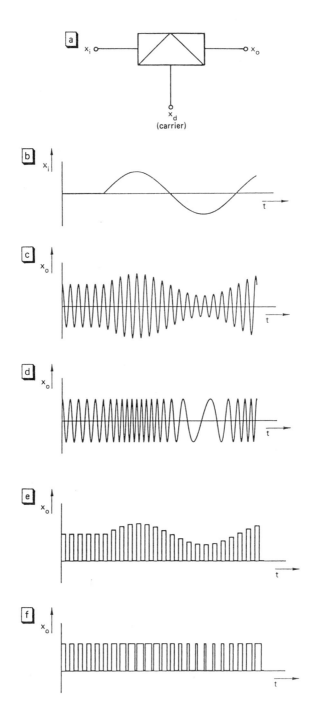

Figure 17.1 (a) Circuit symbol of a modulator; (b) example of an input signal; (c) amplitude modulation; (d) frequency modulation; (e) pulse height modulation; (f) pulse width modulation.

17.1 Amplitude modulation and demodulation

17.1.1 Theoretical background

For the analysis of the properties of modulated signals, we will use a description of the signals in the time domain. We start with a sinusoidal carrier $x_d(t) = \hat{x}_d \cos \omega_d t$. The amplitude of the carrier is varied according to the input signal $x_i(t)$, resulting in a time-varying carrier amplitude $\hat{x}_d(t) = \hat{x}_d(1 + kx_i(t))$. Hence, the modulated output signal is written as:

$$x_o(t) = \hat{x}_d(t) \cos \omega_d t = \hat{x}_d(1 + kx_i(t)) \cos \omega_d t$$

with k a coefficient determined by the modulator. For $x_i(t) = 0$, the output is just $x_d(t)$, the original carrier signal, with constant amplitude. Suppose the input signal is a pure sine wave (with only one frequency ω_i: $x_i(t) = \hat{x}_i \cos \omega_i t$. The modulated signal is:

$$x_o(t) = \hat{x}_d(1 + k\hat{x}_i \cos \omega_i t) \cos \omega_d t$$

Expansion of this expression into separate sine wave components gives:

$$x_o(t) = \hat{x}_d(\cos \omega_d t + m \cos \omega_i t \cos \omega_d t)$$
$$= \hat{x}_d[\cos \omega_d t + \tfrac{1}{2}m \cos(\omega_d + \omega_i)t + \tfrac{1}{2}m \cos(\omega_d - \omega_i)t]$$

with $m = k\hat{x}_i$ the modulation depth, a term that is apparent from Figure 17.2. This figure represents the output waveform for two different values of m. The input signal is still recognized in the envelope of the modulated signal.

Evidently, the modulated signal has three frequency components: one with the carrier frequency (ω_d), one with a frequency equal to the sum of the carrier frequency and the input frequency ($\omega_d + \omega_i$) and one with the difference between these two frequencies ($\omega_d - \omega_i$) (Figure 17.3a).

The modulator produces two new frequency components for each input component, positioned on both sides of the carrier frequency. An arbitrary, aperiodic low frequency input signal has a continuous spectrum as depicted in Figure 17.3b. By modulating the amplitude of a carrier with this signal, the whole frequency band is shifted to a region around the carrier frequency (Figure 17.3c). The bands at either side of the carrier are called the sidebands of the modulated signal. The total bandwidth of an AM signal is thus twice that of the original signal. Each sideband carries the full information content of the input signal.

From the previous discussion, it will be clear that an AM signal does not contain a DC component or low frequency components (if the carrier frequency is high enough), even when the original input signal has indeed such components. Modulated signals can therefore be amplified without being disturbed by offset and drift; these error signals can easily be removed from the amplified output by a high-pass filter.

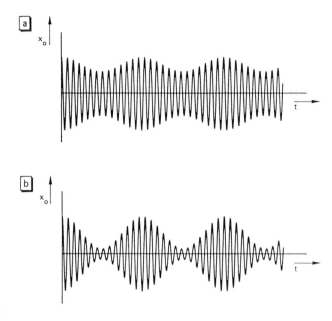

Figure 17.2 A sinusoidal carrier modulated with a sinusoidal input signal, with modulation depth (a) m = 0.25 and (b) m = 0.75.

The new position of the frequency band is just around the carrier frequency. The frequency band of another measurement signal can be shifted to a different position, by modulating a carrier with another frequency (Figure 17.3d). If these and possibly more bands do not overlap, they can all be transported over a single transmission channel (a cable, a satellite link) without disturbing each other. After the reception of the signals, they are separated by demodulation, the conversion of the frequency bands back to their original position.

Example 17.1
The bandwidth of the acoustical signals in a telephone system is limited to a range from 300 to 3400 Hz. The transmission channels have a much wider bandwidth. To achieve efficient signal transport, the various acoustic signals are converted to different carrier frequencies such that many signals can be transported simultaneously over one conductor pair. As both sidebands contain the same information, a special modulation technique is applied (single sideband modulation) to double the channel capacity. Moreover, the carrier frequency is not transported with the signals, to improve further the transmitter efficiency.

The first signal is converted to a band between 12 and 16 kHz, the second to a

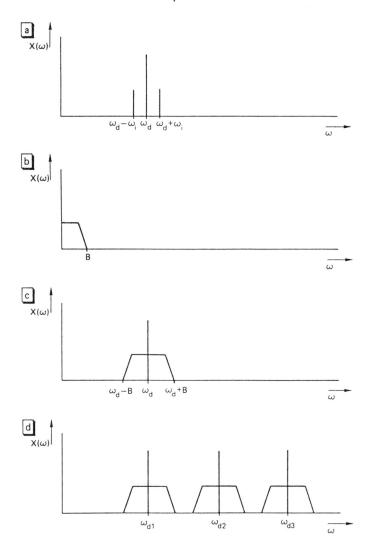

Figure 17.3 The frequency spectra of (a) a sine wave carrier (ω_d) modulated with a sine wave (ω_i); (b) a low frequency input signal with bandwidth B; (c) the corresponding AM signal; (d) a frequency multiplexed signal.

band from 16 to 20 kHz and so on. In this way, one cable composed of 24 conductor pairs can carry 2880 telephone signals. Coaxial cables, with a much wider bandwidth, can transport even more signals: up to 8100 signals over one cable composed of three coaxial conductors.

17.1.2 Methods of amplitude modulation

There are many ways to modulate the amplitude of a carrier signal. One of these methods employs an analog multiplier (Chapter 14). The multiplication of a carrier signal $x_d(t) = \hat{x}_d \cos \omega_d t$ with an input signal $x_i(t)$ results in an output signal $x_o(t) = x_i(t) x_d(t) = x_i(t) \hat{x}_d \cos \omega_d t$ (the scale factor of the multiplier is set to 1 for simplicity). For a sine-shaped input signal $x_i(t) = \hat{x}_i \cos \omega_i t$ the output of the multiplier is:

$$x_o(t) = \hat{x}_i \hat{x}_d \cos \omega_d t \cos \omega_i t$$
$$= \tfrac{1}{2}\hat{x}_i \hat{x}_d [\cos(\omega_d + \omega_i)t + \cos(\omega_d - \omega_i)t]$$

This signal contains only the two sideband components and no carrier (Figure 17.4a).

Figure 17.4b shows this signal in the time domain. Because there is no carrier, it is called an AM signal with suppressed carrier. For arbitrary input signals, the spectrum of the AM signal consists of two (identical) sidebands without carrier.

A second type of modulator is the switch modulator. The input signal is periodically switched on and off, a process that can be described by multiplying the input signal by a switch signal $s(t)$, being 1 when the switch is on, and 0 when the switch is off

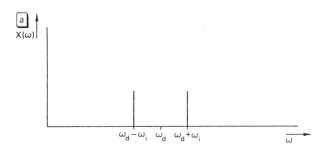

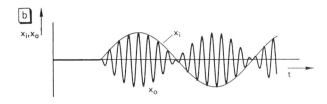

Figure 17.4 (a) The frequency spectrum and (b) the amplitude–time diagram of an AM signal with suppressed carrier.

Amplitude modulation and demodulation

(Figure 17.5). The result is a modulated signal $x_o(t) = x_i(t)s(t)$. To show that this product is indeed a modulated signal, we expand $s(t)$ into its Fourier series:

$$s(t) = \frac{1}{2} + \frac{2}{\pi}\left(\sin\omega t + \frac{1}{3}\sin 3\omega t + \frac{1}{5}\sin 5\omega t + \ldots\right)$$

with $\omega = 2\pi/T$, and T the period of the switching signal. For a sine wave input signal $x_i(t) = \hat{x}_i \cos\omega_i t$, the output signal $x_o(t)$ is:

$$x_o(t) = \hat{x}_i\left(\frac{1}{2} + \frac{2}{\pi}\sum_{n=1}^{\infty}\frac{1}{n}\sin n\omega t\right)\cos\omega_i t \quad (n \text{ even})$$

$$= \frac{1}{2}\hat{x}_i \cos\omega_i t + \frac{2\hat{x}_i}{\pi}\left[\frac{1}{2}(\sin(\omega+\omega_i)t + \sin(\omega-\omega_i)t)\right.$$

$$\left. + \frac{1}{2}\cdot\frac{1}{3}(\sin(3\omega+\omega_i)t + \sin(3\omega-\omega_i)t) + \ldots\right]$$

The spectrum of this signal is depicted in Figure 17.6a; Figure 17.6b shows the spectrum for an arbitrary input signal.

This modulation method produces a large number of sideband pairs, positioned around the fundamental and odd multiples (ω, 3ω, 5ω, ...). The low frequency component originates from the multiplication by the mean of $s(t)$ (here ½). This low frequency component and all components with frequencies 3ω and higher can be removed by a band-pass filter. The resulting signal is just an AM signal with suppressed carrier (Figure 17.6c).

The advantages of the switch modulator are its simplicity and accuracy; the sideband amplitude is determined only by the quality of the switch. A similar modulation can be achieved by periodically changing the polarity of the input signal. This is equivalent to the multiplication by a switch signal with zero mean value; in that case there is no low frequency component as in Figure 17.6.

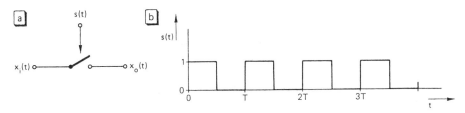

Figure 17.5 (a) An electronic switch, that can act as a modulator; (b) switching signal s(t), serving as carrier.

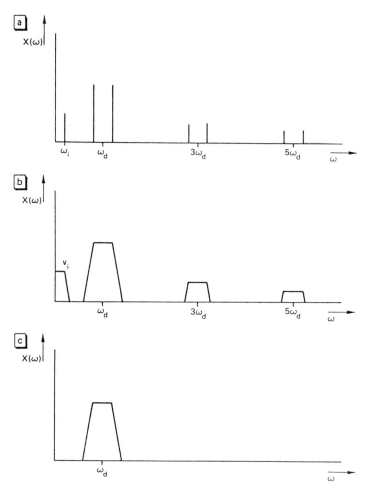

Figure 17.6 Signal spectra for a switch modulator; (a) output spectrum for a sinusoidal input signal; (b) output spectrum for an arbitrary input signal with limited bandwidth; (c) as (b), after low-pass filtering.

We have seen that the absence of DC and low frequency components considerably facilitates the amplification of modulated signals; offset, drift and low frequency noise can be kept far from the signal frequency band. When very low voltages must be measured (hence amplified), it is recommended to modulate them prior to any other analog signal processing that might introduce DC errors. An application of this concept is encountered in the measurement bridge; it can be considered as a third modulation method. The principle is illustrated with the Wheatstone resistance measurement bridge of Figure 17.7a. Such a bridge circuit is widely used for the read-out of resistive sensors whose resistance value changes with the applied physical quantity

Amplitude modulation and demodulation 287

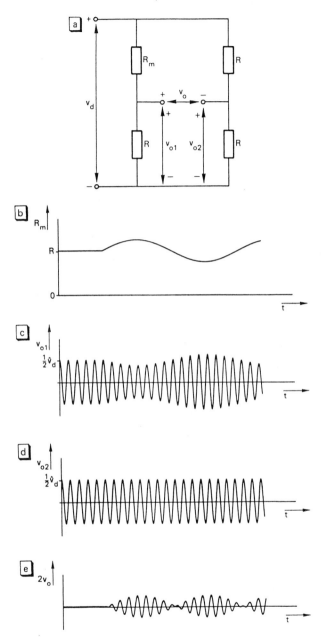

Figure 17.7 (a) Resistance measurement bridge for the measurement of resistance R_m; (b) example of varying resistance value; (c) v_{o1} is an AM signal with carrier $\tfrac{1}{2}v_d$; (d) v_{o2} has a constant value $\tfrac{1}{2}v_d$; (e) bridge output voltage $v_{o1} - v_{o2}$: an AM signal without carrier (scaled by a factor 2 in the figure).

(Section 7.2). The bridge is connected to an AC signal source; the AC signal, usually a sine or square wave, acts as the carrier. In this example, we have just one varying resistance R_m in the bridge (Figure 17.7b). When we consider the bridge as a double voltage divider, the output voltage of the left-hand branch is found to be $v_{o1} = v_d \cdot R / (R + R_m)$ and the output of the right-hand branch is $v_{o2} = \frac{1}{2} v_d$. The latter voltage has a constant amplitude (Figure 17.7d); the amplitude of v_{o1}, however, varies with R_m; it is an AM signal modulated by R_m (Figure 17.7c). It may also be written as $v_{o1} = \frac{1}{2} v_d (1 + f_r(t))$, where $f_r(t)$ varies analogously to R_m. The bridge output signal is $v_o = v_{o1} - v_{o2} = \frac{1}{2} v_d (1 + f_r(t)) - \frac{1}{2} v_d = \frac{1}{2} f_r(t) \cdot v_d$, which is an AM signal without carrier (suppressed by subtracting v_{o2}), as is shown in Figure 17.7e. This output can be amplified by a differential amplifier with high gain; its low frequency properties are irrelevant. The only requirements are a sufficiently high bandwidth and a high CMRR for the carrier frequency, to amplify accurately the difference $v_{o1} - v_{o2}$.

17.1.3 *Methods for demodulation*

The reverse process of modulation is demodulation or (incorrect) detection. Looking at the AM signal with carrier (for instance in Figure 17.2a), we will observe the similarity between the envelope of the amplitude and the original signal shape. An obvious demodulation method would be peak detection (Section 9.2.2). Although this method is applied in radio receivers for AM signals, it is not recommended for instrumentation purposes because of the inaccuracy and noise sensitivity; each transient is considered as a new peak belonging to the signal. A better demodulation method is (double-sided) rectifying and low-pass filtering (Figure 17.8).

The low-pass filter responds to the average of the rectified signal. When properly designed, the output follows the input amplitude.

Obviously, the peak detector and the rectifier detector operate only for AM signals with carrier. In an AM signal without carrier, the envelope is no longer a copy of the input; positive and negative input signals produce equal amplitudes (Figure 17.4b). Apparently, additional information is required with respect to the phase of the input, for a full recovery of the original waveform.

An excellent method to solve this problem, and which has a number of additional advantages, is synchronous detection. This method consists of multiplying the AM signal with another signal that has the same frequency as the original carrier. If the carrier signal is available (as is the case in most measurement systems) this signal can be the carrier itself.

Consider an AM signal with suppressed carrier: $x_o = \hat{x}_o \cos \omega_d t \cdot \cos \omega_i t$. This is multiplied by the synchronous signal $x_s = \hat{x}_s \cos(\omega_d t + \varphi)$, where φ takes into account a possible phase-shift relative to the original carrier. The product of these signals is $x_{dem} = \hat{x}_s \hat{x}_o \cos \omega_d t \cdot \cos \omega_i t \cdot \cos(\omega_d t + \varphi)$. Separation of the frequency components results in:

$$x_{dem} = \frac{1}{2} \hat{x}_o \hat{x}_s \cos \omega_i t \cos \varphi + \frac{1}{4} \hat{x}_o \hat{x}_s [\cos((2\omega_d + \omega_i) t + \varphi) + \cos((2\omega_d - \omega_i) t + \varphi)]$$

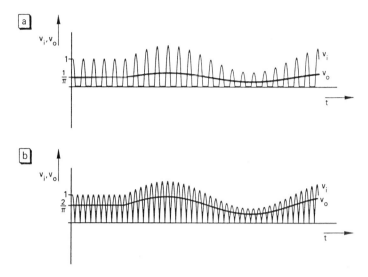

Figure 17.8 (a) Single-sided rectified AM signal v_i, followed by a low-pass filter, responding to the average value v_o; (b) as (a), but now a double-sided rectified signal, producing an average that is twice as high.

With a low-pass filter, the components around $2\omega_d$ are removed, leaving the original component with frequency ω_i. This component has a maximum value for $\varphi = 0$, when x_s has the same phase as the carrier. For $\varphi = \pi/2$, the demodulated signal is zero, and it has the opposite sign for $\varphi = \pi$. This phase sensitivity is an essential property of synchronous detection. Figure 17.9 reviews the whole detection process.

In Figure 17.9b, the spectrum of the AM signal without carrier is depicted, as well as some error signal components. By multiplication by the synchronous signal, a new band is created that coincides with the original one (Figure 17.9c). The low-pass filter removes all components with frequencies higher than that of the original band (Figure 17.9d).

An important advantage of this detection method is the elimination of all error components that are not in the (small) band of the AM signal. If the measurement signal has a narrow band (slowly fluctuating measurement quantities), a low cut-off frequency of the filter can be chosen. Hence, most of the error signals are removed, even with a simple first-order low-pass filter. Synchronous detection permits the measurement of AC signals with a very low signal-to-noise ratio.

17.2 Systems based on synchronous detection

The preceding section showed that synchronous detection is a powerful mechanism in instrumentation to measure small AC signals with a low signal-to-noise ratio. The

290 Modulation and demodulation

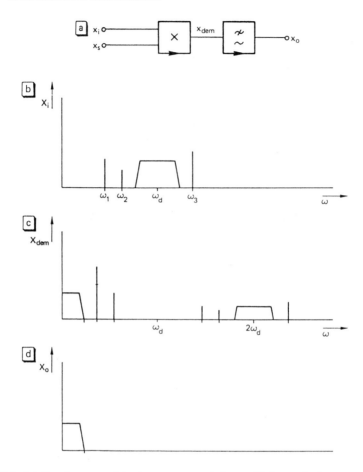

Figure 17.9 (a) Synchronous detector, composed of a multiplier and a low-pass filter; (b) the frequency spectrum of an AM signal without carrier and with some error signals; (c) the spectrum after multiplying with a synchronous signal; (d) the output after the low-pass filter: the original signal component.

measurement bridge from Section 17.1.2 is supplied by an AC source to convert the low frequency sensor signal into an AM signal, in order to bypass the offset and drift problems of high gain amplifiers. Synchronous detection of the amplified bridge output makes the measurement highly tolerant to noise and other interference components (provided that they are outside the signal band). The bandwidth of the detection system is set by a simple low-pass filter of the first order.

Synchronous detection is applied in many measurement instruments, for instance network analysers and impedance analysers. In such instruments the measurement signals are all sinusoidal; the analysis takes place in the frequency domain. Where no synchronous signal is available, it must first be generated. This is achieved by a

17.2.1 Phase-locked loop

A phase-locked loop or PLL is a special electronic system that is able to find the frequency of a sinusoidal measurement signal despite noise and other spurious signals. The PLL consists of a voltage-controlled oscillator (VCO, see Section 16.2.4), a synchronous detector and a control amplifier (Figure 17.10). The output frequency of the VCO is controlled so that it is equal to that of a particular frequency component of the input signal. Due to the high gain of the operational amplifier, the feedback loop forces the amplifier input to be zero. The synchronous detector responds to the frequency difference between the input signal v_i and the output signal v_o of the VCO (due to the multiplication of these signals). At equal frequencies, it responds to the phase difference. If the difference is not zero, the control amplifier controls the VCO via v_s so that the difference is reduced. In the steady state, the input of the amplifier is zero, so the output of the synchronous detector is also zero. This situation corresponds to equal frequencies and a phase difference $\pi/2$ between v_i and v_o.

The synchronous detector is only sensitive to signal frequencies between $f_o \pm B$, with f_o the frequency of v_o, and B the bandwidth of the low-pass filter; other frequencies are ignored. So, the PLL is able to generate a periodic signal whose frequency is exactly equal to that of one particular component of the input signal. This property makes the PLL also suitable as an FM demodulator; the VCO frequency is related to the control voltage v_s. At varying input frequency, v_s varies analogously; v_s is the demodulated FM signal.

A slight extension of the circuit in Figure 17.10 allows the generation of a periodic signal whose frequency equals the sum of two frequencies (Figure 17.11). Assume the frequencies of the input signals v_1 and v_2 are f_1 and f_2, respectively. Synchronous detector 1 measures the difference between f_1 and f_o, the frequency of the VCO. The bandwidth of this synchronous detector is chosen so that $f_o - f_1$ can pass the detector, but not the sum $f_o + f_1$. The second detector measures the difference between the

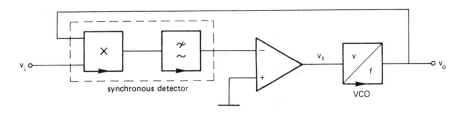

Figure 17.10 Functional diagram of a phase-locked loop (PLL) consisting of a synchronous detector, an amplifier and a voltage-controlled oscillator (VCO).

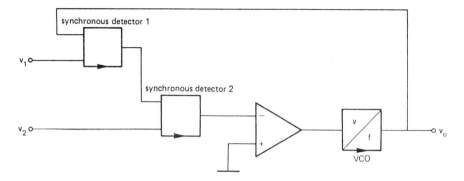

Figure 17.11 Extended PLL, for producing a signal with a frequency equal to the sum of the frequencies of v_1 and v_2.

frequencies $(f_o - f_1)$ and f_2. The feedback makes this signal zero, hence, $f_2 = f_o - f_1$ or $f_o = f_1 + f_2$.

Likewise, the PLL can be used for the generation of multiple frequencies. This is achieved by inserting a frequency divider between the VCO output and the input of the synchronous detector in Figure 17.10. For a frequency division by n, the synchronous detector measures the difference between f_i and f_o/n; this difference is controlled to zero, so $f_o = nf_i$.

The important parameters of a PLL are the lock-in range and the hold range. The lock-in range is that of input frequencies for which the controller locks automatically to that frequency. The hold range is the one over which the input frequency may vary after being locked. The hold range is usually larger than the lock-in range. The hold range depends on the amplitude of the input signal component; the lower this signal, the harder it is for the PLL to control. The lock-in range of a PLL can be shifted by changing the free-running frequency f_o (at $v_s = 0$) of the VCO.

Like many other electronic processing circuits, the PLL is also available as an integrated circuit. This IC contains the analog multiplier, the control amplifier and the VCO. The low-pass filter of the synchronous detector and the components that fix the free-running frequency of the VCO have to be connected externally; such components are hardly integrable. Moreover, the user keeps some freedom for the design. In the following sections, the PLL is considered as a complete system (black box approach).

17.2.2 *Lock-in amplifiers*

A lock-in amplifier is an AC amplifier based on synchronous detection and intended to measure the amplitude and phase of small, noisy, narrow-band measurement signals. A simplified block diagram of a lock-in amplifier is depicted in Figure 17.12. The amplifier has two input channels: the signal channel and the reference channel. The signal channel is composed of AC amplifiers and a band-pass filter. The latter is used to remove part of the input noise prior to detection: the predetection filter. Depending

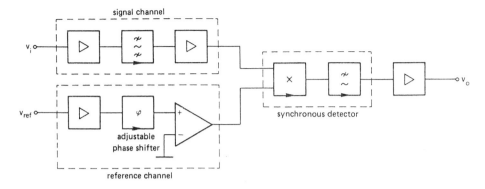

Figure 17.12 A simplified block diagram of a lock-in amplifier for the measurement of the amplitude and phase of small AC signals.

on the type of lock-in amplifier, it is either a manually adjustable or an automatic filter. The reference channel comprises an amplifier, an adjustable phase-shifter and a comparator. An adjustable low drift DC amplifier brings the output to the proper level to display the measurement value. The output signal is a DC or slowly varying signal proportional to the amplitude of the input signal. The adjustable phase-shifter allows maximizing the amplifier's sensitivity; the sensitivity is maximal for $\cos \varphi = 0$ or $\varphi = \pi/2$. With the calibrated phase-shifter, the original phase difference between v_i and v_ref can be determined.

The comparator in the reference channel converts the reference signal into a square wave for a proper control of the switches in the switching synchronous detector. A lock-in amplifier as described here is suitable for the measurement of AC signals of which a synchronous signal is available (like a bridge measurement or a transfer measurement). If there is no synchronous signal, it must still be produced. This is explained by the diagram in Figure 17.13, where a PLL is used for the generation of the synchronous signal.

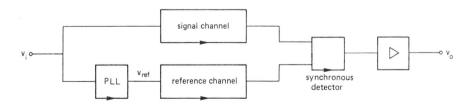

Figure 17.13 Simplified block diagram of a lock-in amplifier with self-generating reference signal.

17.2.3 Chopper amplifiers

A chopper amplifier or chopper-stabilized amplifier is a special type of amplifier for very small DC voltages or low frequency signals. To get rid of the offset and drift encountered in common DC amplifiers, the measurement signal is first modulated using a switch modulator; next it is amplified, then demodulated by synchronous detection (Figure 17.14).

Chopper amplifiers (also called indirect DC amplifiers for obvious reasons) are available as complete modules. Table 17.1 gives a comparison between the properties of an operational amplifier with JFETs (column I) and two types of chopper amplifiers.

In column II, the input voltage is chopped using JFETs as switches. In the varactor amplifier of column III the modulation is performed by varactors, reverse-biased diodes used as voltage-controllable capacitances. The most striking advantage of the chopper amplifier is the combination of a low input offset voltage and a low bias

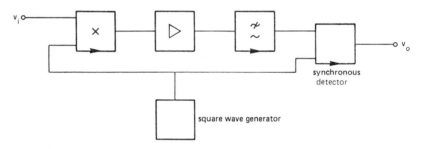

Figure 17.14 Block diagram of a chopper amplifier, for the measurement of very small DC signals.

Table 17.1 Comparison between a JFET operational amplifier and two types of chopper amplifiers

	I JFETs	II JFET chopper	III varactor
A_o	$2 \cdot 10^5$	10^7	10^5
V_{off}	<0.5 mV	<0.02 mV	—
t.c. V_{off}	(adjustable to zero) 2 μV/K	(adjustable to zero) ±0.1 μV/K	(adjustable to zero) 10 μV/K
Supply voltage coefficient	—	±0.1 μV/V	500 μV/V
Stability	—	±1 μV/month	100 μV/month
I_{bias}	10 pA	50 pA	0.01 pA
t.c. I_{bias}	2 pA/K	0.5 pA/K	×2 per 10 K
Bandwidth	1 MHz	3 MHz	2 kHz
R_i	10^{12} Ω	10^6 Ω	10^{14} Ω

current; their non-zero values are due to imperfections of the switches. The varactor amplifier (sometimes called an electrometer amplifier, after its predecessor with electrometer tubes) combines a very high input impedance and an extremely low bias current.

SUMMARY

Amplitude modulation and demodulation

- Amplitude modulation (AM) is a type of signal conversion at which the amplitude of a (usually sinusoidal) carrier signal varies analogously with the input signal. The carrier frequency is high compared to the input frequency. Modulation results in a shift of the input frequency band.
- The reverse process of modulation is demodulation or detection, and results in the original waveform.
- The spectrum of an AM signal consists of two sidebands symmetrically positioned around the carrier frequency. The information content in each of these sidebands is identical to that of the original band.
- Frequency multiplexing is a method of multiplexing in which various signals are converted to different frequency bands by means of modulation. If the bands do not overlap, they can be simultaneously transported over a single transmission line.
- Some methods for amplitude modulation are:
  - analog multiplying of the input signal and the carrier;
  - periodically switching (on/off or $+/-$) the signal; the switching signal acts as the carrier.
- The multiplication of two signals with frequency f_1 and f_2 results in two new frequencies: the sum $f_1 + f_2$ and the difference $f_1 - f_2$.
- Modulated signals are not sensitive to low frequency interference due to the position of the frequency band. Low frequency measurement signals are best modulated prior to any other signal processing, if possible. An example is the measurement bridge with AC supply voltage.
- Demodulation by synchronous detection enables the measurement of signals with very low signal-to-noise ratio. A synchronous detector has a phase-sensitive response.
- Demodulation with a diode peak detector is simple but inaccurate and not applicable to AM signals without carrier.

Systems based on synchronous detection

- Synchronous detection is a type of phase-sensitive amplitude detection, suitable for signals with low signal-to-noise ratio.

- A phase-locked loop (PLL) is an electronic system that generates a periodic signal whose frequency is equal to a particular frequency component of the input signal.
- With a PLL, signals can be generated with frequencies equal to the sum or a multiple of two other frequencies.
- Important parameters of a PLL are the hold range and the lock-in range, defined as the allowable frequency range of the input signal after and before lock-in.
- A lock-in amplifier is an AC amplifier based on synchronous detection. It allows the measurement of the amplitude and phase of very small and noisy AC signals.
- A chopper amplifier is a DC amplifier in which the input signal is first modulated, then amplified as an AC signal and finally demodulated by synchronous detection. Such amplifiers have an extremely low offset voltage and bias currents.

Exercises

Amplitude modulation and demodulation

17.1 What is meant by an amplitude modulated signal with suppressed carrier? What is pulse width modulation?

17.2 A sine wave carrier with frequency 5 kHz is modulated in amplitude by a symmetric triangular signal with fundamental frequency of 100 Hz. Find all frequencies in the modulated signal.

17.3 A frequency multiplex system based on AM should transmit 12 measurement signals having a frequency band of 100–500 Hz each. Find the minimum bandwidth of this system.

17.4 In the Wheatstone bridge below, all resistors are sensors. The bridge is supplied with an AC voltage with amplitude 10 V. The output v_o is amplified by a factor 10^4 and multiplied by a synchronous signal with amplitude 4 V. Calculate the output DC voltage for a relative resistance change $\Delta R/R$ of 10^{-5} and of -10^{-5}.

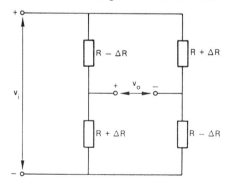

17.5 The same question as in Exercise 17.4. Now, the (amplified) bridge output signal is demodulated with a switched multiplier, whose operation is described by the Fourier series $s(t) = 4/\pi(\cos \omega t + \frac{1}{3}\cos 3\omega t + \frac{1}{5}\cos 5\omega t + \ldots)$; the switching signal is synchronous with the bridge power signal.

Systems based on synchronous detection

17.6 Given a synchronous detector composed of an ideal analog multiplier and a low-pass filter with cut-off frequency of 100 Hz. The frequency of the reference signal is 15 kHz. Find the input frequencies for which this detector is sensitive; make a plot of the frequency characteristic of the sensitivity.

17.7 The synchronous detector from Exercise 17.6 really acts as a band-pass filter. What is the equivalent Q-factor of the detector?

17.8 Given a synchronous detector composed of a switched modulator according to the principle depicted in the figure below, and a low-pass filter with cut-off frequency 200 Hz. The switching period is 0.2 ms. Which frequencies are detected by this system?

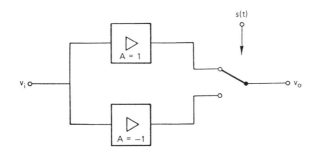

17.9 The figure below shows the block diagram of a system for the testing of a linear signal processing circuit. Which characteristic is displayed?

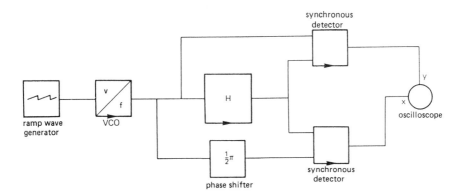

298   Modulation and demodulation

17.10 The output of the comparator in Figure 17.12 is either high or low, corresponding to the polarity of the phase-shifter output. The output levels have accurate values. What are the requirements for the amplitude of the reference voltage v_{ref}, and how do they change when the comparator is omitted?

17.11 Make a sketch of the frequency spectra of all the signals in the lock-in amplifier of Figure 17.12. Assume a narrow band AC input signal.

17.12 Make a plot of the frequency spectra of all the signals in the chopper amplifier of Figure 17.14. Assume a narrow band (quasi-)DC input signal.

18 Digital-to-analog and analog-to-digital conversion

The analog waveform is not suitable for signal processing by a digital processor or computer. Only after conversion into a digital signal can it be so handled. Conversely, many actuators and other output devices require analog signals, so the digital signals from a computer have to be converted into an analog waveform. Analog-to-digital converters (AD converters or ADCs) and digital-to-analog converters (DA converters or DACs) are available as modules or integrated circuits. This chapter starts with a brief description of digital signals and binary codes and continues with the description of the main types of AD and DA converters. The second part of this chapter is devoted to some particular types of converters.

18.1 Parallel converters

18.1.1 *Binary signals and codes*

A digital or binary signal has only two levels, denoted as '0' and '1'. The relation to voltage values or current values depends on the applied components, in particular the technology (bipolar transistors, MOSFETs and possibly others). In digital circuits with for instance bipolar transistors, a '0' corresponds to a voltage below 0.8 V; a '1' represents a voltage above 2 V. Digital signals have a much lower interference sensitivity than analog signals, due to the wide tolerances of the first. The tolerance for interference and noise is paid for by a strong reduction of the information content. An analog voltage of, say, 6.32 V contains much more information than a digital value '0', because in the latter case there are only two possible levels. The minimum amount of information (yes or no; high or low; '0' or '1'; on or off) is called a bit, an acronym for binary digit.

Usually a measurement signal contains much more information than only 1 bit. To represent adequately the information by a binary signal, a group of bits is necessary; such a group is called a binary word. A word consisting of 8 bits is called a byte (an impure acronym for 'by eight'). The term kilobyte stands for 1024 bytes (not 1000!); so, 64 kbytes equals 65 536 bytes or 512 kbits or 524 288 bits. The words byte and bit should not be confused; the notation kb or kB is not clear. In this book, we will not use these abbreviations and write kbyte or kbit.

With n bits, only 2^n different words can be constructed. The number of bits is bound to a maximum, not only for practical reasons, but also due to imperfections of the components in the AD converter that generates the binary words. So, analog-to-digital conversion causes at least one extra error, the quantization error (see also Section 2.1). Typical word lengths of AD converters range from 8 to 16 bits, corresponding to a quantization error of $2^{-8}-2^{-16}$ ($10^{-3}-10^{-6}$). Obviously, it is useless to take an AD converter with many more bits than correspond with the inaccuracy or the resolution of the input signal itself.

Example 18.1
The range of a measurement signal is 0–10 V. There are 10 bits available to represent this signal. The resolution of this representation is $2^{-10} \approx 0.1\%$ or about 10 mV.

Another measurement signal has an inaccuracy of 0.01%. The number of bits, required for a proper representation, is 14, because $2^{14} > 10^4 > 2^{13}$.

A binary word can be written as:

$$G = (a_n a_{n-1} \ldots a_2 a_1 a_0 a_{-1} a_{-2} \ldots a_{-m})$$

where a_i is either 0 or 1 (numbers). The value of G in the decimal number system is:

$$a_n 2^n + a_{n-1} 2^{n-1} + \ldots + a_2 2^2 + a_1 2 + a_0 + a_{-1} 2^{-1} + \ldots + a_{-m} 2^{-m}$$

The coefficient a_n contributes most to G and is therefore called the most significant bit or MSB. The coefficient a_{-m} has the lowest weight and is called the least significant bit or LSB.

The digital signals of AD and DA converters are binary-coded fractions of a reference voltage V_{ref}. The relation between the analog and digital signals of the converter is:

$$V_a = V_{ref}(a_{n-1} 2^{-1} + a_{n-2} 2^{-2} + \ldots + a_1 2^{-n+1} + a_0 2^{-n})$$

$$= V_{ref} \sum_{0}^{n-1} a_i 2^{i-n} \qquad (18.1)$$

The analog voltage is equal to $V_a = G \cdot V_{ref}$, with G a binary number between 0 and 1. Consequently, the MSB of a converter corresponds to a value $\tfrac{1}{2} V_{ref}$, the next bit is $\tfrac{1}{4} V_{ref}$ and so on till, finally, the LSB has a value $2^{-n} \cdot V_{ref}$.

Binary notation is rather inefficient: for the representation of a not-too-small number many bits are required. This explains the use of the hexadecimal notation, which is based on the hexadecimal number system (base 16); the 16 digits are denoted as 0, 1, 2, 3, . . ., 9, A, B, C, D, E and F. The last one, F, has the (decimal) value of 15 or the binary value 1111. The hexadecimal notation is found from the binary

Example 18.2

The binary number 1010011110 can be written as 0010 1001 1110, or 29E hexadecimal. In the decimal number system it is:

$$2 \cdot 16^2 + 9 \cdot 16^1 + 14 \cdot 16^0 = 670$$

This can also be found from the binary notation:

$$1 \cdot 2^9 + 0 \cdot 2^8 + 1 \cdot 2^7 + 0 \cdot 2^6 + 0 \cdot 2^5 + 1 \cdot 2^4 + 1 \cdot 2^3 + 1 \cdot 2^2$$
$$+ 1 \cdot 2^1 + 0 \cdot 2^0 = 670$$

Other codes that are employed are the BCD code and the octal code. The BCD code (binary-coded digit) is structured as follows:

$$\ldots a_2 b_2 c_2 d_2 \; a_1 b_1 c_1 d_1 \; a_0 b_0 c_0 d_0$$

where each group of 4 bits represents the binary-coded decimal digit. This code can be interpreted more easily than the binary code.

Example 18.3

The BCD notation of the decimal number 670 from Example 18.2 is 0110 0111 0000, the binary codes of the decimal digits 6, 7 and 0, respectively.

The octal code, sometimes used in computer programs, has the symbols 0, 1, 2, 3, 4, 5, 6 and 7. The code 10_8 stands for 8_{10} (the index denotes the number system). The octal notation can be derived directly from the binary notation by a partitioning into groups of 3 bits (starting with the LSB), just like the conversion from binary to the hexadecimal notation. The number 670_{10} from Example 18.3 is 1236_8 in octal notation.

Figure 18.1 shows several electronic representations of a binary word. Figures 18.1a and b are dynamic representations (voltages or currents as time signals). In Figure 18.1a, the bits of a word are generated one after another; it is a serial word. The consequence of this form is the relatively long time for each measurement value, which grows in proportion to the number of bits. In Figure 18.1b the bits are generated and transported simultaneously, one line for each bit. There are as many parallel lines as there are bits in a word; it is a parallel word. The information is available all at once, but more hardware is required (cables, connectors, components). Figure 18.1c illustrates a static representation of the same binary word, here with a set of switches. A '0' corresponds to a switch that is off and a '1' to a switch that is on.

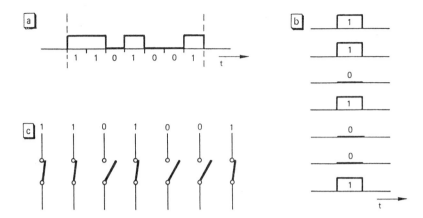

Figure 18.1 Various representations of a binary word: (a) a serial word (dynamic); (b) a parallel word (dynamic); (c) a parallel word (static), using a set of switches.

The time of 1 bit is called the bit time. The transport of a serial n bit word takes n bit times; the transmission of a parallel word is substantially shorter, but takes at least 1 bit time. The conversion process of an analog signal $x(t)$ into a (serial or parallel) word takes time too, which is the conversion time. During that time, the signal may change somewhat. To minimize the uncertainty in both the amplitude and the time of the converted signals, samples are taken at discrete time intervals (Section 1.2). Only the measurement values occurring at those specified moments are converted. The sampled signal should be fixed during the conversion time of the AD converter. This is accomplished by a sample-hold circuit (Section 15.2). When the conversion time is short compared to the actual changes in the input signal, no additional sample-hold circuit is required; the AD converter itself performs the sampling.

18.1.2 *Parallel DA converters*

This section deals with only one type of DA converter, the parallel converter with ladder network. It is the most widely used type for general applications, and is available as an integrated circuit for a very low price.

The first step of the DA conversion is the transfer from the n parallel signal bits to a set of n parallel switches. The binary signals must satisfy certain conditions to activate the switches. Once the digital parallel input signal is copied to the switches, the proper weighting factors must be assigned to each of the (equal) switches; half the reference voltage to the switch for the MSB (a_{n-1}), a quarter of the reference to the next switch a_{n-2} and so on. Figure 18.2 gives an example of this assignment. A current is allocated to each switch corresponding to the weighting factor: $\frac{1}{2}I$ for the MSB, $\frac{1}{4}I$ for the next bit and so on. The current for the last switch (LSB) is $I \cdot 2^{-n}$. In this circuit, the switches have two positions: left and right (1 and 0, respectively). When in

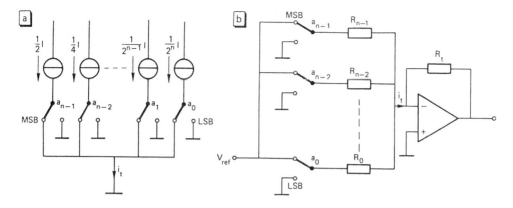

Figure 18.2 Conversion of a binary code into (a) a current; (b) a voltage.

the right position, the current flows directly to ground. In the left position, the current flows to a summing point. The sum of the currents is exactly the analog value that corresponds to the binary input code.

The weighted currents $I/2$, $I/4$ and so on are derived from a single reference voltage (Figure 18.2b). Since the voltage at the non-inverting input terminal of the operational amplifier is zero, the current through each resistor equals V_{ref}/R. The ratio between two subsequent resistance values is exactly 2, so the corresponding currents also differ by a factor of 2. The total current is now:

$$i_t = V_{ref}\left(\frac{a_{n-1}}{R_{n-1}} + \frac{a_{n-2}}{R_{n-2}} + \ldots + \frac{a_1}{R_1} + \frac{a_0}{R_0}\right) = V_{ref}\sum_0^{n-1}\frac{a_i}{R_i}$$

$$= V_{ref}\sum_0^{n-1}\frac{a_i 2^{-n+i+1}}{R_{n-1}}$$

with R_{n-1} the lowest resistance, belonging to the MSB a_{n-1}, and $R_i = R_{n-1}\cdot 2^{n-1-i}$. When we make the feedback resistor R_t equal to $R_{n-1}/2$, the output voltage of the converter is:

$$v_o = -i_t R_t = -V_{ref}\sum_0^{n-1} a_i 2^{-n+i} \qquad (18.2)$$

which, except for the minus sign, is equal to the required equation (18.1). V_{ref} is the full-scale value of the converter: for all $a_i = 1$, $v_o = V_{ref}(-1\,\text{LSB})$.

The configuration of Figure 18.2b is seldom applied because it contains very high and very low resistances, in particular at large numbers of bits. It is rather difficult to make accurate resistors with a high resistance value for the LSB. Moreover, a high resistance, together with the unavoidable parallel capacitance, forms a large time constant, making the converter slow. On the other hand, very low resistances for the

MSB may load the reference source too much, resulting in unacceptable load errors. These obstacles are bypassed by replacing the resistance network with the ladder network of Figure 18.3. A particular property of this network is the input resistance, which is independent of the number of sections, as can easily be seen from Figure 18.3.

Now let us consider the currents through this network. At the first (leftmost) node the input current I splits into two equal parts; one half flows through the resistor $2R$, the other half towards the rest of the network, having an input resistance of $2R$ as well. The latter current, $½I$, splits up again at the second left node in two equal parts, and so on. Consequently, the currents through the resistors $2R$ have values $I/2$, $I/4$, $I/8$ and so on. The network successively divides by two, using only two different resistance values, R and $2R$. This simplifies considerably the design of a DA converter for large numbers of bits.

Figure 18.4 shows how the ladder network is applied in a DA converter circuit. The weighted currents flow either to ground (switch position '0') or to the summing point of the current-to-voltage converter (switch position '1'). The largest current $I/2$ flows to the switch for the MSB (a_{n-1}), the lowest current $I/2^n$ to the switch for the LSB (a_0). The following equations apply for this circuit:

$$I = V_{ref}/R$$

and

$$i_t = \frac{a_{n-1}I}{2} + \frac{a_{n-2}I}{2^2} + \ldots + \frac{a_1 I}{2^{n-1}} + \frac{a_0 I}{2^n} = \sum_0^{n-1} \frac{a_i I}{2^{n-i}}$$

The output voltage is:

$$v_o = -Ri_t = -R \sum_0^{n-1} a_i \frac{I}{2^{n-i}} = -V_{ref} \sum_0^{n-1} a_i 2^{-n+i} \qquad (18.3)$$

which is again equivalent to Equation (18.1).

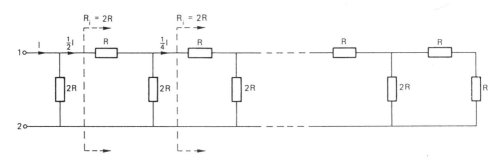

Figure 18.3 A ladder network, containing two resistance values; the input resistance is independent of the number of sections.

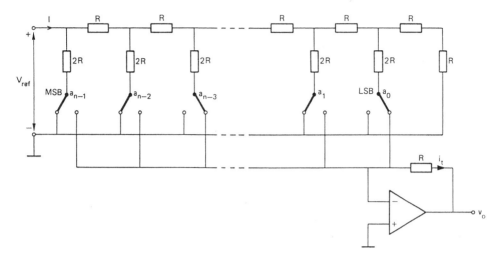

Figure 18.4 A digital-to-analog converter using a ladder network.

Table 18.1 lists the major specifications of an integrated 12 bit parallel DA converter. This type contains an inbuilt reference voltage (with Zener diode, Chapter 9), with an external connection that may be used for other purposes. The ladder network is composed of laser-trimmed SiCr resistors. The switches are made of bipolar transistors. Unlike the circuit discussed before, this DA converter has a current output. The smallest output current step is $2^{-12} \cdot 2\,\text{mA}$, about $0.5\,\mu\text{A}$. This current step corresponds to an input change of 1 bit. Such a step is also called an LSB. The inaccuracy and other precision quantities of a DA converter are usually expressed in terms of this unit, the LSB (see Table 18.1). An inaccuracy of $\pm\tfrac{1}{2}\text{LSB}$

Table 18.1 Specifications of a 12 bit DA converter (ppm = parts per million = 10^{-6})

Input	'1': max. +5.5 V, min. +2.0 V
	'0': max. +0.8 V
Output current	Unipolar −2 mA (all bits '1')
	Bipolar ±1 mA (all bits '1' or '0')
Output offset	<0.05% full scale (unipolar)
Reference voltage	10 V ± 1 ppm/K (unipolar)
Inaccuracy	±¼LSB (25°C), ±½LSB (0–70°C)
Differential non-linearity	±½LSB (25°C)
Monotony	guaranteed (0–70°C)
Settling time	<200 ns (up to ±½LSB)
Power dissipation	225 mW

corresponds (for this 12 bit converter) to $\pm\frac{1}{2} \cdot 2^{-12} \approx \pm 1.2 \cdot 10^{-4}$ of the full scale, or $\pm 0.25\,\mu A$. The differential non-linearity is the maximum deviation from the nominal smallest step (LSB) at the output. Guaranteed monotony means that the output never goes down at an upward change of the input code.

When an output voltage instead of current is desired, the user must add a current-to-voltage converter to the DAC (Figure 18.5). An output voltage range of 0–10 V is achieved by connecting the output of the operational amplifier to terminal 5 of the converter and so using the internal resistor $R = 5\,k\Omega$ as a feedback resistor: $v_o = Ri_o$. The output voltage range can be doubled by making this connection to terminal 4 instead of 5 (dotted line in Figure 18.5). Other ranges can be made by connecting external resistances in series with the internal resistances. By short-circuiting terminals 2 and 3, an extra current of 1 mA is added to the current-to-voltage converter. The input current of the operational amplifier runs from -1 to 1 mA, resulting in a bipolar output voltage with range -5 to 5 V (or from -10 to 10 V).

In the previous discussion, the MSB is denoted by a_{n-1} and the LSB by a_0. Other notations are encountered as well, such as the reverse notation (a_0 for the MSB, a_{n-1} for the LSB), or from 0 (or 1) to n instead of $n-1$: a_0 (or a_1) is the LSB, a_n the MSB or vice versa.

18.1.3 *Parallel AD converters*

The output of an AD converter is a binary code representing a fraction of the reference voltage (or current) that corresponds to the analog input. Evidently, this

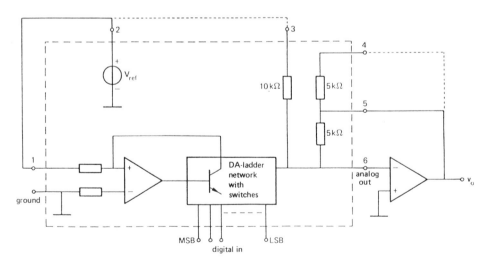

Figure 18.5 Internal structure of an integrated DA converter with current output; a voltage output is obtained with an additional operational amplifier.

input signal must range between zero and the reference, the full scale of the converter. This section is restricted to only one type of AD converter, the successive approximation AD converter. This type of converter belongs to the class of compensating AD converters that all use a DA converter in a feedback loop. The principle is shown in Figure 18.6.

Besides a DA converter, the AD converter contains a comparator, a clock generator and a word generator. The word generator produces a binary code that is applied to the input of the DA converter; it is also the output of the AD converter. The word generator is controlled by the comparator (Section 14.2.1) whose output is '1' for $v_i > v_c$ and '0' for $v_i < v_c$; v_i is the input of the AD converter and v_c is the compensation voltage, identical to the output of the DA converter. When the comparator output is '0', the word generator produces a code that corresponds to a higher compensation voltage v_c; when the output is '1', a new code is generated that corresponds to a lower value of v_c. This process continues until the compensation voltage is equal to the input voltage v_i (a difference of less than 1 LSB). As v_c is the output of the DA converter, its input code is exactly the digital representation of the analog input voltage. In the final state, the (average) input of the comparator is zero; the input voltage v_i is compensated by the voltage v_c.

The clock generator produces a square wave voltage with fixed frequency (the clock

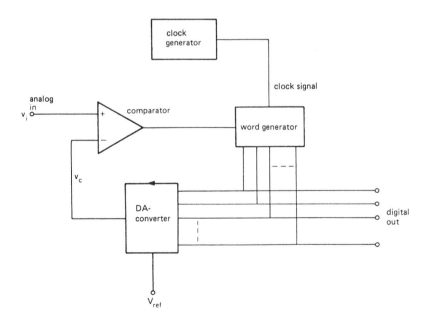

Figure 18.6 Principle of a compensating AD converter using a DA converter in a feedback loop.

frequency). It controls the word generator circuit; at each (positive or negative) transition of the clock, a new word is generated.

Now the question arises how to reach the final state as quickly as possible, that is with the minimum number of clock pulses. As we do not know the input signal, the best estimate is half way along the allowable range, that is $½V_{ref}$. So, the first step of the conversion process is the generation of a code that corresponds to a compensation voltage of $½V_{ref}$ (Figure 18.7). The output of the comparator (high or low) indicates whether the input voltage is in the upper half of the range ($½V_{ref} < v_i < V_{ref}$) or in the lower half ($0 < v_i < ½V_{ref}$). So, the output of the comparator is equal to the MSB a_{n-1} of the digital code we are looking for. If $a_{n-1} = 1$, the compensation voltage is increased by $¼V_{ref}$, to check during the next comparison whether the input is in the upper or lower half of the remaining range. For $a_{n-1} = 0$, the compensation voltage is decreased by that amount. The second comparison results in the generation of the second bit a_{n-2}. The successive comparisons are governed by the clock generator; usually one comparison (so 1 bit) per clock pulse. After n comparisons, the LSB a_0 is known and the conversion process is finished.

The word generator that functions according to the principle explained here is called a successive approximation register or SAR. Most low-cost AD converters are based on this principle. Although the bits are generated successively in time, the bit values are stored in a digital memory (integrated with the converter), to have them available as a parallel binary word. Some converter types have a serial output as well, with the bits available as a series word only during the conversion process.

The conversion time of a successive approximation AD converter is equal to the number of bits n multiplied by the clock period T_c:

$$t_c = n \cdot T_c = n/f_c$$

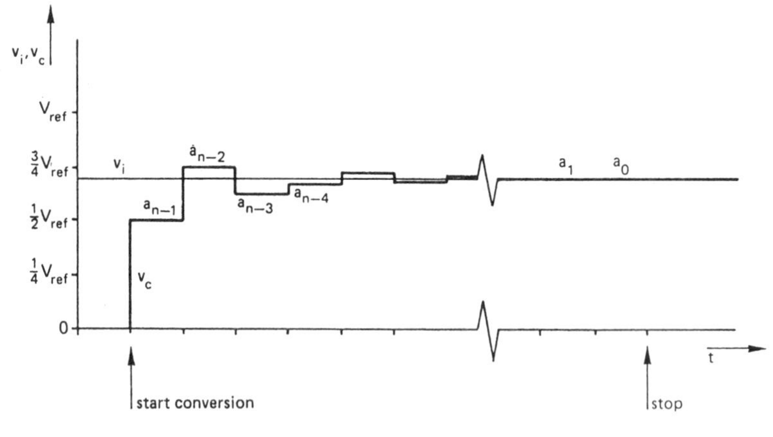

Figure 18.7 Amplitude–time diagram of the compensation voltage in a successive approximation AD converter.

It is independent of the analog input signal. Table 18.2 reviews the main specifications of a 10 bit successive approximation ADC. This type of converter contains a clock generator (500 kHz), a reference voltage source, a comparator and buffer amplifiers at each output. Figure 18.8 shows the internal structure of this fully integrated converter. The functions of the terminals are as follows. The analog input voltage is connected between terminals 1 (ground) and 2 (note the relatively low input resistance). At floating terminal 3, the input voltage range is from 0 to 10 V; at

Table 18.2 Specifications of a successive approximation AD converter

Resolution	10 bit
Analog input	bipolar: −5 V up to +5 V
	unipolar: 0–10 V
Input impedance	5 kΩ
Offset (bipolar)	±2 LSB (adjustable to zero) ±44 ppm/K
Differential non-linearity	$<\pm\tfrac{1}{2}$LSB (25°C)
Conversion time	min. 15 µs; max. 40 µs
Power dissipation	800 mW

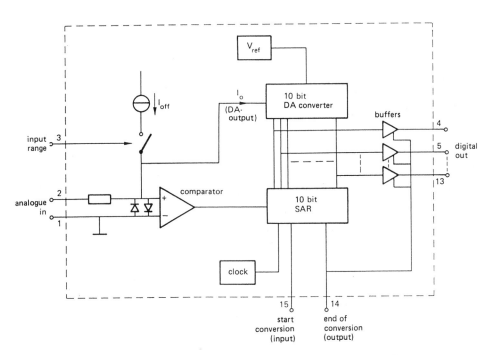

Figure 18.8 Internal structure of an integrated AD converter with successive approximation register (SAR).

grounded terminal 3, an additional current source is connected to the input of the internal DA converter, resulting in a shift of the input range, which is now from −5 to 5 V. Terminals 4 through 13 form the 10 bit parallel digital output. Between the DA converter and the output terminals, 10 buffer amplifiers are connected, called tristate buffers. A tristate buffer has three states: '0', '1' and off; in the off-state there is no connection between the converter and the output terminals. This third state is controlled via an extra input to each buffer. The tristate buffers allow the complete circuit to be electronically disconnected from the rest of the system. This offers the possibility of connecting several devices with their corresponding terminals in parallel, without the danger of mutually short-circuiting the circuits.

As soon as the conversion process is finished, the SAR generates a '0' at terminal 14 (during the conversion this output is '1'). The buffers connect the binary code to the corresponding output terminals. Further, this output signal can be used to let the processor know that the conversion is finished and that the data at the outputs are valid (this terminal is also called the data ready output).

Finally, a binary signal at terminal 15 starts the converter. As long as this input is '1', the converter is in a wait state; as soon as the input is made '0' (for instance by a signal from a computer), the conversion starts.

18.2 Special converters

In the first part of this chapter two popular types of (parallel) AD and DA converters have been described. In this section we introduce several other types: the serial DA converter, two direct parallel AD converters and the integrating AD converter or dual slope converter.

18.2.1 Serial DA converter

The weighting factors of a parallel binary word follow from the position of the bits relative to the 'binary' point. In an ADC or DAC these weighting factors are 1/2 (MSB), 1/4, 1/8 and so on up to $1/2^n$ for the LSB. The allocation of the weighting factors is based on the spatial arrangement of the bit lines or the switches. In a serial word (Figure 18.9a), the allocation is based on the time order of the bits. The first step is the conversion of the active binary quantity (a voltage or a current) into the corresponding switch state (on or off). One side of the switch is connected to a reference voltage V_{ref}; the other terminal is zero when the switch is off, and V_{ref} when the switch is on (Figure 18.9b). The conversion process takes place in a number of sequential phases. Suppose the LSB is in front (the leading bit). The first phase consists of the following actions: division of the voltage $a_0 \cdot V_{ref}$ by 2, and storage of this value in a memory device. In the second phase, the value $a_1 \cdot V_{ref}$ is added to the contents of the memory, the result is divided by 2 and stored again (the old value is deleted). The memory now contains the value $(a_0/4) \cdot V_{ref} + (a_1/2) \cdot V_{ref}$. This process

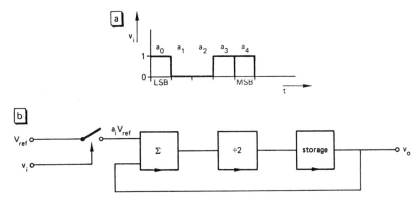

Figure 18.9 (a) An example of a binary serial word, with the LSB as the leading bit; (b) functional diagram of a serial DA converter.

is repeated as long as bits arrive at the input of the converter. At the advent of the MSB (the nth bit), the contents of the memory are:

$$(a_0 2^{-n} + a_1 2^{-n+1} + \ldots + a_{n-2} 2^{-2} + a_{n-1} 2^{-1}) V_{ref}$$

which is the desired analog value GV_{ref}. Figure 18.10 shows a circuit that works exactly like this procedure.

The memory device is a capacitor, which is charged when the switch S_2 is in the left position; when the switch is in the right position, the charge is maintained and can be measured via a buffer amplifier. The sample-hold circuit retains the voltage during the period that C is disconnected from the sample-hold circuit. Both switches, the bit switch S_1 and the switch of the sample-hold circuit, have equal switch rates, equal to the bit frequency. In each bit period, S_2 samples half the refreshed value and transfers this value to the sample-hold circuit.

The structure of this converter does not depend on the number of bits. The inaccuracy is mainly limited to the imperfections of both amplifiers and both memory devices (the capacitor C and the sample-hold circuit).

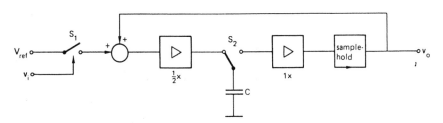

Figure 18.10 A DA converter according to the procedure of Figure 18.9. The switches are controlled by a clock signal.

18.2.2 Direct AD converter

The structure of a direct AD converter is very straightforward (Figure 18.11). The input voltage is compared simultaneously to all possible binary fractions of the reference. For an n bit converter there are 2^n distinct levels. The reference voltage is subdivided into 2^n equally spaced voltages; with the same number of comparators, the input voltage v_i is compared to each of these levels. The number of comparator outputs, counted from the top, is low, the rest is high, depending on the input voltage. A digital decoder combines these 2^n values and generates the required binary code of n bits. This AD converter is characterized by a high conversion speed, limited by the time delay of the comparators and the decoder and a large number of components (and thus a high price).

It is possible to reduce the number of components significantly, at the price of a slower speed (Figure 18.12). Here, the input voltage v_i is compared to $½V_{ref}$, resulting in the MSB. If $a_{n-1} = 1$ (hence $v_i > ½V_{ref}$), v_i is reduced by an amount $½V_{ref}$; otherwise it remains the same. To determine the next bit, a_{n-2}, the corrected input voltage should be compared to $¼V_{ref}$. However, it is easier to compare twice the value with $½V_{ref}$, which is the same. To that end, the voltage $v_i - a_{n-1} \cdot ½V_{ref}$ is multiplied by 2 and compared to a second comparator with half the reference voltage. This procedure is repeated up to the last bit (LSB).

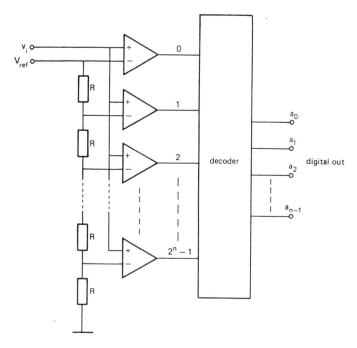

Figure 18.11 An n-bit direct AD converter uses 2^n accurate resistors and 2^n comparators; the conversion speed is very high.

Special converters 313

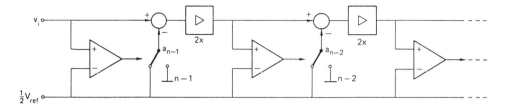

Figure 18.12 Principle of a cascaded AD converter.

As the comparators are now connected in series, this converter is called a cascaded DA converter. It is somewhat slower than the preceding type, because the delay times are in series; the number of components is reduced considerably.

Table 18.3 shows a specification example of the two DA converters described previously. Normally, the conversion time is the inverse of the conversion frequency. However, for very fast cascaded converters this is no longer true. The word bit frequency can be much higher than the conversion speed; the next bit is applied to the first comparator when the preceding bit is in the second comparator and so on (the pipeline effect). This explains why the conversion frequency of the cascaded converter is higher than the inverse of its conversion time.

18.2.3 *Integrating AD converters*

In an integrating AD converter, the input signal is integrated during conversion; the digital output is proportional to the mean value of the input signal. As noise and other interference signals are also integrated, their contribution to the output is low (presuming zero average value of these signal components). For this reason, integrating AD converters are widely used in DC current and voltage measurement systems. Due to the integration of the input, these converters have a slower response compared to the other AD converters.

Table 18.3 Specification example of a direct and a cascaded DA converter

	Parallel	Cascaded
Resolution	8 bit	8 bit
Inaccuracy	±0.15% ±½LSB	±0.15% ±½LSB
Monotony	guaranteed	guaranteed
Differential non-linearity	0.01%	0.01%
Conversion time	35 ns	150 ns
Conversion frequency	20 MHz	11 MHz

314 Digital-to-analog and analog-to-digital conversion

Figure 18.13 shows the principle of a dual-slope integrating AD converter. First the input signal is integrated, then a reference voltage. The conversion starts with the switch S_1 in the upper position. The input signal is connected to the integrator, which integrates v_i during a fixed time period T. At positive constant input and a positive transfer of the integrator, the output increases linearly in time. The comparator output is negative and keeps switch S_2 on. This switch lets a series of pulses with frequency f_0 pass to a digital counter circuit. In digital voltmeters, this is usually a decimal counter. Upon each pulse, the counter is incremented by one count, as long as S_2 is on. When the counter is 'full' (it has reached its maximum value), it gives a command to switch S_1 to disconnect v_i, and to connect the negative reference voltage to the integrator. At the same moment the counter starts again counting from zero. As V_{ref} is negative, the output of the integrator decreases linearly with time. As soon as the output is zero, detected by the comparator, S_2 switches off and the counter stops counting. The content of the counter at that moment is a measure for the integrated input voltage. Figure 18.14 depicts the integrator output for one conversion period and two (constant) different input voltages.

The conversion starts at $t = t_0$; the counter is reset to zero. Switch S_2 goes on as soon as v_c is positive and the counter starts counting. At $t = t_0 + T$, the counter is full; at that moment, the integrator output voltage is:

$$v_{c, t_0 + T} = \frac{1}{\tau} \int_{t_0}^{t_0 + T} v_i \, dt$$

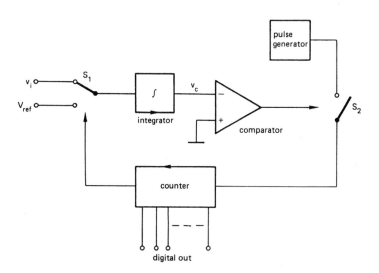

Figure 18.13 The principle of a dual-slope AD converter.

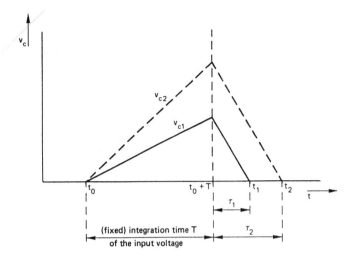

Figure 18.14 The output v_c of the integrator during one conversion period of a dual-slope AD converter, for two different input voltages v_1 and v_2. The integration time for v_1 is τ_1; for v_2 it is τ_2 ($v_1 < v_2$). The integration time τ is proportional to the input voltage v_i.

where τ is the proportionality factor (RC time) of the integrator. At constant input voltage, the slope of the output is v_i/τ. From $t = t_0 + T$, the (negative) reference is integrated, hence v_c decreases with a rate V_{ref}/τ. When $v_c = 0$ (at $t = t_1$), S_2 goes off and the counter stops. The voltage rise during the first integration (of v_i) is fully compensated by the voltage drop during the second integration (of V_{ref}). This maximum voltage is equal to the slope times the integration time, so:

$$\frac{v_i}{\tau} T = \frac{V_{ref}}{\tau}(t_1 - t_0 - T)$$

or

$$v_i = V_{ref}\frac{t_1 - (t_0 + T)}{T}$$

The voltage v_i is transferred to a time ratio. This ratio is measured by the decimal counter. The first integration period T is made so that the counter is full at a power of 10, so 10^n. As the counter restarts with the second integration at $t = t_0 + T$, the number of pulses during the second integration is equal to $(t_1 - t_0 - T) \cdot 10^n/T = (v_i/V_{ref}) \cdot 10^n$. When v_i is not constant during the integration period, the number of pulses at the end of the second integration period equals:

$$\frac{10^n}{V_{ref}} \int_{t_0}^{t_0+T} v_i dt$$

316 Digital-to-analog and analog-to-digital conversion

At first sight this conversion method seems rather laborious. However, the resulting output code depends solely on the reference voltage (and of course the input voltage) and not on other component values. The only requirement for the integrator and the frequency f_0 is that they should be constant during the integration period.

The method also allows a strong reduction of interference from the mains (50 or 60 Hz spurious signals). To suppress 50 Hz interference, the integration time is chosen as a multiple of 20 ms. The average of a 50 Hz signal is zero in that case.

Digital voltmeters based on the dual-slope technique have an inaccuracy of less than 10^{-4} to 10^{-5}. The converters are available as a module or integrated circuit.

SUMMARY

Parallel converters

- A binary signal has only two levels, denoted as '0' and '1'. The information content is restricted to 1 bit at a time; a binary signal is highly fault tolerant.
- A binary word is a group of bits. A byte is a group of 8 bits; a kilobyte is 1024 bytes. The bit with the greatest weight is the most significant bit or MSB; that with the lowest weight is the least significant bit or LSB.
- The finite number of bits limits the resolution of the measurement quantity; the error due to the AD conversion is at least the quantization error, $\pm\frac{1}{2}$LSB.
- Besides the decimal and the binary number systems also the hexadecimal and octal systems are used (bases 16 and 8, respectively). The BCD code is a combination of decimal and binary code.
- Binary words are materialized dynamically (by voltages or currents) or statically (by switches), either as parallel words or serial words. A serial word can be transmitted along a single line, a parallel word needs as many lines as there are bits.
- Conversion from the analog to the digital signal form requires sampling and quantization. Both steps can introduce additional errors.
- The digital signal of DA and AD converters is the binary code of a fraction G of the reference (voltage): $V_a = GV_{ref}$, with $0 < G < 1$. The reference is the full scale. The weight of the MSB is $\frac{1}{2}V_{ref}$.
- The inaccuracy parameters of AD and DA converters are expressed in units of LSB. The LSB of an n bit converter with reference voltage V_{ref} is $V_{ref}/2^n$.
- The ladder network in a parallel DA converter generates a series of currents which differ by a factor of 2 successively; they are the weighting factors of the bits.
- The differential non-linearity of an AD or DA converter is the maximum deviation from the nominal step of 1 LSB. If the differential non-linearity is more than 1 LSB, monotony is no longer guaranteed.

- Compensating AD converters use a DA converter in a feedback loop. A successive approximation is a widely used type converter. With n bits, the conversion is performed within n steps (comparisons; clock pulses).

Special converters

- A serial DA converter converts a binary-coded signal directly upon reception of the bits, using a capacitor as a memory device and a sample-hold circuit.
- An n bit direct AD converter has 2^n comparators, but is very fast. The derived cascaded converter is also fast, but has a reduced number of components (n comparators in series).
- An integrating AD converter responds to the integral (or average) of the analog input signal. These converters are slow but accurate, because of the high noise and interference immunity.
- The twofold integration (once the input, once the reference) of the dual-slope integrator makes the system insensitive to component tolerances. It is therefore widely used in accurate voltage and current measurement systems.

Exercises

Parallel converters

18.1 Complete the table shown below.

Binary	1010111					
Octal		577				111
Decimal			257		111	
Hexadecimal				8F	111	

18.2 The reference voltage of a 10 bit DA converter is 10 V. Calculate the output voltage at an input code 1111100000 (MSB first).

18.3 The reference voltage of a 12 bit DA converter has a temperature coefficient of ±2 ppm/K. Find the inaccuracy in the output voltage over a temperature range from 0 to 80°C, expressed in LSB.

18.4 What is meant by the differential non-linearity of a DA converter? What is monotony?

18.5 Integrated digital circuits employ the binary number system, rather than, say, a number system with base 4, where a value gives us a choice of 1 out of 4. Why this rather inefficient number system?

18.6 The clock frequency of a 10 bit successive approximation AD converter is 200 kHz. Find the (approximated) conversion time of this converter.

18.7 Explain the term 'multiplying DAC' for a DA converter with external reference.

18.8 What is the function of the two diodes connected in antiparallel, at the input of the integrated circuit of Figure 18.8?

Special converters

18.9 A serial binary signal with bit frequency of 1 Mbit/s is applied to a serial DAC of Figure 18.10. Find the conversion time for a 14 bit serial word.

18.10 The input signal of the DAC in Figure 18.10 is the 3 bit word 101. Make a plot of the output signal versus time. The capacitor is uncharged for $t<0$.

18.11 The specifications of a 3 bit cascaded converter of Figure 18.12 are:
 - reference voltage V_{ref} = 5.000 V;
 - offset voltage of the comparators: ±6 mV;
 - offset voltage of the amplifiers: ±6 mV;
 - inaccuracy of the gain factor: ±0.5%.

Calculate the maximum error in the digital output, due to each of these specifications, expressed in LSB, for an input voltage of 0.630 V.

18.12 Explain why the pulse frequency is not important in the dual-slope converter.

18.13 The integration period of an integrating AD converter is 100 ms ±1 µs. Determine the maximum conversion error due to a 50 Hz interference signal whose rms value is 1 V.

19 Digital electronics

This chapter deals with integrated circuits for the processing of digital signals. Section 19.1 starts with an introduction to the Boolean algebra and proceeds with a functional description of some digital components. Section 19.2 deals with several widely used digital circuits: multiplexers, adders, counters and shift registers.

19.1 Digital components

19.1.1 *Boolean algebra*

Analog signals and transfer functions are adequately described by time functions and frequency spectra. Such descriptions are useless for digital signals and processing circuits. For this purpose, we employ a particular mathematical method which was first used for the description of logic processes in 1847 by George Boole.

Logic statements have two values: 'true' (T) and 'false' (F). We will illustrate this with some electronically oriented examples. Figure 19.1 shows a circuit of a lamp and a switch. We can pose several statements with respect to this system, for instance: ℓ: 'the lamp is on' and s: 'the switch is on'. These two statements can both be true or false, but, if ℓ is true, s is true too. The relations between various statements can be described by, for instance, a truth table. Table 19.1 is a truth table for the circuit of Figure 19.1, with the statements already given. We can of course define other statements; the truth tables will change accordingly, but the physical operation does not change. Table 19.2 is another truth table for the same circuit in Figure 19.1, but with different statements.

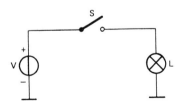

Figure 19.1 If the statement 'the switch is on' is true, then the statement 'the lamp is on' is true too.

Table 19.1 Truth table of the circuit in Figure 19.1. s = 'the switch is on'; ℓ = 'the lamp is on'

s	ℓ
False	False
True	True

Table 19.2 Truth table for the circuit in Figure 19.1. s = 'the switch is on'; ℓ = 'the lamp is off'

s	ℓ
False	True
True	False

Suppose we have a lamp circuit with two switches in series (Figure 19.2). The truth table is longer: there are four possible combinations of the two switches. Table 19.3 gives the truth table for the same statements as before, now denoted as a and b: 'the switch is on' and ℓ: 'the lamp is on'. For simplicity, we use the symbols T and F for true and false, respectively. As can be expected from the figure, the lamp is only on (ℓ = T) when both switches are on ($a = b =$ T). This is an example of the logic operation AND: $\ell = a$ AND b; the statement ℓ is only true if both statements a and b are true. There are several notations for this operation: AND, $\wedge$, $\cap$ or $\cdot$; in this book we use the symbol $\cdot$, or the operator notation is left out: $\ell = a \cdot b = ab$.

Another way to control a lamp by two switches is drawn in Figure 19.3; the corresponding truth table is given in Table 19.4. In this case, at least one of the two switches must be on, to light the lamp. This operation corresponds with the logic operation OR: $\ell = a$ OR b. Possible notations are OR, $\vee$ or +. In this book we will use the + symbol. From the truth table it follows that this is an 'inclusive OR': either a or b or a and b let the lamp light, including this third possibility.

The 'exclusive OR' (or EXOR) is an operation characterized in Table 19.5: only one of the statements a and b must be true to let ℓ be true. This operation is denoted as

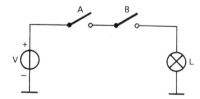

Figure 19.2 A circuit with two switches in series is described by the logic AND operation.

Table 19.3 The truth table for the circuit in Figure 19.2

a	b	ℓ
F	F	F
F	T	F
T	F	F
T	T	T

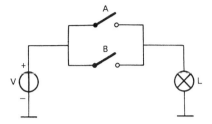

Figure 19.3 A circuit with two switches in parallel is described by the logic OR operation.

Table 19.4 The truth table for the circuit of Figure 19.3.
a = b = 'the switch is on';
ℓ = 'the lamp is on'

a	b	ℓ
F	F	F
F	T	T
T	F	T
T	T	T

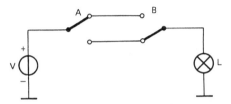

Figure 19.4 A two-way switch, described by an EXOR operation.

Table 19.5 The truth table of the exclusive OR operation

a	b	ℓ
F	F	F
F	T	T
T	F	T
T	T	F

EX-OR, EXOR, $\veebar$ or $\oplus$; in this book we use the symbol $\oplus$, hence $\ell = a \oplus b$. A physical realization of the exclusive OR operation is given in Figure 19.4, the two-way switch.

The statements given before are not applicable for this type of switch. Instead we use the statement 'the switch is in the rest position' which means the switch position (that is, the direction of the bold bar) is just as it is shown in the figure.

A further logic operation is the inversion or negation, denoted as NOT a or $\bar{a}$: we use the latter notation. If a is a statement, for instance 'the lamp is on', then, if a is true, the lamp is indeed on. The lamp is also on when $\bar{a}$ (NOT a) is false. The truth table of the inversion is given in Table 19.6.

In digital electronics other symbols for F and T are used: '0' and '1', respectively. Note that the symbols '0' and '1' are just symbols for the logic variables false and true, and must not be confused with the binary numbers 0 and 1. The symbol choice is, in principle, arbitrary. Physically, the logic variables can be presented in various ways, for instance a low and a high voltage, current or no current, high or low frequency and so on. In digital circuits, a '0' is usually a voltage below 0.8 V and a '1' is a voltage above 2 V. Data transmission (telephone lines) uses two different frequencies.

From now on we use the symbols 0 and 1 (without quotation marks). Table 19.7 shows the basic logic operations in a truth table, now with the symbols 0 and 1. The

Table 19.6 The truth table for the logic negation or inversion

a	$\bar{a}$
F	T
T	F

Table 19.7 The truth table for $a \cdot b$; $a + b$; $a \oplus b$; $\bar{a}$ and $\bar{b}$

a	b	$a \cdot b$	$a + b$	$a \oplus b$	$\bar{a}$	$\bar{b}$
0	0	0	0	0	1	1
0	1	0	1	1	1	0
1	0	0	1	1	0	1
1	1	1	1	0	0	0

symbols a, b and ℓ are logic variables; they have only two values: 0 or 1. In composite logic equations, like $z = x + y\bar{w}$, the order of the logic operations is: inversion – logic AND – logic OR. Another order is indicated by brackets, just as in normal algebraic equations: $z = (x + y)\bar{w}$. The Boolean algebra has a number of rules that facilitate the calculations with logic variables. These rules can be proved by making the complete truth table; for all possible combinations, the left-hand and right-hand sides of the logic equations must be equal.

- The following rules are valid for the logic values 0 and 1:

$$\begin{aligned}
0 \cdot 0 &= 0 & 0 + 0 &= 0 \\
0 \cdot 1 &= 0 & 0 + 1 &= 1 & \bar{0} &= 1 \\
1 \cdot 0 &= 0 & 1 + 0 &= 1 & \bar{1} &= 0 \\
1 \cdot 1 &= 1 & 1 + 1 &= 1
\end{aligned} \tag{19.1}$$

Next is a list of some rules for general logic variables.

- Law of equality:

$$\begin{aligned}
a \cdot a \cdot a \cdot \ldots &= a \\
a + a + a + \ldots &= a
\end{aligned} \tag{19.2}$$

- Commutative laws for addition and multiplication:

$$\begin{aligned}
a \cdot b &= b \cdot a \\
a + b &= b + a
\end{aligned} \tag{19.3}$$

- Associative laws for addition and multiplication:

$$(a \cdot b) \cdot c = a \cdot (b \cdot c)$$
$$(a + b) + c = a + (b + c) \tag{19.4}$$

- Distributive laws:

$$a \cdot (b + c) = ab + ac$$
$$a + (b \cdot c) = (a + b)(a + c) \tag{19.5}$$

- Modulus laws:

$$0 \cdot a = 0 \qquad 1 \cdot a = a$$
$$0 + a = a \qquad 1 + a = 1 \tag{19.6}$$

- Negation laws:

$$a \cdot \bar{a} = 0$$
$$a + \bar{a} = 1 \tag{19.7}$$
$$\bar{\bar{a}} = a$$

$$\left. \begin{array}{l} \overline{a \cdot b} = \bar{a} + \bar{b} \\ \overline{a + b} = \bar{a} \cdot \bar{b} \end{array} \right\} \text{De Morgan's theorem} \tag{19.8}$$

- Absorption laws:

$$a \cdot (b + a) = a$$
$$a + a \cdot b = a \tag{19.9}$$

$$a \cdot (\bar{a} + b) = a \cdot b$$
$$a + \bar{a} \cdot b = a + b \tag{19.10}$$

As can be seen from these formulas, each logic equation has a counterpart, the dual equation. This dual form is found by replacing each 0 by 1 and vice versa, and replacing each $+$ by $\cdot$ and vice versa. The laws are used to simplify logic expressions and, hence, to simplify logic circuits that are described by these equations.

Digital operations can also be visualized with Venn diagrams. In Figure 19.5, some of these diagrams are given, to illustrate the operations $A \cdot B$, $A + B$, $\bar{A}$ and $A \oplus B$. This method is not useful for complex logic operations.

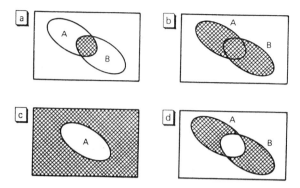

Figure 19.5 Representation of logic operations with Venn diagrams: (a) A·B; (b) A + B; (c) $\bar{A}$; (d) A⊕B.

19.1.2 *Digital components for combinatory operations*

Digital circuits are either combinatory or sequential circuits. The output of a combinatory circuit is determined exclusively by the combination of the actual input signals. The output of a sequential circuit can also depend on earlier input values; a sequential circuit has memory properties. Figure 19.6 shows an overview of the most commonly used logic elements, their symbols and the corresponding Boolean equations. The official IEC (International Electrotechnical Commission) symbols and the American symbols are given.

The elements in Figure 19.6 are called logic gates; they can have more than two inputs but only one output. The following is a short description of the gates in Figure 19.6.

- AND gate
  The output is 1 if all inputs are 1. It functions like a set of switches in series (Figure 19.2).
- OR gate
  The output is 1 if one or more inputs are 1. The OR gate functions as a set of parallel switches (Figure 19.3).
- Inverter (NOT)
  The output is the complement of the input. This element is usually combined with other gates. It is symbolized by a small inverted circle at the input or output (see for instance the following NAND gate).
- NOT-AND or NAND
  The output is only 0 if all inputs are 1. It is an AND gate in series with an inverter. The function is similar to a series of switches parallel to the load (Figure 19.7); the lamp is on as long as not all switches are on.

Digital components

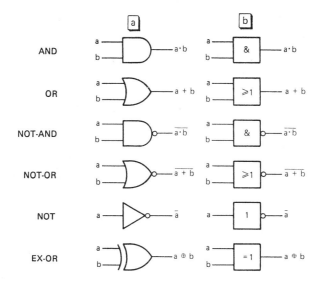

Figure 19.6 (a) The American and (b) the European symbols for the logic elements.

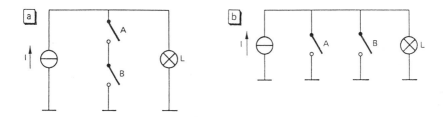

Figure 19.7 (a) A circuit with a NAND function, the dual form of the circuit in Figure 19.3; (b) a circuit with a NOR function, the dual form of the circuit in Figure 19.2.

- **EX-OR or EXOR**
  This is an OR gate in the literal sense: the output is only 1 if either input a, or input b (or c, etc.) is 1, if only one of the inputs is 1. The two-way switch in Figure 19.4 is an example of a circuit with an EXOR function.

Logic gates are available as integrated circuits. Usually, there are two or more gates in a single IC, depending on the number of inputs, for instance four gates with only two input terminals (a 'quad'), three gates with three inputs each (a 'triple'), two gates with four inputs (a 'dual') and an eight input gate in one encapsulation. Inverters are available as a group of six in one encapsulation (a 'hex'). Figure 19.8 shows the internal structure of several digital ICs.

326 Digital electronics

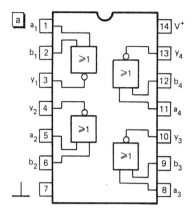
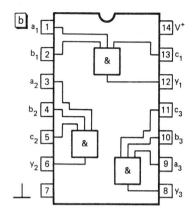
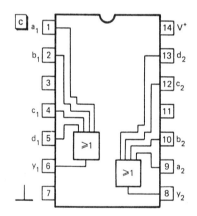
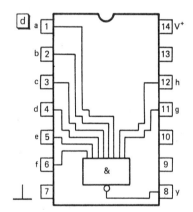

Figure 19.8 Internal structure of various logic gate ICs: (a) quad 2-input NOR gate; (b) triple 3-input AND gate; (c) dual 4-input OR gate; (d) 8-input NAND gate.

There are also integrated circuits with a combination of gates for a particular function (adders, multiplexers); these circuits are described in the second part of this chapter.

Gates are composed of active electronic components like transistors. They need to be powered for proper operation. Most gate ICs require a voltage supply of 5 V. These are the transistor–transistor logic (TTL) circuits composed of bipolar transistors, diodes and resistances. Figure 19.9a shows an example of a TTL NAND circuit.

Also gates composed of MOSFETs (Section 11.1.2) operate at a supply voltage of 5 V. Components that are made in the CMOS technology (complementary MOS:

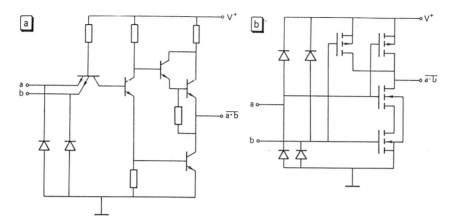

Figure 19.9 Internal structure of a NAND gate, made in (a) a bipolar TTL process; (b) in CMOS technology.

circuits with both p-channel and n-channel MOSFETs) operate at supply voltages between 3 and 15 V. Particular properties of CMOS components are the very high input impedance ($\approx 10^{12}\,\Omega$) and a remarkably low power dissipation. The power consumption of CMOS-integrated circuits is extremely low; they can easily be battery powered. Batteries are discharged faster by their own leakage current than by the CMOS circuits. Figure 19.9b shows the internal structure of a CMOS NAND gate.

19.1.3 *Digital components for sequential operations*

Sequential circuits produce an output that depends on the actual input combination and on earlier combinations; they exhibit a memory function. The most important representative of this category is the flipflop, a digital component with two or three inputs and two complementary outputs; it is composed of a combination of gates. We will describe only two types: the *SR* flipflop and the *JK* flipflop.

SR flipflop

The simplest *SR* flipflop is constructed with two NOR gates (Figure 19.10a). The element has two inputs, called the s (set) and r (reset) inputs. The two outputs are q and $\bar{q}$. When $sr = 10$ (this means $s = 1$ and $r = 0$), the output is 1 ($q = 1$, $\bar{q} = 0$). For $sr = 01$, the output is 0. When we start with one of these input combinations (10 or 01) and change to $sr = 00$, the output remains unchanged; the flipflop remembers which of the inputs (r or s) was 1. This is illustrated in Table 19.8, the truth table of the flipflop. The combination $sr = 11$ should be avoided; the two outputs are not complementary (both are 1). Further, when changing from this state to $sr = 00$, the output will be either 0 or 1, depending on which gate has the faster response; the output state cannot be foreseen.

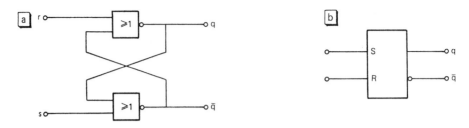

Figure 19.10 (a) A circuit configuration and (b) the symbol of an RS flipflop.

Table 19.8 The truth table of the SR flipflop in Figure 19.10. The symbol '–' means: 0 or 1, a 'don't care'; q_n means the output prior to a clock pulse, q_{n+1} indicates the output after that clock pulse

s	r	q_n	q_{n+1}	$\bar{q}_{n+1}$	
0	0	–	q_n	$\bar{q}_n$	Store
0	1	–	0	1	Reset
1	0	–	1	0	Set
1	1	–	0	0	
1→0	1→0	0	?	?	Not allowed

The symbol of the RS flipflop is given in Figure 19.10b.

The output of the flipflop changes upon a change in r or s (when applicable). In circuits with many flipflops, it is often necessary to activate all flipflops at the same moment. To that end, the flipflop is extended with two AND gates with a common input, the clock input. The clock signal is a square wave voltage used to synchronize digital circuits. In this case, the clock controls the transfer of the input values r and s to the inputs R and S of the flipflop. This is illustrated in Figure 19.11a. For zero clock ($c = 0$), the outputs of the AND gates are zero, irrespective of r and s; the flipflop is in the hold mode. Immediately $c = 1$, the outputs of the AND gates are equal to r and s ($R = r$, $S = s$): the flipflop is either set ($rs = 01$) or reset ($rs = 10$) (or remains unchanged when $rs = 00$). Figure 19.11b shows the circuit symbol of the clocked RS flipflop, with C the clock input.

When two RS flipflops are connected in series (Figure 19.12a), binary information (1 bit) is transferred from the first flipflop to the second, with the clock as a command signal. In Figure 19.12a, the clock input of the second flipflop is inverted, as is indicated by the little circle. The first flipflop responds to the positive edge of the clock (from 0 to 1), the second one when the clock changes from high to low. This

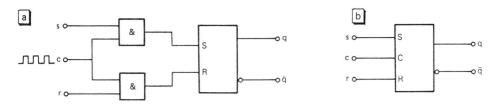

Figure 19.11 (a) An RS flipflop with clock input 'c', to synchronize several flipflops in a circuit; (b) the circuit symbol of a clocked RS flipflop.

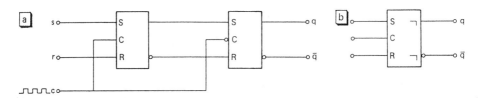

Figure 19.12 (a) A master–slave flipflop, composed of two clocked SR flipflops. One flipflop responds to the rising edge of the clock, the other to the falling edge; (b) circuit symbol of the master–slave flipflop, indicated by the ⌐ sign.

combination is called a master–slave flipflop. At $c = 1$, the first flipflop (the master) transfers the information from the input to the output; the second flipflop (the slave) is in the hold mode. At the transition from $c = 1$ to $c = 0$ ($\bar{c} = 1$) the information is transferred to the output of the slave, while the master is in the hold mode. So, a master–slave flipflop transfers the input in two phases to the output. The advantage of such a two-phase action is the time separation between the changes at the input and at the output. This allows the direct connection of flipflop outputs to the inputs of other flipflops. Time delay and transients have no impact on the operation. The symbol of the master–slave flipflop is given in Figure 19.12b.

JK flipflops

The *JK* flipflop is a master–slave flipflop, controlled by a clock signal. It is undoubtedly the most widely used flipflop. The two inputs are called *J* and *K*, and the outputs are q and $\bar{q}$. The four possible input combinations each have a different effect on the output. The truth table (Table 19.9) shows the four modes of operation.

The truth table indicates how the output q_n after the nth clock pulse changes to q_{n+1} one clock pulse later. For $jk = 00$, the flipflop is in the hold mode, for $jk = 10$, the flipflop is set ($q = 1$ after the next clock pulse) and for $jk = 01$, the flipflop is

Table 19.9 Two types of truth tables for a *JK* flipflop; the symbol '—' means 0 or 1; q_n is the output after the *n*th clock pulse, q_{n+1} the output after the (*n* + 1)st clock

j	k	q_{n+1}	q_n	q_{n+1}	j	k
0	0	q_n	0	0	0	—
0	1	0	0	1	1	—
1	0	1	1	0	—	1
1	1	$\overline{q}_n$	1	1	—	0

reset: $q = 0$. So far, the operation is similar to that of the *SR* flipflop. At the combination $jk = 11$, the output is inverted at the next clock pulse (it 'toggles'). In this mode, the flipflop performs as a frequency divider; a clock signal with frequency f results in a square wave output q with half the clock frequency (Figure 19.13a).

The circuit symbol of the master–slave *JK* flipflop is given in Figure 19.13b.

Figure 19.14 shows the pin connection diagram of a commercial-type *JK* flipflop. The integrated circuit contains two totally independent flipflops in a 16-pin encapsulation. The flipflops in this IC have, besides the *j* and *k* inputs, two other inputs: *s* (set or preset) and *r* (reset or clear). These inputs control the output independently of the clock, and are called asynchronous inputs, in distinction to the *j* and *k* inputs, which are synchronous inputs. The truth table (Table 19.10) shows the various modes of this flipflop.

There is a wide variety of this kind of flipflop, for instance flipflops with only one asynchronous input or types where one or more inputs are internally inverted. Some ICs contain two *JK* flipflops with a common clock and asynchronous inputs, to save pins. To select the right type of flipflop, not only the number and nature of the inputs must be taken into account, but also the maximum power consumption and the maximum clock frequency.

Flipflops from the TTL series operate at clock frequencies up to 100 MHz; the power consumption is about 40 mW (dual flipflop). The power consumption of CMOS components depends on the clock frequency. At 30 MHz (roughly the maximum for CMOS types), the power consumption is about 50 mW. At lower frequencies this

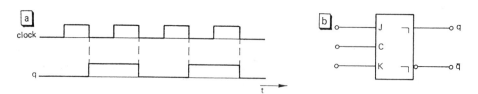

Figure 19.13 (a) A JK flipflop can act as a digital frequency divider for jk = 11: the output frequency is half the clock frequency; (b) the circuit symbol of a master–slave JK flipflop.

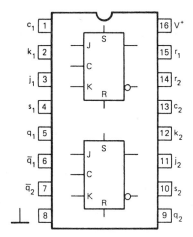

Figure 19.14 The internal structure of a dual JK flipflop with asynchronous set and reset inputs.

Table 19.10 The truth table of a *JK* flipflop with asynchronous set and reset. The sign ↓ means that the flipflop triggers at the negative edge of the clock (from 1 to 0). '–' means 0 or 1 (don't care); q_n is the output after the *n*th clock pulse, q_{n+1} the output after the $(n + 1)$st clock

Modes	Inputs					Outputs	
	s	r	c	j	k	q_{n+1}	$\bar{q}_{n+1}$
Asynchronous reset (clear)	0	1	–	–	–	0	1
Asynchronous set	1	0	–	–	–	1	0
Not allowed	1	1	–	–	–	1	1
Hold mode	0	0	↓	0	0	q_n	$\bar{q}_n$
Synchronous reset (0)	0	0	↓	0	1	0	1
Synchronous set (1)	0	0	↓	1	0	1	0
Synchronous inversion	0	0	↓	1	1	$\bar{q}_n$	q_n

decreases dramatically; at 10 kHz it is less than 10 μW. Power dissipation and clock frequency vary strongly from type to type. The user is referred to the data books of the manufacturers. An example of the full specification of a digital component (a dual *JK* flipflop) is given in Appendix A2.2.

19.2 Logic circuits

This part of the chapter on digital electronics contains some examples of circuits composed of logic elements. We will discuss the digital multiplexer, the logic adder,

332 Digital electronics

counters and shift registers. The chapter ends with an illustrative example, the design of a counter circuit for a particular application.

19.2.1 *Digital multiplexer*

A digital multiplexer has essentially the same function as an analog multiplexer. The difference is that the digital multiplexer has binary input and output signals. This multiplexer can be built up from combinatory elements (logic gates). Figure 19.15 shows the circuit of a digital multiplexer with eight inputs d_0 to d_7. With the selection inputs s_0, s_1 and s_2, one of the eight signal inputs is selected; the value of the selected d input (0 or 1) is transferred to the output y. With the enable input e, the output can be fixed at 0, irrespective of the selected input (compare the enable input of the analog multiplexer in Section 15.2). The truth table of this multiplexer, available as an integrated circuit, is shown in Table 19.11.

19.2.2 *Digital adder*

Logic variables are either true or false, denoted before by 0 and 1. A group of logic variables can be represented by a string of ones and zeros, or even as a binary number with bits 0 and 1, corresponding to the logic values 0 and 1. The logic

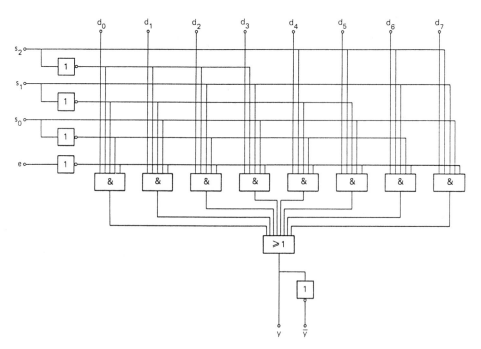

Figure 19.15 A digital multiplexer with eight inputs d_0–d_7. The inputs s_0–s_2 are for the channel selection; e is the enable input.

Table 19.11 The truth table of the 8-channel digital multiplexer of Figure 19.15; '—' means don't care

e	s_2	s_1	s_0	Y
1	—	—	—	0
0	0	0	0	d_0
0	0	0	1	d_1
0	0	1	0	d_2
0	0	1	1	d_3
0	1	0	0	d_4
0	1	0	1	d_5
0	1	1	0	d_6
0	1	1	1	d_7

variables are now considered as the value of a particular bit of a binary number. With this in mind, it is easy to understand the principle of a digital summing circuit, based on binary numbers.

Figure 19.16 shows a combinatory circuit for the addition of two binary numbers A and B, each of only 1 bit. The sum is a 2 bit number, as the maximum sum is $1_{10} + 1_{10} = 2_{10} = 10_2$. From the truth table (Table 19.12) we conclude that the LSB s_0 of the sum is the result of an EXOR operation on the variables a and b, whereas the MSB (here s_1) is the result of an AND operation on a and b.

The circuit of Figure 19.16 is called a half-adder; the variable s_0 is the sum bit, and s_1 is the carry bit. For the summation of two binary numbers of 2 bits each, the circuit is extended to that of Figure 19.17.

This circuit consists of a half-adder for the determination of the LSB (s_0) and the first carry bit (c_0). This carry bit determines, together with the MSB of A and B (a_1 and b_1), the next carry bit, which in this case is the third bit and hence the MSB of the

Table 19.12 The truth table of the binary adder in Figure 19.16

A	B	s_1	s_0	Decimal
0	0	0	0	0 + 0 = 0
0	1	0	1	0 + 1 = 1
1	0	0	1	1 + 0 = 1
1	1	1	0	1 + 1 = 2

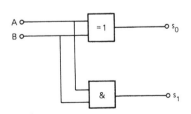

Figure 19.16 The circuit of a half-adder, for the arithmetic summation of two 1 bit numbers.

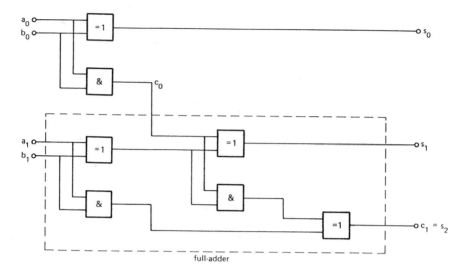

Figure 19.17 The circuit for the arithmetic addition of two 2 bit numbers. The circuit can be extended with more full-adders for the addition of larger numbers.

sum. The lower part of this circuit is a full-adder. The circuit can be extended by more such full-adders for the addition of larger binary numbers. The truth table (Table 19.13) shows the various binary values in the circuit, for each possible combination of the two input numbers.

19.2.3 Digital counters

A counter is a digital circuit that enables the counting of pulses, zero-crossings or periods of a periodic signal. No fixed frequency is required. Figure 19.18a shows a simple counter circuit consisting of a chain of JK flipflops connected in series. The q output of each flipflop is connected to the clock input of the next flipflop. All j and k inputs are 1, so all flipflops act as a toggle (Table 19.9 and Figure 19.13). Each flipflop halves the frequency of the clock input, as can be seen in Figure 19.18b.

The group of binary outputs corresponds to the binary-coded number of clock pulses that has passed the first flipflop. The counter is incremented 1 bit at each clock pulse; it is an up counter. A down counter is realized likewise, but now the $\bar{q}$ outputs must be connected to the clock inputs of the next flipflop; the q outputs correspond to the output codes of the counter. Starting from the situation where all flipflop outputs are 0, the circuit counts as follows: 0000–1111–1110–. . .–0001–0000–1111, and so on. When drawing the time diagram of such a down counter, it must be realized that the information at the j and k inputs is stored at the positive edge of the clock pulse,

Logic circuits

Table 19.13 The truth table for the adder in Figure 19.17

A		B		S			A + B = S Decimal
a_1	a_0	b_1	b_0	s_2	s_1	s_0	
0	0	0	0	0	0	0	0 + 0 = 0
0	0	0	1	0	0	1	0 + 1 = 1
0	0	1	0	0	1	0	0 + 2 = 2
0	0	1	1	0	1	1	0 + 3 = 3
0	1	0	0	0	0	1	1 + 0 = 1
0	1	0	1	0	1	0	1 + 1 = 2
0	1	1	0	0	1	1	1 + 2 = 3
0	1	1	1	1	0	0	1 + 3 = 4
1	0	0	0	0	1	0	2 + 0 = 2
1	0	0	1	0	1	1	2 + 1 = 3
1	0	1	0	1	0	0	2 + 2 = 4
1	0	1	1	1	0	1	2 + 3 = 5
1	1	0	0	0	1	1	3 + 0 = 3
1	1	0	1	1	0	0	3 + 1 = 4
1	1	1	0	1	0	1	3 + 2 = 5
1	1	1	1	1	1	0	3 + 3 = 6

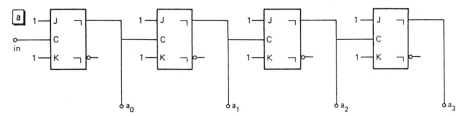

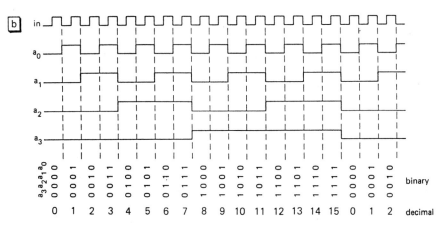

Figure 19.18 (a) A 4 bit binary counter, composed of four master–slave JK flipflops each in the toggle mode; (b) time diagram of an upcounter with the corresponding binary and decimal output codes.

and transferred to the output q at the negative edge of the clock, in agreement with the master–slave principle.

Due to time delay, each flipflop will trigger at a small time interval later than its predecessor (not shown in Figure 19.18): the flipflops do not trigger at the same time which explains the name 'asynchronous counter' or 'ripple counter'; the bits change one after another. This introduces serious timing errors, in particular in complex digital circuits. Synchronous counters do not have this problem. An example of a 4 bit synchronous counter is depicted in Figure 19.19. All clock inputs are connected to each other, so the flipflops trigger at the same moment. The jk inputs control the flipflops. The truth table (Table 19.14) shows for which condition $jk = 11$ (toggle). The first flipflop acts as a toggle for output a_0; the second (output a_1) must toggle for $a_0 = 1$. So the jk inputs of this flipflop must satisfy $j = k = q_0$. The third flipflop (output a_2) must toggle for $a_0 = 1$ and $a_1 = 1$: hence, j and k must be equal to $q_0 \cdot q_1$. The last flipflop (with output a_3) should toggle only if $a_2 a_1 a_0 = 111$, or $j = k = q_0 q_1 q_2$.

Obviously, some extra gates have to be added to the string of flipflops, to realize the synchronous operation. Synchronous counters are available as complete integrated circuits.

There is a wide variety of counter types on the market, such as counters with a reset input (which sets all bits to zero), an enable input (which stops the counter), the possibility to load the counter with an arbitrary value (preset) and counters that can count both up and down. Some 4 bit counters count from 0000 to 1001 (0 to 9) and not from 0000 to 1111 (0 to 15). These binary counters usually have an extra output that indicates the counter state 9.

To make the proper choice for a particular application, the designer should consult the data books for the various types, their specifications and also the maximum frequency and power dissipation.

19.2.4 *Shift registers*

A shift register can store digital information and transfer this information on command of a clock pulse. Shift registers are encountered in all kinds of computers and digital signal processing equipment, such as for arithmetic operations.

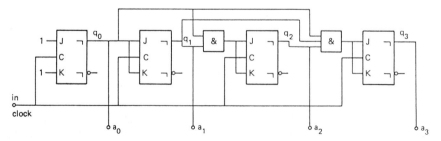

Figure 19.19 A 4 bit counter, running synchronously with the clock.

Table 19.14 One counting cycle of the synchronous 4 bit counter in Figure 19.19

	q_3 / a_3	q_2 / a_2	q_1 / a_1	q_0 / a_0	Outputs that must change at the next clock pulse
0	0	0	0	0	
1	0	0	0	1	q_1
2	0	0	1	0	
3	0	0	1	1	$q_1\,q_2$
4	0	1	0	0	
5	0	1	0	1	q_1
6	0	1	1	0	
7	0	1	1	1	$q_1\,q_2\,q_3$
8	1	0	0	0	
9	1	0	0	1	q_1
10	1	0	1	0	
11	1	0	1	1	$q_1\,q_2$
12	1	1	0	0	
13	1	1	0	1	q_1
14	1	1	1	0	
15	1	1	1	1	$q_1\,q_2\,q_3$
0	0	0	0	0	
1	0	0	0	1	q_1
	.				
	.				
	.				

Example 19.1

The decimal number 45 (101101 in binary form) is doubled by shifting the bits one position to the left (putting a 0 at the free most-right position). The result is 1011010_2, which is equal to 90_{10}. Division by 2 corresponds to shifting the bits one position to the right (filling the empty place with 0). The result is $010110_2 = 22_{10}$. The missing LSB corresponds to an arithmetic rounding-off.

Like a counter, the shift register is composed of a string of flipflops, but now the q and $\bar{q}$ outputs are connected to the j and k inputs of the next element. The clock inputs are all connected to each other; the circuit operates synchronously (Figure 19.20).

Due to this particular coupling, each flipflop is loaded either by a 1 ($jk = q\bar{q} = 10$) or a 0 ($jk = q\bar{q} = 01$). If at the first clock pulse the input is 1 and for the rest of the time 0, then this 1 shifts through the register, at each clock pulse one position to the right; at the end it disappears (Figure 19.20b). The shift register of Figure 19.20 can be loaded serially; a binary word shifts bit after bit via the left flipflop into the register; the readout can be done in parallel ($a_0 - a_3$) but also serially (via a_3). There are shift registers that can be parallel loaded as well, which is faster than serial loading.

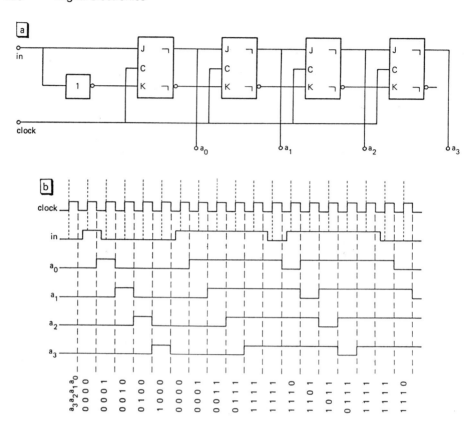

Figure 19.20 (a) A 4 bit shift register composed of four master–slave JK flipflops; (b) corresponding timing diagram with a shifting 1 and a shifting 0: ...... = input change; ---- = output change.

With respect to loading and readout, we can distinguish four types of shift registers: serial in–serial out; serial in–parallel out; parallel in–serial out and parallel in–parallel out. Some registers can combine serial and parallel operation. Most available shift registers have 4 or 8 bits, some can shift to the left as well as to the right.

Except for arithmetic binary operations, shift registers are also used for communication between digital instruments, for instance between a computer and a terminal (a monitor). The link is a serial data path, whereas the computer and the terminal operate with parallel words. To transmit a binary word, it is first loaded (parallel) into a shift register, from which the bits are transported serially. At the receiver, the bits are loaded serially into the shift register and the word is read out in parallel. To the instruments the communication seems parallel; only the (lower) transmission speed points to serial data transport.

Figure 19.21 shows the pin connections of an integrated 4 bit bidirectional universal shift register. This shift register can shift to the left as well as to the right. Other features are parallel loading, parallel readout, reset (all outputs 0) and hold (the output remains unchanged). Table 19.15 is the function table for this shift register. This integrated circuit is available in TTL and in CMOS technology.

19.2.5 An application example

This section shows how to use digital components for a particular application. We consider bottles on a conveyor belt that must be positioned in a crate. The goal is the design of a digital circuit giving a sign when 12 bottles have passed (corresponding to a

Table 19.15 The function table of the shift register in Figure 19.21. q_{ij} means: q_i at time j. '–' means 0 or 1

Function	Inputs					Outputs			
	c	mr	s_1	s_0	d_i	$q_{0,n+1}$	$q_{1,n+1}$	$q_{2,n+1}$	$q_{3,n+1}$
Reset (clear)	–	1	–	–	–	0	0	0	0
Hold	–	0	0	0	–	$q_{0,n}$	$q_{1,n}$	$q_{2,n}$	$q_{3,n}$
Shift left	↑	0	1	0	–	$q_{1,n}$	$q_{2,n}$	$q_{3,n}$	d_{sl}
Shift right	↑	0	0	1	–	d_{sr}	$q_{0,n}$	$q_{1,n}$	$q_{2,n}$
Parallel load	↑	0	1	1	d_i	d_0	d_1	d_2	d_3

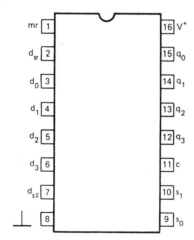

Figure 19.21 The pinning diagram of a 4 bit shift register. mr = master reset; d_{sr} = serial data input for shifting to the right; $d_0 \ldots d_3$ = parallel inputs; d_{sl} = serial data input for shifting to the left; ⊥ = ground; V^+ = power supply voltage; $q_0 \ldots q_3$ = parallel outputs; c = clock input; s_1, s_0 = selection inputs (Table 19.15).

full crate); furthermore, the number of passed bottles must be indicated on a display. This requires a counter that counts from 1 to 12 and starts at 1 again. Every time a bottle passes a certain point, the counter is incremented by one step. A sensor (for instance a photodetector, see Section 7.2) gives a binary signal that is 0 for a bottle and 1 for an empty place. This signal is used as a clock signal for the counter circuit.

We need a counter, a display and a circuit to control the display. After having consulted several data books, we chose (for instance) the digital components shown in Figures 19.22 and 19.23.

The circuit in Figure 19.22 is a synchronous decimal counter (Section 19.2.3) with the possibility of synchronous parallel loading. Table 19.16 shows the functional operation of this counter.

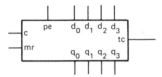

Figure 19.22 A synchronous decimal counter with possibility for synchronous parallel loading of a 4 bit number, according to Table 19.16.

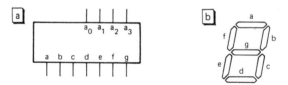

Figure 19.23 (a) A decoder for the conversion from the binary code to the seven-segment code; (b) seven-segment display, consisting of seven independent light sources.

Table 19.16 Functional operation of the decimal counter in Figure 19.22; '(a)' means 1, only for count 9; '–' means 0 or 1, don't care

Function	Inputs				Outputs	
	mr	c	pe	$d_{i,n}$	$q_{i,n+1}$	tc
Reset	1	–	–	–	0	0
Parallel load	0	↑	1	0	0	0
	0	↑	1	1	1	(a)
Count	0	↑	0	–	Count	(a)

Logic circuits 341

The binary number to be loaded is connected to the inputs d_0–d_3; if pe (parallel enable) is 1, the counter output $q_0q_1q_2q_3$ becomes equal to $d_0d_1d_2d_3$ at the next upgoing clock pulse. With the input mr (master reset), the counter is asynchronously reset to 0000. The output tc (terminal count) is 1 when the counter output is 1001_2 (decimal 9): this enables the triggering of a second counting section for the tens.

The decoder for the conversion of a binary number to the seven-segment code is given in Figure 19.23a. The output signals a–g each activate one segment of the display, as is shown in Figure 19.23b.

The truth table of this decoder is given in Table 19.17. Most seven-segment decoders can be connected directly or via resistors to an LED display.

The design thus far is depicted in Figure 19.24. The counter outputs are connected to the inputs of the decoder which controls the LEDs of the display. The seven LEDs have a common terminal that must be connected to ground; the resistors in series with the other terminals limit the control current to a maximum value (a precautionary measure).

This circuit has the following counting sequence: 0, 1, 2, ..., 8, 9, 0, ... The system must count to 12, so an extra display element is required. We chose a display for only a 1, which can simply be controlled by a JK flipflop and a buffer. The flipflop should generate a 1 when the clock goes up during a counter output of 9. For this purpose the tc output is used. The flipflop must be of the master–slave type; the information about the tc is stored at the negative edge of the clock pulse and appears at the output at the positive edge of the clock. This operation is seen in Figure 19.25, which is the counter circuit of Figure 19.24 with the extension for counting up to 19.

Table 19.17 The truth table for the binary to seven-segment decoder in Figure 19.23a

a_3	a_2	a_1	a_0	a	b	c	d	e	f	g	Display
0	0	0	0	1	1	1	1	1	1	0	0
0	0	0	1	0	1	1	0	0	0	0	1
0	0	1	0	1	1	0	1	1	0	1	2
0	0	1	1	1	1	1	1	0	0	1	3
0	1	0	0	0	1	1	0	0	1	1	4
0	1	0	1	1	0	1	1	0	1	1	5
0	1	1	0	1	0	1	1	1	1	1	6
0	1	1	1	1	1	1	0	0	0	0	7
1	0	0	0	1	1	1	1	1	1	1	8
1	0	0	1	1	1	1	1	0	1	1	9
1	0	1	0	0	0	0	0	0	0	0	
1	0	1	1	0	0	0	0	0	0	0	
1	1	0	0	0	0	0	0	0	0	0	
1	1	0	1	0	0	0	0	0	0	0	
1	1	1	0	0	0	0	0	0	0	0	
1	1	1	1	0	0	0	0	0	0	0	

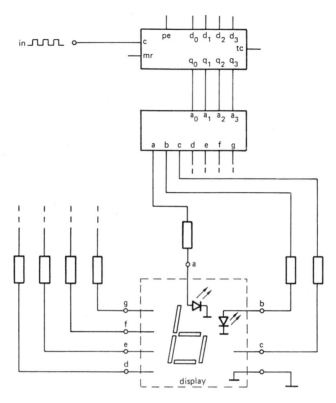

Figure 19.24 A decimal counter with display, to count from 0 to 9.

From the timing diagram in Figure 19.25b, it appears that the circuit first counts from 0 to 19; as the flipflop is not reset, the next cycles run from 10 to 19. We must switch off the display for the tens after count 12. This is accomplished by resetting the flipflop at count 12, similar to how it is set at count 9. Looking at the time diagram, we see that at count 12 there is a unique combination: $q_2 = 1$ and $q_A = 1$. So, $q_2 \cdot q_A$ is a suitable signal for switching off the display (see the AND gate in Figure 19.26). With this additional circuit, the counting sequence is 0, . . ., 11, 12, 3, 4, ... To reset the counter to 1 (after count 12), we utilize the parallel load feature of the counter IC. To that end, we make the inputs $d_0 d_1 d_2 d_3$ equal to 1000 (d_0 is the LSB). When $pe = 1$, this number is loaded into the counter at the next clock pulse. This may only happen after count 12. We already have a signal that satisfies this condition: the k input of the flipflop. Unfortunately, this signal cannot be used for the pe without introducing timing problems. For a proper operation of sequential circuits their inputs must be stable during a specified time interval before and after the clock pulse. The k input changes as soon as the counter output is no longer 12, whereas the pe signal must be stable during the time that the upgoing clock changes the counter output from 12 to 1.

Logic circuits   343

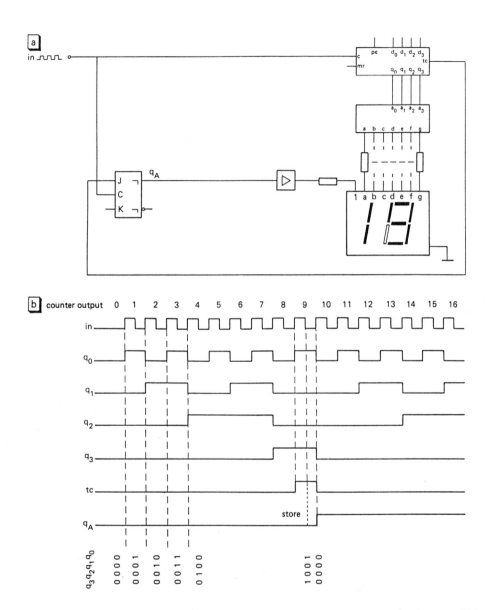

Figure 19.25 (a) Counter circuit of Figure 19.24, extended with a display for the tens. This circuit counts to 19; (b) the corresponding time diagram, showing the control of the extra flipflop.

A signal that satisfies these conditions is derived from an extra flipflop which triggers on the negative edge of the clock pulse (not a master–slave flipflop). The j input is connected to a signal that is 1 during a counter output of 12 whereas the k input is connected to a signal that is 1 for a counter output of 1, and 0 at 12, for instance $\bar{q}_A$ (see timing diagram in Figure 19.26b).

This pe signal can eventually be used to indicate a full crate; another suitable signal for that purpose is $z = q_2 q_A$ (see the timing diagram).

To reset the counter at an arbitrary moment, the reset inputs of both flipflops and the decimal counter are connected to a reset switch (a push button); as long as this switch is on the counter output remains zero: the reset is an asynchronous signal.

The designed counter is a fully synchronous circuit (except for the reset). It is quite possible to make an asynchronous design that might even be simpler. The advantage of a synchronous operation, however, is that all signal transitions occur at fixed moments; this simplifies the design work, in particular for very complex systems.

SUMMARY

Digital components

- The description of digital signals and processing systems is based on the Boolean algebra for logic variables; a logic variable can have only two values: true or false, T or F, 1 or 0.
- A relation between two or more logic variables can be represented by a logic equation or a truth table.
- Four basic logic operations are AND, OR, NOT and EXOR. Digital circuits that realize such operations are logic gates.
- The output of a combinatory digital circuit depends only on the combination of the actual (logic) inputs.
- The output of a sequential digital circuit not only depends on the actual input combination but also on earlier input values (memory function). A flipflop is an example of such a circuit.
- An SR flipflop has three states: set ($sr = 10$, output 1), reset ($sr = 01$, output 0) and hold ($sr = 00$, output remains unchanged).
- A JK flipflop has, besides the three operation modes of the SR flipflop, a fourth mode: inversion or toggle ($jk = 11$); in this mode, the flipflop behaves as a frequency divider (factor 2).

Logic circuits

- Digital circuits can operate in a synchronous or in an asynchronous way; in synchronous circuits all flipflops trigger at the same moment, on command of a clock pulse; in asynchronous circuits this is not the case.

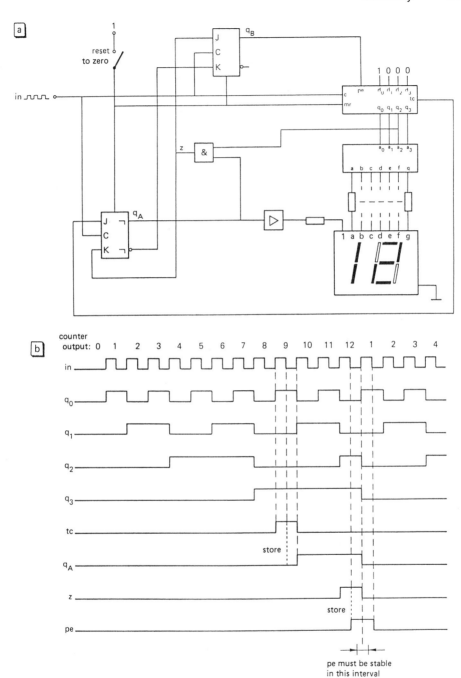

Figure 19.26 (a) The complete circuit of the digital counter that counts from 1 to 12; (b) the corresponding time diagram, showing all the relevant control signals.

- Examples of circuits composed of combinatory elements (logic gates) are the digital multiplexer and the binary adder.
- A digital counter is a circuit composed of flipflops which can count numbers of clock pulses. There are synchronous and asynchronous counters. Decimal counters are binary counters that count up to 10.
- A shift register performs the division or multiplication by 2, by shifting the chain of bits to the right or the left, respectively. It is also used for the conversion of parallel words into serial words and vice versa.
- For the control of a seven-segment display, special digital circuits are available, called seven-segment decoders.

Exercises

Digital components

19.1 Simplify the following logic equations.

(a) $x + x \cdot y$
(b) $\bar{x} + x \cdot y$
(c) $\bar{x} + \overline{x \cdot y}$
(d) $x(x \oplus y)$
(e) $x(y + z) + \overline{xyz}$

19.2 Make a truth table for the following logic relations.

(a) $(a \oplus b)(\overline{a \cdot b})$
(b) $(a + \bar{b} + \bar{c})(\bar{a} + \bar{b} + c)(b + c)$

19.3 Given the combinatory circuit of the figure below, make a truth table for this circuit.

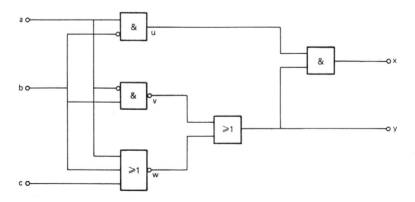

19.4 The following truth table belongs to a *D* flipflop. Give a circuit diagram for such a flipflop with only one *JK* flipflop and some logic gates.

d_n	q_n	q_{n+1}
0	–	0
1	–	1

Logic circuits

19.5 Make a truth table for the multiplexer in Figure 19.15; the output *y* is a function of the inputs d_n, the control inputs s_0, s_1 and s_2, and the enable input *e*.

19.6 Logic gates exhibit time delay. Discuss the effect of time delay in full-adders for large binary numbers.

19.7 The output of the counter depicted in the following figure is $a_2 a_1 a_0 = 011$. Find the counting sequence.

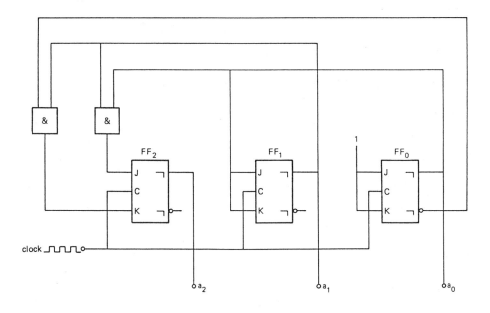

20 Microprocessor systems

A microprocessor system is a digital signal processing system where the processing operations are performed by a microprocessor, which is a processing unit on a single integrated circuit (IC). Besides this processor, the system contains various other digital ICs, necessary for a proper operation of the microprocessor. An essential part of a microprocessor system is the memory. The first part of this chapter is devoted to various types of integrated semiconductor memories. The second part of the chapter deals with the basic structure of microprocessor systems and the microprocessor IC.

20.1 Semiconductor memories

20.1.1 *The function of memories in a microprocessor system*

A general structure of computers and microprocessor systems is depicted in Figure 20.1. Such a structure is called the 'von Neumann architecture', after the American mathematician who had a strong influence on the development of the first computers. The arrows in the figure indicate the direction of the information transfer. A series of instructions describes the successive operations that must be performed on the data by the central processing unit or CPU. The list of instructions is the program, which is stored in the program memory. The instructions are carried out one after another by the CPU. Both data and instructions are represented as binary signals.

The data enter the system via the input interface, a circuit that converts the data structure as it exists outside the system to a structure that can be handled by the CPU. Examples of interface circuits are AD converters (Chapter 18) and parallel-to-serial converters (Section 19.2.4). The incoming data are stored in an internal data memory until they are needed for further processing. Other functions of the data memory are the storage of interim results and of output data that cannot be sent immediately to the output interface. The output interface converts the processed data into a suitable form for further processing outside the processor system.

As data and instructions are both binary-coded signals, the data memory and the program memory are usually combined in a single memory.

Usually, a microprocessor system contains two types of memories. The first kind is utilized for the storage of information that changes continuously, like signal samples and interim results of processing operations. For this purpose, read–write memories

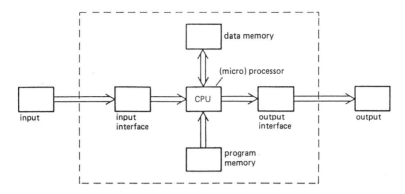

Figure 20.1 Basic structure of a computer and a microprocessor system, based on the von Neumann structure.

are employed (sometimes called RAM, see Section 20.1.2). Most types of RAM lose their contents when the power supply is switched off. The storage of permanent data (as for look-up tables and permanent programs) requires a type of memory that keeps its contents even when the power supply is switched off. Such a non-volatile memory is the read-only memory (ROM, see Section 20.1.3).

Besides these external memories RAM and ROM, a CPU has an internal memory, with a limited storage capacity. Usually, such memories consist of flipflops and shift registers (Chapter 19). Figure 20.2 shows the structure of a microprocessor system with read-only and read–write memory. The CPU is a single integrated circuit, the microprocessor. The external memory consists of one or more ICs, depending on the required storage capacity. The data transport between the memory and the CPU is usually in parallel, word by word, indicated in Figure 20.2 with the double-lined data links.

The selection of a particular data word in the memory is achieved by putting a unique address word, corresponding to the location in the memory, on the address lines. The selected data word is connected to the (parallel) data lines. Besides data lines and address lines, the system also has some control lines, for instance the clock and a line to indicate whether data should be written to or read from the selected memory location (the read–write line).

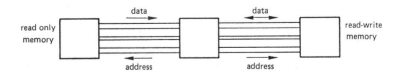

Figure 20.2 The most simple structure of a microprocessor system consists of a CPU and external memory, subdivided into read-only and read–write memory.

The reading process of a memory involves the following actions. First, the processor generates the address code for the desired memory location. At the same time, the read–write line is activated to enable the read operation. Next, the bit values on the selected memory location are connected to the data lines and, finally, transferred to the appropriate location in the internal memory of the CPU.

When writing to a memory (setting the individual memory cells to 0 or 1), the processor generates the proper address code, connects the data bits to be stored to the data lines and activates the read–write line. Then the data lines are connected to the selected memory cells where the bits are stored.

20.1.2 *General aspects of semiconductor memories*

For the storage of a large amount of information, special ICs are used, comprising the external memory of the processor system. One memory IC consists of a matrix of binary cells (small electronic circuits), each capable of storing 1 bit of information.

A memory whose cells can be addressed immediately is called a random-access memory (RAM). Its counterpart is the serial-access memory (SAM). A shift register is an example of a SAM. The information is not directly accessible; the contents can only be read (and written) bit after bit, like a magnetic tape. We have seen that some shift registers can be parallel loaded and read; these shift registers can be considered as both a RAM and a SAM.

Memories are also distinguished with respect to the possibility of writing (new) data. A memory with permanent, unchangeable contents is called a read-only memory (ROM). They are used to store data and other information that must be held permanently, like tables for mathematical formulas.

Although the name RAM refers to the accessibility of the memory locations, memories that can be (re)written directly from the processor are also called RAM. A compact disk is an example of a ROM but it is also a RAM. In this book we use the names read-only memory (ROM) and read–write memory.

An important aspect of a semiconductor memory is the internal organization, the way the bits are arranged and grouped in the memory chip. Memories for long words (many bits per word) require fewer addresses, for the same memory capacity.

Example 20.1
Memories with a storage capacity of 16 kbits are available as circuits with 16 384 words of 1 bit, 4096 words of 4 bits or 2048 words of 8 bits each. Usually this is denoted as $16\,k \cdot 1$, $4\,k \cdot 4$ and $2\,k \cdot 8$, respectively.

Another important specification of a semiconductor memory is its access time, that is the time interval between addressing the memory and the appearance of the data at the output. The access time ranges from less than 10 ns for the fast memories to several hundred nanoseconds for the slower types.

The internal structure of a semiconductor memory is illustrated in Figure 20.3, for a

Semiconductor memories 351

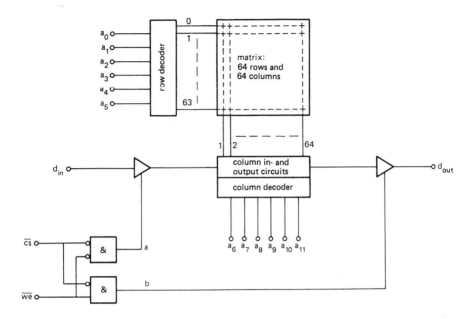

Figure 20.3 Basic structure of a memory circuit of 4096 bits. Each node of a row and a column of the matrix contains one memory cell of 1 bit.

4 k · 1 memory. The device has 18 pin connections in total: 12 address lines $a_0 \ldots a_{11}$, the input d_{in}, the output d_{out}, the chip select cs, the write enable we and two connections for the power supply (ground and 5 V). The memory cells are located at the nodes of the 64 rows and 64 columns in the matrix. The address lines a_0 to a_5 select 1 row from the 64 rows; the corresponding 64 bits are connected to the column lines. At the same time, the address lines a_6 to a_{11} select one of these 64 columns, so 1 cell out of the 4096 cells is connected to the output.

The chip select signal (cs) has the following function: if it is 1, the outputs a and b of the AND gates down left in Figure 20.3 are 0 (the inputs are inverted prior to the AND action). The tristate buffers (Section 18.1) at the input and output of the memory device are switched off, disconnecting the input and output terminals from the external circuits. The memory can be neither written nor read (see the truth table of the memory in Table 20.1). The we signal, which is active only when the chip is turned on by the sc signal, determines which of the operations (read or write) is executed. For a we signal 0, the output of the AND gate b is also 0 and output a is 1. The output buffer disconnects the output, whereas the input buffer connects the input to the memory circuit. Apparently, this corresponds to a write action. When we is inverted, the input buffer is disconnected whereas the output buffer lets pass the selected bit to the output terminal d_{out}.

Table 20.1 Truth table of the memory circuit in Figure 20.3. '–' means 0 or 1, don't care

Function	$\overline{cs}$	$\overline{we}$	a	b	Output
Not selected	1	–	0	0	Disconnected
Write	0	0	1	0	Disconnected
Read	0	1	0	1	d_{out}

Note that in the figure the control signals are indicated with their inversed values: $\overline{cs}$ and $\overline{we}$. This means that the chip is selected for an externally applied select signal that is high (1), hence $\overline{cs} = 1$ (and $cs = 0$). The cs function (and here also the we function) is said to be active-low (as against active-high).

The memory in Figure 20.3 has only one data input and output, as the words consist of only 1 bit. Memories for data words with more bits have parallel inputs and outputs. Some memory chips for words with many bits have common read and write outputs (they are multiplexed), which is allowed because reading and writing will not usually occur at the same time (Figure 20.4).

Another method to save pins is the shared use of data and address lines or, at many addresses, the separation of the address into two parts. For instance, the first 8 bits of a 16 bit address are applied, then the remaining 8 bits. The 16 bits are internally concatenated to a complete 16 bit address. The user or designer has to look carefully into the data books of the manufacturers of memory chips, to make the proper selection.

20.1.3 *Read-only memories*

There are many kinds of ROM circuits on the market. We will consider only three commonly used types: the mask ROM, the fuse-link programmable ROM and the erasable programmable ROM (or EPROM). A mask ROM consists of a matrix of field-effect transistors (MOSFETs, Section 11.1.2), arranged in rows and columns (Figure 20.5). Some gates are connected, some not, depending on the information that has to be stored. The device is called a mask ROM because during the last fabrication step a mask is used for the deposition of the connection pattern.

Each column line in Figure 20.5 is connected via a resistor to the positive power supply voltage; the state is high (1), unless one of the FETs in the column is short-circuited to ground. This happens only for those FETs with a gate connection belonging to the selected row. The row selection is effected by applying a high voltage (1) to the row line, so all FETs of that row with a gate connection are on (conducting).

The programming of this ROM type is done during the fabrication; this is only efficient for large series of ROMs with identical information contents.

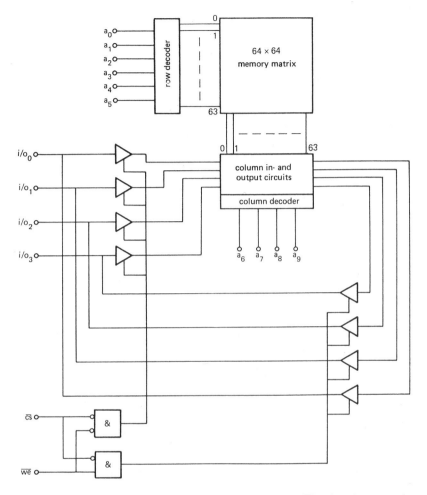

Figure 20.4 A 4096 bit memory chip for 1024 words of 4 bits. The data inputs and outputs share the terminals, by multiplexing.

Figure 20.6 shows the principle of a fuse-link programmable ROM (or fuse-link PROM). This type of ROM can be programmed by the user, by a special instrument, a PROM programmer. The switches in the figure are the fuse-links, comparable with safety fuses. The PROM programmer can selectively burn fuses by a high current pulse through the selected link. Those switches are permanently off. Fuse-link PROMs are used for smaller series where mask ROMs are not remunerative.

A third type of ROM is the erasable programmable ROM (the EPROM), a user-programmable memory. The stored information can be erased either by ultraviolet light or electrically: the electrically erasable or alterable PROM (EEPROM or EAPROM).

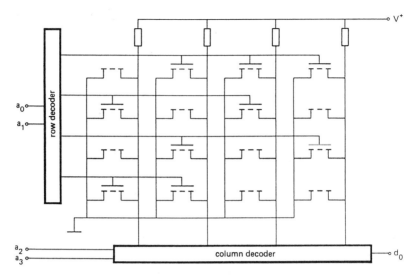

Figure 20.5 Basic structure of a 16 × 1 bit mask ROM.

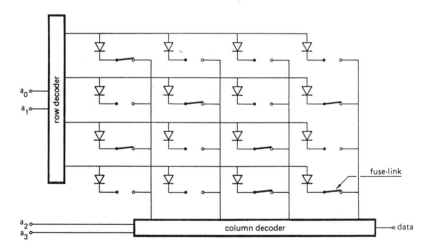

Figure 20.6 Basic structure of a 16 × 1 bit fuse-link PROM.

The principle of the EPROM and EAPROM is based on special MOSFETs whose gates are fully isolated from the environment; there are no external or internal connections. Nevertheless, the gates can be charged by applying a high electric field; the stored charge is maintained for a year or more, without the need for power supply. A charged gate corresponds to a conducting channel. This way of programming requires a much higher voltage than available in most electronic circuits; to rewrite the chips, they must be removed from the circuit. Erasing an EPROM is performed by ultraviolet light; at sufficient intensity, the gates are all discharged. To that end, the chip is illuminated by a UV light source via a small window in the encapsulation (Fig. 20.7). The EAPROM is erased by applying a high reverse electric field; this can be done for each individual bit cell. EPROMs and EAPROMs are applied in those situations where the information content must be retained after switching off the power, but whose information must be changed from time to time.

20.1.4 *Read–write memories*

The internal organization of a read–write memory has already been described in Section 20.1.2. With respect to the principle of data storage, we can distinguish two types: static and dynamic memories. Both types lose the stored information when the power is switched off.

Static memories are composed of flipflops as memory cells (Section 19.1.3). The dynamic types are composed of two components: a capacitor (the memory) and a MOSFET (the selector switch) (Figure 20.8). The cell is selected by applying a voltage on the gate. The information (charge or no charge on the capacitor) is available as a low or a high drain voltage. The capacitor loses its charge rather quickly, because its capacitance is made small, to save space on the chip surface. So, the information must be periodically refreshed. Each memory cell is refreshed every 2 ms, by an internal circuit. The user has to supply the address and an additional signal to activate the refresh operation.

Dynamic memories are available that perform the refreshment automatically; they behave externally as a static memory. Dynamic memories are very compact and are applied for the storage of many bits on a single chip.

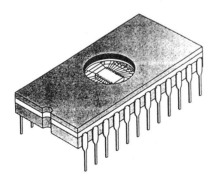

Figure 20.7 An example of the encapsulation of an EPROM. When the chip is illuminated with a UV light beam, the memory is erased completely.

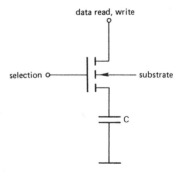

Figure 20.8 One cell of a dynamic memory. The small capacitance discharges rather quickly, so the cell requires periodical refreshment.

20.2 Structure of microprocessors

A microprocessor is a collection of digital circuits on a single silicon chip. Most of these digital circuits have already been discussed in Chapter 19. A major but not essential distinction between a microprocessor and the digital circuits discussed so far is its complexity; a microprocessor chip contains many logic circuits with various functions. Another, more essential difference is the programmability of a microprocessor; the successive function modes can be assigned beforehand and the corresponding operations are executed automatically one after another.

There are many different types of microprocessors on the market. They differ mainly with respect to the internal structure (the architecture) of the processor and the number of different functions. We will confine our discussion to the essential aspects of the microprocessor. For more details on how to use microprocessor chips, the reader is referred to the instruction manuals, which contain detailed information about the pin connections and the programming of the processor.

20.2.1 *Bus structures*

The number of internal connections in an electronic circuit grows significantly with increased number of components. In the case of a microprocessor, the number of connections would be unacceptably high if made in the conventional way. A solution to this interconnection problem is found in a completely different connection structure: the bus. A bus is a group of parallel lines (conductors) along which the data transfer takes place. Each circuit intended for communication with other circuits is connected to this bus (Figure 20.9). Apparently, all systems are connected in parallel. Data reception can take place by several of the connected systems at the same time because of the high input resistance; the transmission of data is, however, restricted to only one system at a time, due to the low output resistance. Hence, the bus connections have to be disconnected electronically; this is, for instance, done by tristate buffers at the outputs of the circuits (Figures 18.8 and 20.4).

Most microprocessor systems have three distinct buses. The data bus transports binary-coded signals (data words or instruction codes). The number of data lines is a characteristic parameter of the microprocessor. An 8 bit microprocessor means a

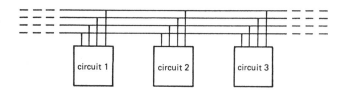

Figure 20.9 A bus structure reduces the number of interconnections between the system parts. As all systems are in parallel, only one system at a time can put the data on the bus.

processor with a data bus consisting of 8 parallel lines. The address bus carries the binary-coded addresses of the memory locations. With a 16 bit address bus (16 parallel lines), 2^{16} different memory locations can be selected (so 65 536 locations). The control bus consists of various lines for the control of the digital circuits, the clock (for synchronization), the read write line (to indicate whether to read or to write) and the *cs* signals (to activate individual chips). Figure 20.10 is an example of a microprocessor system in which this bus structure is applied. The bus structure offers a high flexibility; it is easy to add or replace subsystems without changing the structure of the system. For instance, the memory capacity can be extended simply by connecting one or more additional memory chips to the bus lines.

The bus system has a disadvantage too. Only one word at a time can be transported along the bus. This seriously limits the transmission speed and the execution time of the program.

There are microprocessors with a different structure, to avoid this bottle-neck. However, up to now the common types of microprocessors have used the bus structure given in Figure 20.3.

20.2.2 *Internal organization of a microprocessor*

The internal organization of a microprocessor differs from type to type. Fortunately, there are also common characteristics which will be discussed briefly in this section. One guideline for the discussion is the simplified diagram of a microprocessor (Figure 20.11).

To execute a program, the microprocessor successively transfers the instruction codes from the program memory (where the program is stored beforehand) to an internal memory circuit and executes these instructions. A special register in the CPU, the program counter, keeps count of the instructions. When the instructions are stored in the right order in the memory, it is sufficient to increment the program

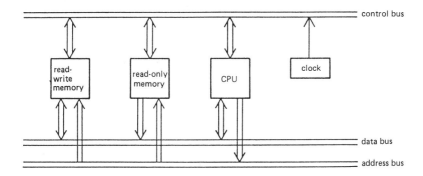

Figure 20.10 The three buses in a microprocessor system: address bus, data bus and control bus.

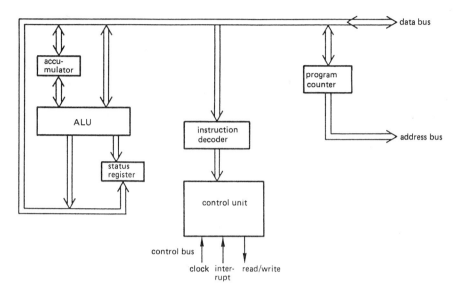

Figure 20.11 Simplified diagram of a microprocessor.

counter by 1 after the execution of each instruction. The program counter output is just the address of the location where the next instruction code is stored.

The program counter output can also be changed intentionally; the program continues with the instruction stored at the new address pointed to by the program counter. In this way, it is possible to make a jump in the program. The microprocessor has special instructions for that purpose, the jump instructions.

Example 20.2
When switching on a microprocessor, the program counter starts at 0000 (hexadecimal notation). When this memory location contains the code for a jump instruction to jump to address 2000, the processor proceeds, after the execution of this jump instruction, with the instruction stored at the location 2000.

The translation from the coded instructions to a real operation by the CPU is performed by the instruction decoder and the control unit. The instruction decoder interprets the data coming from the data bus and the control unit generates the proper signals for the internal logic operations. The control unit also synchronizes all the digital circuits by means of the clock. The representation of the control unit in Figure 20.11 as a separate block is not quite true; it must be considered as a network of circuits distributed over the whole system.

The execution of the basic operations is accomplished by the arithmetic and logic unit (ALU) which contains at least an adder for two binary words. Usually, there are some extra functions, like subtraction, logic operations (AND, OR, EXOR), shifting and counting. The result of an operation executed by the ALU appears in a special register, the accumulator, which serves as a source for the ALU as well. In another register some particular properties of the result are stored; this is the status register. It is composed of the condition flipflops: one for a positive or negative result, one for zero or non-zero accumulator contents, and some others. These bits are called flags and are used for the execution of conditional instructions; depending on the status of a particular flag bit (as a result of the last operation), the processor either proceeds with the next instruction or skips it.

Example 20.3
Suppose a certain part of a program must be repeated 600 times. The number 600 is stored somewhere in the memory, and is used as a 'counter'. Assume the subprogram starts at address 3000 (hexadecimal). This subprogram contains an instruction that puts the counter number into the CPU, decrements the number by 1 and places it back at the original location. As long as the result of decrementing (subtraction) is not zero, the zero-status flag is zero. When finally, after 600 cycles, the counter is zero (hence the result of the last subtraction), the flag goes to 1. The last instruction of the subprogram is a conditional jump; the program returns to address 3000 if the zero flag is 0.

20.2.3 *Interfacing*

Interface circuits perform the matching between the CPU and other system parts. This matching is necessary for a number of reasons.

- Signal structure
  The data bus of the processor uses fixed voltages to represent 0 and 1 (for instance 0 and 5 V). Outside the CPU, other signal structures may be used, like non-electrical signals (for instance optical), frequency-modulated signals or currents and voltages with other values.
- Synchronization
  When two systems communicate with each other, the timing of the signals is usually not equal. Synchronization and matching of the transmission speeds are required for proper data communication.
- Number of bus lines
  The number of lines in a microprocessor bus is fixed, depending on the type of microprocessor. Other systems may have different numbers of bus lines and different bus line definitions.
- Codes
  There are many ways to encode a signal: serial, parallel, with or without certain test codes for fault correction, and so on. A microprocessor needs a particular code; conversion to that type of code is necessary.

When communicating with peripheral instruments that have a low data processing speed, the much faster computer has to wait for the proper data coming from these instruments. To avoid unnecessary waiting, the processor has special interrupt inputs. When an instrument is ready for communication with the CPU, it activates one of the interrupt inputs, that are continuously checked by the CPU. Upon an interrupt, the processor finishes the current instruction and starts with a subroutine, a subprogram that (in this case) performs the communication with the instrument that has activated the interrupt. The result of this procedure is that the processor only communicates with peripherals when they are ready to do so. The interrupt mechanism is used for instance for the readout of the keys of a keyboard. Many processors have various interrupt inputs; some (or all) can be temporarily switched off (by the program), to avoid interrupts at an undesired moment (priority).

There are special interface circuits for bidirectional data transport with the CPU: these are called input/output ports or IO ports. Many peripheral instruments have such an IO circuit. The IO circuits are connected to the data and address buses of the processor, like memory circuits. An IO IC also must be selected, like a memory IC. The IO port has a special input, the chip select input (compare the *cs* of a memory). It is connected to the address bus via a special circuit, the address decoder or selector. An example of a simple address selector is given in Figure 20.12. It consists of a series of digital comparators, which compare the state of the address lines with the particular address word of the device; this word can be chosen freely by, for instance, setting a series of switches. If the address on the data bus corresponds with the address word of the device (the IO circuit), then the output of the selector is 1 and activates the chip select input of the IO IC, upon which the CPU transfers the data to that device.

This way of addressing an IO IC is similar to that for a memory location. The method is therefore called memory-mapped IO. There are other methods of addressing IO circuits, but these are not discussed further here.

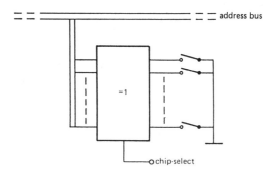

Figure 20.12 A digital comparator, to compare a parallel word (adjustable with the switches) with the data on the address bus. The output acts as a chip select signal that activates the circuit with the corresponding address.

To illustrate the interfacing of a microprocessor, we consider now the operation of a particular IO interface circuit in detail. In this example we have an IO circuit for the transmission of parallel words, parallel input–output interface (PIO), widely used for connecting the CPU to other peripheral circuits. Figure 20.13 shows the pin connection diagram of this IC. The PIO has three parallel ports (A, B and C), that can be connected to the data bus of the CPU. All three ports can act either as input port or as output port, depending on the control signals. Table 20.2 (the truth table of the PIO) clarifies the various operation modes.

The control of the PIO is as follows: with the address lines a_0 and a_1 one of the three ports is selected; with the remaining address lines the chip is selected (via an address decoder as shown in Figure 20.12). The read–write signal of the control bus indicates whether data are input or output.

ICs are also available for serial data transport. In particular, for the transmission over a long distance, it is more efficient to transmit the data bit after bit; this requires only a single conductor pair. Of course, the transmission speed is much lower. Serial interface ICs convert parallel words from the data bus into serial words for the transmission line and vice versa. Such a circuit is called a UART or USART (universal

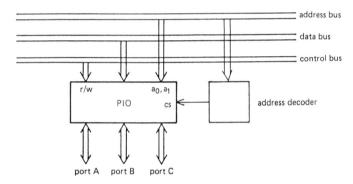

Figure 20.13 An interface circuit able to connect three parallel peripherals to the microprocessor.

Table 20.2 The truth table of the PIO in Figure 20.13; '–' means 0 or 1, don't care

cs	a_1	a_0	r/w	Function
1	0	0	0	Port A is input for the data bus
1	0	1	0	Port B is input for the data bus
1	1	0	0	Port C is input for the data bus
1	0	0	1	Port A is output for the data bus
1	0	1	1	Port B is output for the data bus
1	1	0	1	Port C is output for the data bus
0	–	–	–	Tristate

synchronous/asynchronous receiver/transmitter). These ICs are used in modems (modulator/demodulator) for the transmission of computer data over a telephone line or a special data link.

20.2.4 *Programming*

We can distinguish various levels of abstraction in the programming of a microprocessor (Table 20.3).

In Section 20.2.2 we met some instructions on the machine (CPU) level: conditional jump; decrement; add. The codes for the machine instructions consist of one or more binary words. In an 8 bit microprocessor, these words contain 8 bits (1 byte). With an 8 bit instruction code $2^8 = 256$ different machine codes can be represented.

> **Example 20.4**
> The contents of an auxiliary register in the CPU (for example register D) must be transferred to the accumulator (register A). For such an operation the processor has a special instruction. This instruction is executed when the word 01111010 (or 7A hexadecimal) is applied to the instruction decoder. Of course, this code may differ for different types of microprocessors. This is an example of a 1 byte instruction.

> **Example 20.5**
> The contents of the memory location with address F0A5 (hexadecimal) must be transferred to the accumulator (register A). According to the specifications of the manufacturer, the machine code is 00000110–11110000–10100101 (or 06F0A5 hexadecimal). This instruction requires three 8 bit words, stored at three successive locations in the memory.

When there is no instruction for a certain operation, the programmer must find an adequate combination of permitted instructions.

A microprocessor system contains at least a number of ICs (the CPU and a read-only memory), as well as a start program. This program is stored in a ROM. When switching on the system, the CPU executes this start program; it begins with the instruction stored at address 0000. The contents and the nature of the start program depend on the particular application.

Table 20.3 The levels of abstraction in programming

Program language	Description
Machine language	Binary control codes for the CPU
Assembler language	Description of the instructions in machine codes by simple words or abbreviations
Abstract computer language	Application-oriented language (no relation with the processor or structure)

The realization of the starting program and the other programs is done with a development system which is, in turn, also a processor-based system, offering facilities for the development and testing of the programs, with the use of keyboard and monitor.

The machine code consists of mere binary codes; working with such codes is very cumbersome and easily leads to errors. Programming at assembler level is much easier; the codes are translated to simple functional names, mnemonics, that can be easily interpreted by the programmer. This is the assembler language in Table 20.3.

Example 20.6
The instruction in Example 20.4 is, in assembler: LD A,D (load register A with the contents of register D). The mnemonics of the instruction in Example 20.5 is LD A,A5F0.

From this example it appears that the assembler language is much easier to use; from the mnemonics it is immediately clear what is meant by a certain instruction. The assembler language is coupled to the type of microprocessor. This is not the case for application-oriented languages as Pascal and C. There are program translators (compilers) available to translate programs written in, for instance, Pascal into the assembler or machine language of a particular type of microprocessor. In the next example, we compare an instruction written in three languages: Pascal, assembler and machine code.

Example 20.7
Pascal:
IF Y ⩾ Z THEN Y:=2Y   (if Y equal to or larger than Z, Y must be multiplied by 2)

Assembler: (Y is at memory location 4000, Z at 4001)
LD A,(4000)	transfer Y in 4000 to the accumulator
LD B,A	put Y in register B
LD A,(4001)	transfer Z from 4001 to the accumulator
SUB A,B	subtract Z from Y; result is in A
JMP 9000	if the contents of A are negative, go to address 9000 (to proceed with another routine)
LD A,(4000)	transfer Y from 4000 to the accumulator
RLA	rotate contents of the accumulator 1 position to the left (Section 19.2.4)
LD(4000),A	transfer Y from the accumulator back to location 4000

Machine code (hexadecimal notation):
... 3A 00 40 02 3A 01 40 90 FA 00 90 3A 00 40 17 32 00 40 ...

Machine code (contents of the program memory):
... 00111010 00000000 01000000 00000010 00111010 00000001 01000000
10010000 11111010 00000000 10010000 00111010 00000000 01000000
00010111 00110010 00000000 01000000 ...

After having developed and tested the program, it is transferred to the microprocessor system. A fixed program (like the start program) is stored via a PROM programmer to the system ROM, positioned in the microprocessor system.

SUMMARY

Semiconductor memories

- Semiconductor memories are composed of memory cells, arranged in columns and rows, forming a matrix. Each cell can store only 1 bit of information.
- Memory devices are distinguished with respect to their accessibility into serial access (SAM) and random access (RAM), with respect to the ability of writing into read-only (ROM) and read–write memories (referred to as RAM).
- Important specifications of memory devices are the memory-cell organization, the number of words (addresses), the word length (in bits), the access time and the power consumption.
- The most important ROM types are the mask ROM, the fuse-link ROM, the EPROM and the EAPROM or EEPROM.
- Read–write memories are divided into static and dynamic memories; a dynamic memory requires periodic refreshment of the stored information.

The structure of microprocessor systems

- A microprocessor is a programmable digital signal processing circuit, consisting of an arithmetic and logic unit (ALU), several auxiliary registers and control circuits, together with the CPU, integrated on a single silicon integrated circuit.
- The list of instructions to be executed by the processor is the program. This program is stored in one or more memory ICs. Fixed programs are stored in read-only memories (ROM), variable programs are stored in read–write memory (RAM).
- The bus structure is a concept that facilitates interconnections in complex electronic systems. The bus consists of a group of parallel lines; the circuits

can be electronically connected to the bus lines. A microprocessor has a data bus, an address bus and a control bus. The number of bits of the data bus is a characteristic parameter of a microprocessor.
- For the internal organization of the microprocessor (the CPU), a number of auxiliary circuits is available. The major registers are the accumulator, the status register (with the flags) and the program counter. An important part is the instruction decoder for the translation of the instruction codes.
- A running program can be interrupted externally, via the interrupt inputs. The processor terminates the current instruction and starts at the beginning of the interrupt program.
- Data exchange takes place via interface circuits. Interface devices for bidirectional transport are called IO ports. There are interface circuits for parallel and for serial data communication.
- An assembler is a program that translates an assembly program (with mnemonics) into machine code. A compiler is a program that translates a user-oriented program into the language of the processor.

Exercises

Semiconductor memories

20.1 Explain why the output buffers of a semiconductor memory circuit are tristate buffers and not normal-type buffers.

20.2 In Figure 20.6, what is the function of the diodes in series with the fuse-links?

20.3 Is it possible to create an EPROM with bipolar transistors instead of FETs?

20.4 Discuss the advantages of a static CMOS memory compared to a non-static memory and a non-CMOS memory.

20.5 A 64 kbit memory is organized in words of 1 bit.
   (a) How many memory cells does this memory device contain?
   (b) How many distinct addresses are necessary?
   (c) How many bits are necessary for the addressing?

20.6 As the number of bits increases in a memory, the number of pins for the data and the addresses grows accordingly. Describe a method that reduces the number of pins, while keeping full access to all memory locations.

20.7 Explain why the address lines of various memory ICs can be connected simultaneously to the bus, but not the data lines.

The structure of microprocessor systems

20.8 What is the function of these CPU registers: accumulator, program counter and status register?

20.9 A particular type of 8 bit microprocessor has a 16 bit address bus. How many bytes are needed for the instructions for:
   (a) the transfer of the contents of an arbitrary memory location to the accumulator;
   (b) an EXOR operation on the contents of a CPU register;
   (c) subtraction of the number 3A (hexadecimal) from the contents in the accumulator;
   (d) a conditional jump instruction to an arbitrary memory location?

21 Measurement instruments

Measurement systems are available for almost any electric quantity. All these instruments are based on the electronic principles and circuits described in the preceding chapters. The first part of this chapter gives a review of several commonly used measurement instruments: multimeters, oscilloscopes, plotters, signal generators and analysers. The second part of this chapter deals with computerized measurement systems.

21.1 Electronic measurement instruments

21.1.1 *Multimeters*

Multimeters are the most frequently used electronic instruments. They are low cost and easy to use. Multimeters are suitable for the measurement of voltages, currents and resistances. The input voltage is converted into a digital code and displayed on a liquid crystal display (LCD) or an LED display. Input currents are first converted into a voltage; resistance measurement is performed by applying an accurately known current (generated in the instrument) to the unknown resistance and measuring the voltage across it.

Electronic multimeters have a high input impedance, in the order of $1\,M\Omega$ (in the voltage mode). Most multimeters have autoscaling and autopolarity features; the polarity and the units are displayed next to the measurement value. Autoranging and autopolarity are realized by comparators and electronic switches (reed switches or FETs).

21.1.2 *Oscilloscopes*

An oscilloscope is an indispensable measurement instrument for the testing of electronic circuits. It visualizes the otherwise invisible electronic signals, in particular periodic signals. Some oscilloscopes can display non-periodic signals (transients) as well.

The display section of the oscilloscope consists of a cathode-ray tube, two pairs of deflection plates and a phosphorescent screen (Figure 21.1). The cathode (a heated filament) emits a beam of electrons towards the screen and produces a light spot where the electrons hit the screen. Without deflection the spot is just in the middle of the screen. The beam can be deflected independently in horizontal or vertical directions (x and y directions), by the application of a voltage on the deflection plates. The position of the light spot changes in proportion to the voltage on the plates. To produce an amplitude–time diagram on the screen, the x plates are connected to a ramp voltage generated internally in the instrument. This time-base signal sweeps the spot from left to right with constant speed, resulting in a horizontal line on the phosphorescent screen. The speed can be varied stepwise, to obtain a calibrated time-scale. The signal to be observed is connected, via an amplifier with adjustable gain, to the y plate, letting the spot move in the vertical direction. The combination of horizontal and vertical displacement results in an image of the amplitude–time diagram on the screen.

When the beam arrives at the end of the scale, the time-base signal returns to the left in a very short time; while returning, the beam is switched off (blanking), to make it invisible to the observer. This process is repeated periodically. Due to the inertia of the human eye and the afterglow of the phosphor, the observer gets the impression of a full picture of (part of) the periodic signal.

If the frequency of the periodic input signal is an exact multiple of that of the time-base, the image on the screen is stable (Figure 21.2). When the number of periods does not fit in the interval of the ramp, the picture is running over the screen, which hampers observation. To obtain a stable picture, irrespective of the input frequency, the time-base signal is triggered on command of the input. A comparator (Section 14.2.1) compares the input signal with an adjustable reference level and starts the time-base when the input passes that level. The result is depicted in Figure 21.3, for a sine wave signal and various trigger levels and time-base frequencies.

Most oscilloscopes have an automatic trigger facility; it automatically finds a proper trigger level. For composite signals the user can switch a filter to trigger only on the low or the high frequency components. Other features are external triggering (on a separate signal applied to an extra terminal) and delayed trigger (the trigger is delayed over an adjustable time interval).

Figure 21.4 shows a simplified block diagram of an oscilloscope; the blanking circuit is left out for simplicity.

A capacitor in series with the input of the y amplifier/attenuator allows the user to eliminate the DC component of the input signal, for instance when the AC component to be observed is much smaller than the average value. Most instruments have the

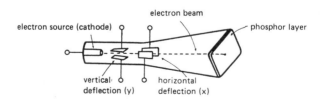

Figure 21.1 Basic structure of a cathode-ray tube in an oscilloscope.

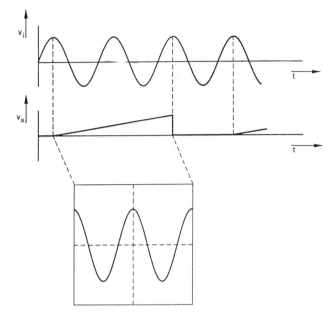

Figure 21.2 Image on an oscilloscope of a sinusoidal signal v_i. The picture is stable when the input frequency is just a multiple of the time-base frequency.

possibility of an external x input signal (instead of the internal time-base). This allows the visualization of, for instance, the input–output transfer (x–y characteristic) of an electronic network.

There are oscilloscopes for more than one input signal, including the dual or multichannel oscilloscopes (up to 12 channels). With such an instrument, several signals can be observed simultaneously. There are two approaches to obtain multichannel operation with a single electron beam. The first is called alternate mode; the (two) signals are written on the screen alternately, during one complete period of the time-base (Figure 21.5a). For high signal frequencies (and thus high time-base frequencies), the images appear ostensibly simultaneously on the screen. For low frequencies the pictures appear visibly one after another; in that case the chopped mode is preferred. In this mode, the electron beam is switched at a relatively high speed between the two input channels (Figure 21.5b). Each of the signals is displayed in small parts (chopped) but, due to the high chopping frequency, this is not visible to the observer.

Other important specifications of an oscilloscope are the sensitivity of the y input(s) and the maximum speed of the trace on the screen (hence the maximum input frequency). Oscilloscopes for general use have a sensitivity of about 1 mV/cm (better: 1 cm/mV) and a bandwidth of 15 MHz. The time-base of such oscilloscopes is adjustable from about 50 ns/cm to roughly 0.5 s/cm. There is a wide variety of

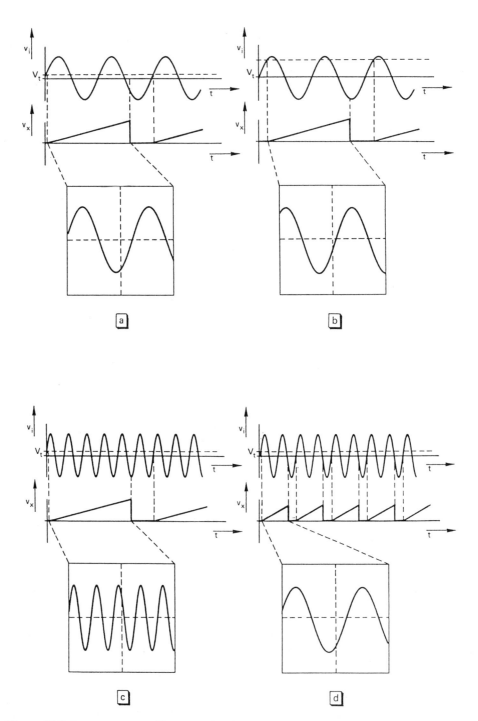

Figure 21.3 Image on an oscilloscope of a sine wave input signal for (a) a low trigger level v_t; (b) a high trigger level; (c) a high input frequency, with time-base and trigger level as in (a); (d) a high frequency with adjusted time-base frequency.

Electronic measurement instruments 371

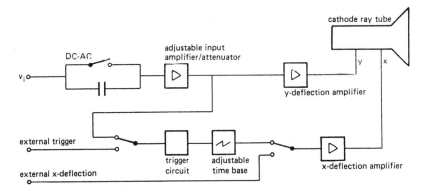

Figure 21.4 Simplified block diagram of an oscilloscope.

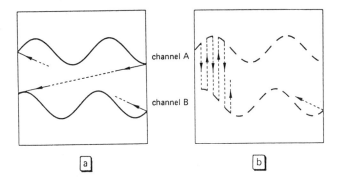

Figure 21.5 Two ways to view several signals with a single electron beam: (a) alternately writing the input during a time-base period; (b) chopped: each signal is displayed during a very short time interval of the time-base period.

oscilloscopes for particular applications, like very high frequencies (up to 1 GHz), low frequencies (using long-persistence phosphors) and single events; with special techniques, an almost permanent image is obtained using a storage oscilloscope. Unlike the oscilloscopes discussed so far, operating in real-time mode, there are also oscilloscopes that sample (part of) the input signals, after which the stored data are presented on the screen in a user-defined way (digital or sampling oscilloscope).

The oscilloscope is provided with one or more probes, connected to the oscilloscope by flexible cables, for better accessibility to the measurement spot. However, the capacitance of the probe cable reduces the bandwidth of the oscilloscope. The input impedance of the oscilloscope is modelled by a resistance R_i

and a capacitance C_i in parallel (Figure 21.6a). Furthermore, the cable capacitance is C_k and the source resistance is R_g. The voltage transfer equals:

$$V_i/V_g = \frac{Z_i}{Z_i + Z_g} = \frac{R_i}{R_i + R_g + j\omega R_i R_g (C_i + C_k)}$$

Normally, $R_g \ll R_i$, so the transfer can be approximated as:

$$V_i/V_g = \frac{1}{1 + j\omega R_g (C_i + C_k)}$$

Obviously, this corresponds with the transfer of a low-pass filter with $-3\,\text{dB}$ frequency $\omega = 1/\{R_g(C_i + C_k)\}$ (see Sections 6.1.1 and 8.1.1).

Example 21.1
Suppose an input impedance of 1 MΩ // 100 pF (typical values for a normal oscilloscope), a cable capacitance of 100 pF and a source resistance of 1 kΩ. The bandwidth of the measurement system is now limited to 1.3 MHz (the oscilloscope is assumed to have a much wider band). So, the image of a sine wave with frequency 650 kHz is 10% too small.

For this reason the probe is provided with an adjustable attenuator (voltage divider, Figure 21.6b). The signal is attenuated but the frequency transfer characteristic can be made fully independent of the frequency. The transfer of the system in Figure 21.6b is:

$$V_i/V_g = R_i(1 + j\omega RC)/\{R_i + R + R_g + j\omega[(R + R_g) \\ \times (C_i + C_k)R_i + (R_i + R_g)RC] - \omega^2 R_g R R_i C(C_i + C_k)\}^{-1}$$

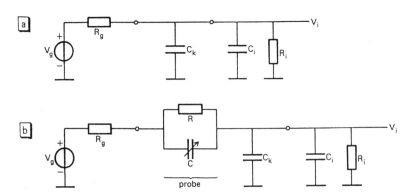

Figure 21.6 (a) The model of a voltage source and the oscilloscope with probe cable capacitance C_k and input impedance $R_i//C_i$; (b) as (a), extended with a probe attenuator.

In normal cases, $R_g \ll R$ and $R_g \ll R_i$, hence:

$$V_i/V_g \approx R_i(1+j\omega RC)/[R_i + R + j\omega RR_i(C + C_k + C_i) - \omega^2 R_g RR_i C(C_i + C_k)]^{-1}$$

$$= \frac{R_i}{R_i + R}(1+j\omega RC)\left(1 + j\omega \frac{RR_i}{R_i + R}(C + C_k + C_i)\right.$$

$$\left. - \omega^2 \frac{RR_i}{R_i + R} R_g C(C_i + C_k)\right)^{-1}$$

Neglecting the term with ω^2, which is usually allowed, the transfer has the real value $R_i/(R_i + R)$ (and so independent of the frequency) under the condition $RC = RR_i(C + C_k + C_i)/(R_i + R)$ or $RC = R_i(C_k + C_i)$.

Example 21.2
A typical value for the probe attenuation is a factor of 10. With an input resistance $R_i = 1$ MΩ (as in Example 21.1), the series resistance R in the probe must be 9 MΩ. The corresponding value of C is about 13 pF. The input impedance of the total measurement system (oscilloscope with probe) is now 10 MΩ // 12 pF.

The probe attenuator allows the measurement of voltages from a source with rather high input impedance, without affecting the frequency response of the system. Prior to the measurement the user must adjust the probe capacitance C for frequency-independent transfer. The adjustment is performed using a calibrated square wave test signal available at an extra terminal on the front of the instrument.

21.1.3 *Plotters*

A plotter is used to present and register single events as transients and for the registration of a measurement signal over a long time. The plotting is performed by a movable pen; in some plotter types, the paper moves or both the pen and the paper move.

There are two main groups of plotters: *x–y* plotters and *x–t* plotters. In *x–y* plotters, the pen is moving in two directions *x* and *y*, in proportion to the input signals v_x and v_y, respectively. In large computer-controlled plotters, the pen moves in a horizontal direction whereas the paper moves in a perpendicular direction.

An *x–t* plotter registers the amplitude–time diagram of a signal. The pen moves in one direction in proportion to the input signal and the paper moves with a constant speed in a perpendicular direction.

In many types of plotters pen control is performed by a feedback control loop, to achieve high position accuracy (Figure 21.7). The pen motor is controlled by a differential amplifier that responds to the difference between the input signal and the voltage on the slider of an accurate linear displacement sensor (Section 7.2), a

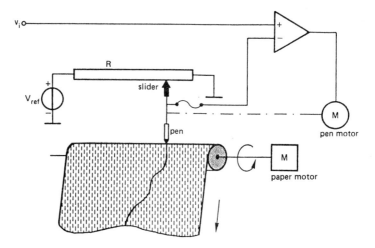

Figure 21.7 Principle of a servo-plotter. The pen motor is controlled such that the input voltage is forced to zero.

precision potentiometer. The motor moves until the difference is zero and the pen is in the position that corresponds with the input voltage.

The bandwidth of the system is limited by the inertia of the mechanical system and ranges from 0 up to several tens of hertz. The sensitivity can be very high, up to $1\,\mu\text{V/cm}$. In x–t plotters, the time-scale is adjustable from several mm per hour up to several metres per second.

There are multipen plotters (up to six channels) for the simultaneous registration of several signals to study their relation in the time domain. For the registration of even more signals at the same time, the plotter is used in time multiplex (Section 1.1), where the signals are distinguished by various colours of ink.

Computer-controlled plotters plot the stored signals sequentially, exchanging the pens to give each trace a different colour. The pens are positioned in the pen carousel of the plotter.

21.1.4 *Signal generators*

Signal generators are indispensable instruments for testing and error tracing. There is a wide variety of types, ranging from simple sine wave oscillators (Section 16.1) to very complex, processor-controlled systems for various signal shapes and signal parameters.

DC signal sources produce an accurately adjustable DC voltage or current, derived from a Zener reference source (Section 9.1). The DC voltage or current is used either as a test signal or as a (stabilized) power supply voltage.

Sine wave generators produce a sinusoidal voltage, mainly as a test signal (for instance to measure the frequency response of an analog signal processing system). They are available for frequencies ranging from very low values (down to 10^{-4} Hz) up to several GHz. The basic principles are discussed in Chapter 16.

A pulse generator produces periodic square wave and pulse wave voltages with adjustable frequency, pulse height and pulse width or duty cycle (Section 16.2.1). There are special types generating pulses with a very short rise (and fall) time, down to 0.1 ns, used for the investigation of wide band systems.

Periodic signals, like ramp, triangle and square wave signals, are generated by a function generator. Some of these instruments have multiple outputs (for instance a triangle and a square wave with equal frequency). They are versatile instruments and widely used in the laboratory.

The frequency stability of the signals is determined by the stability of circuit components such as resistors and capacitors (Chapter 16). If a much higher stability is required, a crystal oscillator is used. A piezoelectric element (Section 7.2.5) is forced to resonate mechanically by applying an AC signal. The resonance frequency is determined almost exclusively by the dimensions of the crystal, so the output frequency is as stable as the crystal. Other frequencies are derived from the resonance frequency by frequency division (with flipflop circuits, Section 19.1).

Some generators have a voltage-controlled output frequency: they are called voltage-controlled oscillators (VCO; Section 16.2.4). In combination with a second generator, the frequency can be swept linearly or logarithmically over a predefined range (Figure 21.8). Such a sweep generator is used in instruments for the analysis of frequency-dependent transfers and impedances.

Finally, we mention noise generators. These instruments generate a stochastic signal, usually derived from a noise-producing element, for instance a resistor that produces white noise (Sections 2.1.3 and 5.2.2). The frequency spectrum or the bandwidth of the noise is varied using electronic filters. Noise generators based on digital principles generate quasi-stochastic signals; the binary signal is periodic, and within each period the transitions occur randomly. Noise signals are often utilized as a test signal; the advantage over a sinusoidal test signal is that noise contains a continuous range of frequency components.

All voltage generators should have a low output impedance for a proper matching (Section 5.1.3). A standard value is 50 Ω, a value that is employed in high frequency

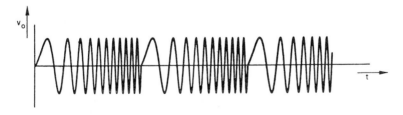

Figure 21.8 An example of an output signal of a sweep oscillator.

systems as well (characteristic impedance, Section 5.1.3). The output amplitude is adjustable, continuously from 0 to 10 V or more. Some generators have an adjustable DC level (offset).

Low-cost generators have manually controlled potentiometers for continuous adjustment of the signal parameters. In most processor-based generators the frequency, amplitude and offset are set by buttons. In general, these instruments are far more expensive, but they have better accuracy and stability, and are more flexible than analog instruments.

21.1.5 *Counters, frequency meters and time meters*

Instruments for the measurement of pulses, frequency and time intervals are all based on counting signal transitions. The input signal is applied to a comparator or Schmitt trigger (Sections 14.2.1 and 14.2.2) which transfers the analog signal into binary form, with preservation of the frequency. This binary signal is applied directly to a digital counter circuit (Figure 21.9), which can be reset by an externally controllable reset button. The number of input pulses is therefore counted. The same circuit can be used for the measurement of the input frequency; the comparator output is only applied to the counter during an accurately known time interval. The counter counts the number of pulses or zero crossings per unit of time, that is the frequency. The result is displayed directly in hertz. The required reference time is derived from a stable, accurate (crystal) oscillator. The range of the frequency counter depends on the range of the counter but also of the time interval. The latter can be varied stepwise with an adjustable frequency divider. By changing the range in factors of 10, the display can simply change accordingly by shifting the decimal point.

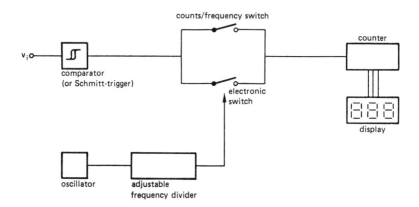

Figure 21.9 Principle of an electronic counter. As a frequency-measuring instrument, the number of pulses is counted during a defined time interval, adjustable with the frequency divider.

Electronic measurement instruments

The measurement time of low frequencies would become too long using this principle. In such cases, it is more convenient to measure the period of time rather than the frequency. This can be done by simply interchanging the position of the comparator and the reference oscillator in Figure 21.9. The counter counts the number of pulses from the reference oscillator during half the period of the input signal. Similarly, the range can be changed by using the frequency divider.

The measurement range of frequency meters can be as high as 100 MHz; special types count even to several GHz. The input impedance has a high value, typically 1 MΩ for the common types and 50 Ω for the high frequency instruments.

The input signal should always exceed a specified minimum value, otherwise a correct conversion to a binary signal is not guaranteed.

21.1.6 *Spectrum analysers*

A spectrum analyser is a measurement instrument for the visualization of the frequency spectrum of a signal. Figure 21.10 shows the basic structure of the instrument. The voltage-controlled band-pass filter (VCF) has a very narrow band. The position of the filter in the frequency band is varied electronically by a control voltage, generated by a ramp generator. During the filter sweep, the amplitude of the output signal is determined continuously, for instance with a peak detector (Section 9.2.2). A spectrum is obtained on the screen when the amplitude information is applied to the vertical deflection plates of the cathode-ray tube and the ramp voltage to the horizontal deflection plates.

The peak detector is not suitable for small signals (Section 9.2.2), so the amplitude is always detected by synchronous detection (Section 17.1.3), as is depicted in Figure 21.11.

The synchronous detector behaves as a narrow band-pass filter with a central frequency equal to the frequency of the reference signal (Section 17.2).

21.1.7 *Network analysers*

A network analyser measures and displays the frequency characteristic (Bode plot) of a two-port network. Figure 21.12 shows the simplified structure of a network analyser.

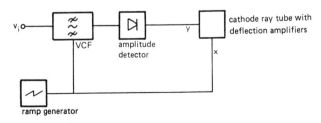

Figure 21.10 Basic structure of a spectrum analyser.

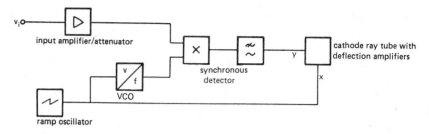

Figure 21.11 Basic structure of a spectrum analyser with synchronous detector as band-pass filter and amplitude detector.

The VCO generates a sinusoidal voltage with controllable frequency; in manual mode the frequency can be adjusted by the user; in the sweep mode, the instrument varies the frequency by a ramp voltage on the control input of the VCO. At constant (calibrated) VCO output, the output of the amplitude detector is a measure for the amplitude transfer of the network. The phase transfer is determined by synchronous detection. A synchronous detector responds to the cosine of the phase difference and the respective amplitudes (Section 17.1). A linear-phase response is achieved by conversion of both input and output signals to binary signals (square wave signals) with fixed amplitude. The output after synchronous detection is linearly proportional to the phase difference between the input and output of the system under test, and independent of the signal amplitude.

By controlling the frequency of the VCO over a certain frequency range, the Bode plot (and eventually the polar plot, Section 6.2) can be projected on the screen.

21.1.8 *Impedance analysers*

An impedance analyser measures the complex impedance of a two-pole network at discrete or continuously varying frequency.

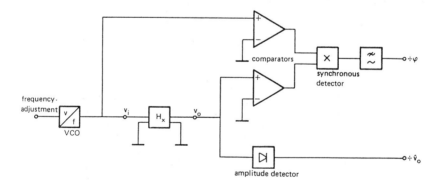

Figure 21.12 Simplified structure of a network analyser for the measurement and the display of both the amplitude transfer and the phase transfer of a two-port network.

Impedance analysers usually have a digital display or a cathode-ray tube for a complete picture. Figure 21.13 shows the simplified structure of an instrument that displays the impedance on a numerical display.

The VCO voltage is applied to the impedance under test. The current through the impedance is converted into a voltage (Section 12.1.1). The amplitude and phase of this voltage are measured in a similar fashion to the network analyser of the preceding section. The instrument is equipped with AD converters and a microprocessor to calculate from the amplitude and phase other impedance quantities such as its real and imaginary parts (Section 4.1.1), depending on the model of the network; the user can choose a model that fits best to the actual circuit.

21.1.9 *Logic analysers*

During the design of digital circuits, it is very convenient and sometimes necessary to be able to observe a large number of digital signals simultaneously on a screen. Oscilloscopes are not suitable for that purpose because of the limited number of channels. For this application, a logic analyser has been developed which allows the presentation of a large number of binary signals. The name logic analyser is, however, misleading; the instrument performs no analysis at all, but only displays the digital signals over a certain time interval, such as the time chart of Figure 19.18. The user analyses this picture and draws the conclusions. This does not mean that a logic analyser has no other uses.

To observe the fast, non-periodic level transitions, the logic analyser stores part of the signals during a certain time interval. These signals are sampled periodically and the results are stored in the memory of the system. Later, the results can be displayed on the screen in a form that is suitable to the user.

Most logic analysers have trigger facilities; the user can start the sampling at an arbitrary moment, which is useful for fault searching in a program.

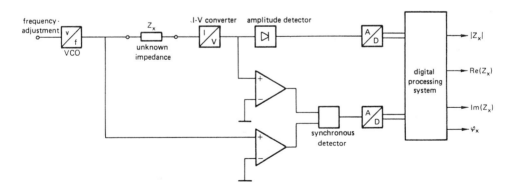

Figure 21.13 Simplified structure of an impedance analyser with microprocessor for the calculation of various impedance parameters.

Another feature in some instruments is the ability to make histograms, to observe how often and how long certain program routines are used, and to improve the efficiency of developed programs. A logic analyser is an almost indispensable tool in the development of circuits with microprocessors.

21.2 Computer-based measurement instruments

21.2.1 *Computer-based instrumentation*

Measurement systems controllable by a computer do not differ essentially from other measurement instruments. They have special interfaces for communication with a computer. The task of the computer is controlling the instrument and further processing of the measurement data. As a computer has a large processing capacity, several instruments can be connected to the computer simultaneously (Section 1.1). As in computers and processors, a bus structure (Section 20.1.3) is applied to reduce the wiring and the number of interface circuits (Figure 21.14). The instruments are all connected in parallel to the bus. To transport the data in an orderly fashion, each instrument is equipped with a more or less intelligent interface.

Several organizations for standardization have tried to define and introduce a standard bus structure suitable for measurement instruments from various manufacturers. Two organizations have succeeded in producing an acceptable standard for instruments, the IEC (International Electrotechnical Commission) and the IEEE (Institute of Electrical and Electronics Engineering). The standards are the IEC-625 and the IEEE-488 bus, respectively. As these bus systems are widely used in instrumentation, we will discuss them in this book. The two bus standards are almost identical; in the next sections, we will only refer to the IEC bus.

21.2.2 *The IEC-625 instrumentation bus*

The IEC bus is restricted to a maximum of fifteen measurement instruments (or in certain circumstances thirty-one). These instruments can be of various kinds, such as multimeters, frequency counters, analysers, plotters, signal generators and so on.

Two instrument functions are distinguished: 'listen' and 'talk'. An instrument that functions as a listener can only receive messages from the bus; a talker, on the contrary, can only transmit messages. Most instruments, however, are designed for

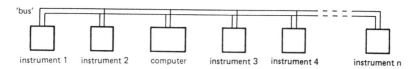

Figure 21.14 An example of a bus system for the connection of several measurement instruments to a single computer system.

both transmitting and receiving messages, for instance for the control of the measurement range and for the transmission of the data. No instrument can act as a listener and a talker at the same time. Several instruments can function simultaneously as listeners, but only one instrument can talk at a time.

To regulate the information stream along the bus, one of the instruments connected acts as a controller or supervisor; this usually is a computer. The controller determines which instrument is the talker and which the receiver. Each instrument has a unique address (set with a series of small switches, usually located on the back plane of the instrument). Only the instrument with the address that corresponds to the address transmitted by the computer receives the message. It is even possible to let two instruments communicate without the intervention of the computer.

The IEC-625 bus consists of sixteen lines; eight lines are reserved for parallel transmission of binary data, the other eight lines are for control purposes and will be described later.

The data transport is bit parallel and word serial; the data are transmitted byte after byte. The data consist of measurement data or commands for the selected instruments. One message may comprise several bytes.

To guarantee a proper data transport, irrespective of the differences in response times of the instruments, the bus contains three special control lines, the handshake lines. These lines have the following functions:

- NRFD (Not Ready For Data): all instruments indicate by this line whether they are ready to accept data. The talker must wait until all listeners are ready, which is indicated by a false NRFD. Logically speaking, this signal is the OR operation on all connected NRFD lines: NRFD = NRFD(1) + NRFD(2) + ...
- DAV (Data Valid): this signal is made true by the instrument assigned as talker, just when all active listeners are ready to receive data (NRFD is false). The receiving instruments know that the data on the data lines contain relevant information.
- NDAC (Not Data ACcepted): the listeners make this signal true as soon as data transmission must take place (which is only possible for DAV true). NDAC remains true as long as the last instrument has received the data from the data lines. The talker must maintain the data on the data lines at least until that moment. A possible next byte can only be put on the data lines when NDAC is false. The NDAC is a logic OR function of the individual NDAC of the instruments: NDAC = NDAC(1) + NDAC(2) + ...

Figure 21.15 shows the timing diagram of the handshake procedure followed for each byte transmitted.

The data transport over the bus is asynchronous; due to the handshake procedure, instruments with diverse processing times can be connected to the bus, so the transport speed is determined by the slowest of the connected instruments.

Some bus lines are connected in a wired OR manner (Figure 21.16). The bus lines (here for the NRFD line) are all connected to the positive power supply voltage via a

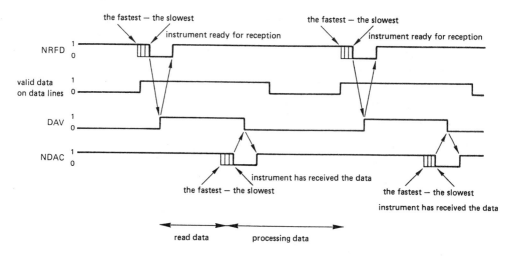

Figure 21.15 The time diagram of the handshake procedure in the IEC bus.

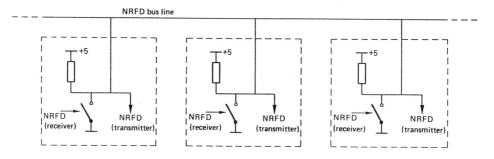

Figure 21.16 Wired OR lines, here for the NRFD line. When the listener is not ready to accept data, the switch is off; the line voltage is low (0) and the talker is waiting.

resistor. Each instrument contains a switch that can connect the lines to ground. The line voltage is only high if none of the instruments has short-circuited the line; the line is low (0) if at least one instrument has put the switch on. In this way, the OR function between the NRFD lines is realized without the necessity of separate OR gates (which would introduce complex wiring).

The remaining five control lines are the management lines. They have the following functions:

- IFC (InterFace Clear): the controller can reset the bus; all connected instruments are set in an initial state and wait for further commands.

- ATN (ATtentioN): the controller can switch from command mode to data mode. In the data mode, the data lines are used to transport measurement data; in the command mode signals for the control of the instruments are transported over the data lines.
- REN (Remote ENable): the controller takes over the control of the instruments; the control elements at the front plate are switched off.
- SRQ (ServiceReQuest): one or more instruments can give a sign to the controller for the execution of a particular program.
- EOI (End Or Identify): in the data mode (ATN = false), the EOI means that the current byte is the last byte of the message. In the command mode the EOI refers to the execution of a service request.

Normally, the user of the bus system has nothing to do with the bus interfaces. Special ICs take care of the communication, including the handshake procedure. Instruments that are provided with an IEC bus connection can be connected mutually with a special cable (via piggy-back connectors). Evidently, the user has to write a program for the controller to execute the desired measurement sequence.

21.2.3 *An example of a computer-based measurement system*

To illustrate the versatility of computers in instrumentation, this section considers an example of a computer measurement system for a particular application.

The example refers to the automatic testing of a specific type of humidity sensor. We will discuss two aspects: automated measurements during the production of the sensors and automatic measuring in the development phase.

We have chosen an Al_2O_3 humidity sensor. This sensor consists of an aluminium substrate of which the top is anodically oxidized, forming a porous layer (Figure 21.17).

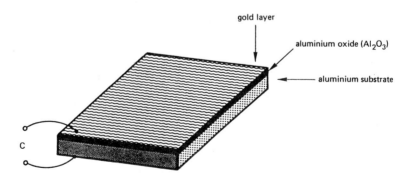

Figure 21.17 A capacitive humidity sensor. The top layer of gold is very thin and permeable to water molecules. The capacitance changes with the amount of absorbed water.

On top of the oxide, a very thin gold layer is deposited, permeable to water vapour molecules. Water vapour from the air (gas) enters the pores in proportion to the humidity content. The sensor behaves as a capacitor whose capacitance changes according to the amount of absorbed water, hence the relative humidity.

The formation of the porous layer is a random process. So, all sensors have different absorption characteristics, even when fabricated under identical environmental conditions. Also, the production parameters can vary (the anodization process is easily affected by the temperature and the composition of the electrolyte). All these differences result in different absorption characteristics of the produced sensors: the capacitance as a function of the humidity. This means that all sensors need to be calibrated individually, if the manufacturer wants to give precise specifications. In that case, the sensors are delivered with a unique calibration chart or a PROM with a calibration table. The calibration can be performed automatically (Figure 21.18).

For reasons of efficiency, a large number of sensors are put together in a measurement cell (climate chamber) with controllable temperature and humidity. A multichannel impedance analyser (Section 21.1.8) measures the capacitance of each sensor, whereas the measurement results are plotted or printed on a piece of paper.

Each measurement sequence starts with the initialization of the measurement instruments; the climate chamber is set at the proper temperature and humidity, the impedance analyser at the desired frequency. Next, the computer-controlled multiplexer in the analyser scans the connected sensors and the corresponding measurement data are stored in the computer memory. Then the computer sets the climate chamber at another temperature or humidity; after a certain time (the climate chamber and the sensors have a rather long response time), the sequence is repeated. In this way, the whole humidity range can be covered. At the end, the measurement results are printed or a complete characteristic is plotted for each individual sensor. The chart is provided with additional data, such as the measurement frequency, the date and time of calibration and the type and series number of the sensor. Furthermore, the computer searches for sensors that fall beyond the specified tolerances.

This automatic measurement system can also be used during the research phase of the sensors. For instance, the optimal measurement frequency can be investigated.

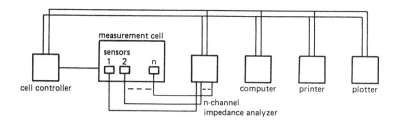

Figure 21.18 An example of a measurement set-up for the automatic calibration of humidity sensors. The calibration curves and other specifications can be plotted automatically on the paper.

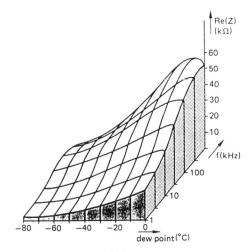

Figure 21.19 An example of a set of characteristics, for the capacitive humidity sensor in Figure 21.17, as can be obtained by a computerized measurement system; the real part of the sensor impedance is plotted as a function of both frequency and humidity.

To that end, a three-dimensional plot can be produced by the system, by measuring at a series of different frequencies (Figure 21.19).

These pictures give quick insight into the behaviour of the sensor as a function of several parameters. It is also possible to plot a number of other parameters, such as the real and imaginary part of the impedance, the modulus and the argument, all versus frequency. As the sensor behaviour can hardly be predicted, it is possible that other output parameters are a more suitable measure for the humidity. Having collected that information, the optimal output parameter and measurement frequency can be determined.

With the same measurement system, much more information can be retrieved, such as the temperature sensitivity of the sensor, its response time and the effect of process parameters on the characteristic properties of the sensor. Another program allows the measurement of other quantities, with almost the same instruments.

Measurement sequences that are repeated many times justify the purchase of these expensive automatic measurement instruments.

SUMMARY

Electronic measurement instruments

- A multimeter is an instrument for the measurement of various electrical quantities, usually voltage, current and resistance.

- With an oscilloscope one or more electrical signals can be visualized as a function of time. Periodic signals can be represented continuously as a stable picture. The horizontal scale (time-base) and the vertical scale (signal amplitude) are adjustable.
- A plotter produces a picture of a signal on paper. The signal frequency is restricted to several tens of hertz.
- Signal generators produce electrical signals as, according to the type of instrument, DC signals, sine waves, triangular waves, ramp and pulse signals and noise. Function generators produce various periodic signals, some of them simultaneously. The output amplitude and the frequency are adjustable over a wide range.
- A frequency counter is suited to the measurement of the frequency of periodic signals, their period time or the number of pulses or zero crossings.
- With a spectrum analyser, the frequency spectrum of an electrical signal is visualized over an adjustable range of frequencies.
- A network analyser allows the investigation of the impedance of two-port networks over a wide frequency range. Depending on the type of instrument, Bode plots (amplitude characteristic and phase characteristic) can be visualized.
- An impedance analyser is suitable for the measurement of various impedance parameters of a (passive) two-pole network, as a function of frequency.
- With a logic analyser, the time diagram of a number of digital (binary) signals is imaged simultaneously, over adjustable time intervals.

Computer-based measurement instruments

- Computers speed up the measurement time and facilitate data presentation. The computer is used for the control of the connected measurement instruments and for the storage and presentation of the measurement data.
- The various instruments in an automatic measuring system are connected via a bus structure. A widely used standardized bus is the IEC-625 bus (or IEEE-488 bus).
- The IEC instrumentation bus is based on asynchronous data transport. To be independent of the processing speed of the connected instruments, the data transfer is controlled by a handshake procedure.

Exercises

Electronic measurement instruments

21.1 Three voltmeters are calibrated for rms values. Meter A is a true rms meter; meter B measures the average of the modulus (double-sided rectified signal) and meter C measures the average of the signal clamped with its negative peaks to zero (see Section 9.2.3). The three meters are calibrated correctly for sinusoidal input signals. The meters are used to measure three different signals (table below), all symmetrical, average zero and peak value of 10 V. What do these instruments indicate?

meter	A	B	C
input voltage			
sine			
square			
triangle			

21.2 Why do some multimeters have four terminals for a resistance measurement instead of only two?

21.3 Describe the principles of the alternate mode and the chopped mode in a multichannel oscilloscope. What is an easy check to see whether an oscilloscope is in the alternate mode?

21.4 Make a picture of the amplitude transfer characteristic V_o/V_i in Figures 21.6a and b (measurement with and without probe). The components have the following values: $R_g = 250$ kΩ; $R_i = 1$ MΩ; $C_i = 26$ pF; $R = 9$ MΩ; $C = 24$ pF; $C_k = 190$ pF. The resistance R_g may be negligible compared to R.

21.5 The following figure shows three possible pictures of the square test signal on the screen of the oscilloscope, during the adjustment of the probe. Which of the pictures corresponds to a correct adjustment, and why?

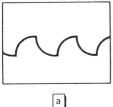
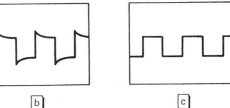

21.6 The oscilloscope can produce Lissajous figures. To that end, a sinusoidal voltage $y(t) = \sin \omega t$ is connected to the Y input of the oscilloscope, and a voltage $x(t) = \sin(\omega t + \varphi)$ to the X input (Figure 21.4). Suppose $\omega = 825$ rad/s. Make a plot of the picture on the oscilloscope for the following values of φ: 0°; 90°; 180°; 270°; 300°.

21.7 At the input of an x–t plotter, a pure linear ramp signal is applied. The plotter writes a curved line. Which of the following effects may cause this non-linearity?
(a) input offset voltage;
(b) non-linearity of the input amplifier;

388 Measurement instruments

(c) non-linearity of the driving motor;
(d) non-linearity of the potentiometer;
(e) irregular paper feed;
(f) drift of the reference voltage;
(g) input offset voltage drift.

21.8 Digital noise generators are based on digital logical circuits. What does the output of such a generator look like?

21.9 The oscillator in the following network analyser (shown in the figure below) produces a sine wave signal $v_1 = A \sin \omega t$. The amplitude of the square wave output signals of the comparators is B; the transfer of the network is $C = |C| e^{j\varphi}$. Make an amplitude–time diagram of all other signals in this system.

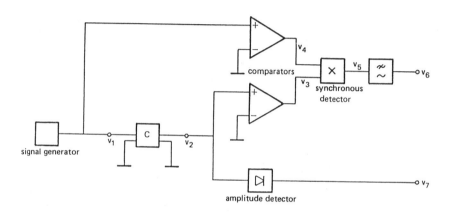

Computer-based measurement instruments

21.10 Given a computer-based measurement system based on the IEC-625 bus, with this system, the Bode plot of the transfer of arbitrary two-port networks must be plotted. The requirements are plotter resolution: 0.4 mm; lengths of the axes: 40 cm for the frequency axis and 20 cm for both amplitude and phase axes; frequency range: 10 Hz–100 kHz; amplitude range: 40 dB; phase range: 2π rad.
(a) Give a plot of the structure of such a measurement system;
(b) Find the relative frequency steps;
(c) Calculate the required resolution (in dB) of the amplitude detector;
(d) Calculate the required resolution (in radians or degrees) of the phase meter.

22 Measurement errors

22.1 Types of measurement errors

It is impossible to measure a physical quantity with infinite accuracy. The presentation of measurement results must contain an indication of the inaccuracy of the measured values. Without such an indication, the measurement data are of little meaning. The most simple way to indicate the inaccuracy is with the number of significant digits. As a rule, only the last (least significant) digit is not accurate. The inaccuracy interval is, in this case, plus or minus half the unit of the last digit, unless otherwise specified.

Example 22.1
The measured value of a voltage is given as 10 V. The true value is between 9.5 and 10.5 V. When the voltage is given as 10.0 V, the actual value lies between 9.95 and 10.05 V. The notation for a more precise tolerance interval (if possible to specify) is, for instance, 10.00 ± 0.02 V. Never use notations like 10 ± 0.02 V or 10.000 ± 0.02 V.

22.1.1 Types of errors

Three types of measurement errors can be distinguished: mistakes, systematic errors and random errors. Mistakes are human errors, such as miscalculations and wrong reading of a display or the settings of an instrument. These errors will not be considered further here. They can be avoided by working conscientiously. A systematic error is an error that remains the same when repeating the measurement under identical conditions. Possible causes of such errors are wrong calibration, offset and impedance mismatch (see Section 5.1.3).

Example 22.2
The short-circuit current of a voltage source is measured using a current meter with input resistance R_M. The measured value must be multiplied with $(R_g + R_M)/R_g$ to find the true value of the short-circuit current; without correction for the load error, the measurement result shows a systematic error.

Systematic errors can be traced or eliminated by repeating the measurement with other instruments or changing the measurement method by (re)calibration, careful inspection and analysis of the measurement system and the measurement object (correct modelling), or by correction factors (as in the preceding example).

Random errors have different values when repeating the measurement. Possible causes are interference from outside, noise of the measurement system or the measurement object. As the value of the error cannot be predicted, it is described in terms of probability (Section 2.2.4). In most cases, the random errors show a normal distribution or Gaussian distribution (Figure 22.1).

The probability of the value being within the interval $[x, x + \Delta x]$ is equal to $\int_x^{x+\Delta x} p(x)\,dx$ (the shaded area in Figure 22.1). The mean or average value has the maximum probability of occurring. This average $x_m = \int_{-\infty}^{\infty} x p(x)\,dx$ is the best estimation of the measurement value. The actual values are scattered around the average. A measure for the dispersion of the measurement values is the variance, defined as $\sigma^2 = \int_{-\infty}^{\infty} (x - x_m)^2 p(x)\,dx$. The variance σ^2 is nothing else than the average of the squared deviation from the mean value x_m. The square root σ is called the standard deviation. At known probability density function, it is possible to calculate the probability of the true measurement value being outside the interval $[x, x + \Delta]$. Table 22.1 shows this probability for various intervals around the mean of the normal distribution function.

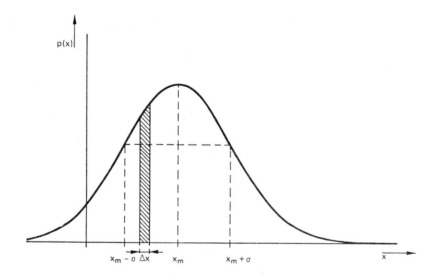

Figure 22.1 The probability density function p(x) of a stochastic parameter with normal or Gaussian distribution. The shaded area is a measure for the probability of the measurement value being within the interval Δx.

Table 22.1 The probability of the true value having a deviation more than $n\sigma$ from the average of a normally distributed quantity

Interval	Probability of being outside the interval
$x_m \pm 0.6745\sigma$	0.5000
$x_m \pm \sigma$	0.3172
$x_m \pm 2\sigma$	0.0454
$x_m \pm 3\sigma$	0.0028

It can be considered as the reliability of the estimation of the measurement value. It is not possible to specify the maximum value of a random error, as the probability of a larger error is never zero. Sometimes the maximum error is specified anyway; usually the 3σ error is meant in that case; the probability of the true value being within the $\pm 3\sigma$ interval is 0.9972 (Table 22.1).

So far we have considered the measurement quantity as a continuous variable. Obviously, there is only a limited number of measurement data and the intervals Δx have finite width. In that case, the probability density function is actually a histogram (Figure 22.2).

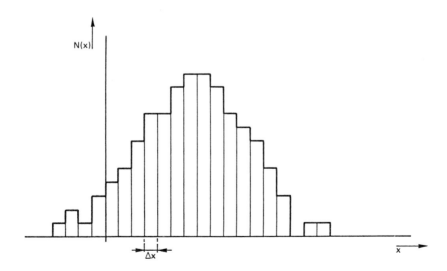

Figure 22.2 A histogram: the number of measurements N(x) whose results lie within the interval Δx, as a function of x.

For a finite number of measurement values the mean value is given as $x_m = (1/n)\Sigma_{i=1}^{n} x_i$, with n the number of measurements and x_i the measured values. The variance is $\mathrm{var}(x) = \sigma^2 = [1/(n-1)]\Sigma_{i=1}^{n}(x_i - x_m)^2$; here also, the standard deviation is the square root of the variance.

22.1.2 Error propagation

Often the value of a physical quantity or parameter is calculated from the measurement of several other quantities. The errors of those measurements will affect the inaccuracy of the calculated value. In this section we will present several rules to find the error in the final measurement result.

In Section 1.2, we introduced two ways of specifying the inaccuracy: the absolute and the relative inaccuracy. The absolute error is the difference Δx between the true value x_0 and the measured value x: $\Delta x = |x - x_0|$. The relative error is defined as $|\Delta x/x| = |(x - x_0)/x|$. Both the absolute and relative errors are positive, unless explicitly specified otherwise. The true value is supposed to be within the interval $[x - \Delta x, x + \Delta x]$. For example, the absolute error of a resistor with resistance value of $80 \pm 2\,\Omega$ is $2\,\Omega$, whereas its relative error is 1/40 or 2.5%. Evidently, the actual deviation might be less.

Assume a quantity z being a function of several measurement parameters: $z = f(a, b, c, \ldots)$, each with an absolute inaccuracy $\Delta a, \Delta b, \ldots$. The maximum error in z is found to be:

$$\Delta z = \left|\frac{\partial f(a, b, c, \ldots)}{\partial a}\right| \Delta a + \left|\frac{\partial f(a, b, c, \ldots)}{\partial b}\right| \Delta b$$

$$+ \left|\frac{\partial f(a, b, c, \ldots)}{\partial c}\right| \Delta c + \ldots$$

The total error is the sum of the positive errors of the individual measurement errors: the maximum absolute error.

The true error is smaller, because some of the errors are positive and others negative, so they might partly cancel. From the expression of the total absolute error, the total relative error is derived:

$$\left|\frac{\Delta z}{z}\right| = \left|\frac{\partial f(a, b, c, \ldots)}{\partial a} \cdot \frac{a}{f(a, b, c, \ldots)} \cdot \frac{\Delta a}{a}\right|$$

$$+ \left|\frac{\partial f(a, b, c, \ldots)}{\partial b} \cdot \frac{b}{f(a, b, c, \ldots)} \cdot \frac{\Delta b}{b}\right|$$

$$+ \left|\frac{\partial f(a, b, c, \ldots)}{\partial c} \cdot \frac{c}{f(a, b, c, \ldots)} \cdot \frac{\Delta c}{c}\right| + \ldots$$

From these two expressions, some practical rules for the calculation of the absolute and relative errors can be derived; they are given below without further proof.

Suppose $z = f(a, b)$, with $z = z_0 \pm \Delta z$, $a = a_0 \pm \Delta a$ and $b = b_0 \pm \Delta b$:

- for addition ($z = a + b$) and subtraction ($z = a - b$), the absolute errors add: $|\Delta z| = |\Delta a| + |\Delta b|$;
- for multiplication ($z = a \cdot b$) and division ($z = a/b$) the relative errors add: $|\Delta z/z| = |\Delta a/a| + |\Delta b/b|$;
- for power functions ($z = a^n$), the relative error is multiplied by the modulus of the exponent n: $|\Delta z/z| = |n| \cdot |\Delta a/a|$.

Example 22.3
The measured current through a resistance of $50 \pm 1\ \Omega$ is 0.40 ± 0.02 A. The nominal voltage across the resistance is $V = IR = 20$ V, with a relative error equal to $1/50 + 0.02/0.40 = 2\% + 5\% = 7\%$. Hence, the absolute error equals ± 1.4 V. The dissipated power is $P = I^2 R = 8$ W, with a relative error $2 \cdot 5\% + 2\% = 12\%$. The calculated power is given as 8 ± 1 W.

It is useless to specify the maximum absolute error when the errors are random. In such cases, we will specify the mean value and the variance. The mean value of $z = f(a, b, \ldots)$ is $z_m = f(a_m, b_m, c_m, \ldots)$ and the variance is:

$$\sigma_z^2 = \left(\frac{\partial f(a, b, c, \ldots)}{\partial a} \right)^2 \sigma_a^2 + \left(\frac{\partial f(a, b, c, \ldots)}{\partial b} \right)^2 \sigma_b^2$$

$$+ \left(\frac{\partial f(a, b, c, \ldots)}{\partial c} \right)^2 \sigma_c^2 + \ldots$$

In the last expression for σ_z^2 the mean values of the variables a, b, etc., must be substituted. With these expressions, the mean and variance of the final measurement result with random errors can be calculated. These rules also apply for quantities with a distribution function other than the normal distribution.

22.2 Measurement interference

One of the major causes of measurement errors is interference or noise, provoked by the interaction of the environment on the measurement system. Often these errors can be avoided by proper measures. Some of the causes of interference and the remedies are discussed in this section.

22.2.1 Causes of interference

Errors due to interference are usually random; they can easily be recognized as such by their fluctuating character. We will briefly review the major causes of interference in electronic measurement systems.

- The capacitance of a cable connecting the measurement system with the measurement object may show fluctuations due to mechanical vibrations (Figure 22.3).
  The input voltage V_i of the measurement system equals $V_i = V_g \cdot R_i / (R_i + R_g + j\omega R_g R_i C_k)$, so when at constant source voltage the cable capacitance C_k changes, the input voltage varies accordingly. The effect can be reduced or even avoided by using high-quality and properly fixed cables.
- The signal transfer of a measurement system may vary with temperature due to the temperature coefficient of the components (Section 1.1); small DC error signals are generated due to the thermoelectric effect (Section 7.2.4), in particular when the input connections, possibly consisting of different materials, are subject to temperature gradients.
- Electrical signals can be induced capacitively into the measurement system via stray capacitances, for instance the capacitance between the system input terminals and the mains (Figure 22.4).

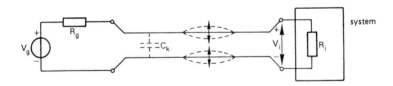

Figure 22.3 Capacitance fluctuations due to mechanical vibrations.

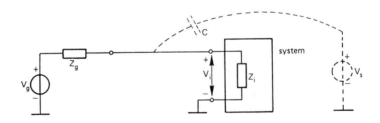

Figure 22.4 Capacitive signal injection via stray capacitances.

The mains voltage (50 Hz) and other signals from, for instance, fuel engines (the ignition), electric motors (the commutator switching) and thyristor circuits, cause capacitively injected interference signals. These systems produce high voltage peaks that can easily enter the measurement system, even via a very small capacitance. Using the model of Figure 22.4, the input voltage due to the error signal alone is found to be $V_{i,s} = V_s(Z_g//Z_i)[Z_g//Z_i + 1/(j\omega C)]^{-1}$. Obviously, in situations where both Z_i and Z_g have high values, the error input signal can be considerably high. The effect is called hum, because of the noise it produces in audio instruments (where the same effect may occur).

- Error signals can also enter the measurement system via magnetic induction (Section 7.1.3). The situation is depicted in Figure 22.5.

Magnetic stray fields are produced by currents flowing through conductors (for instance the mains) and other power instruments (for instance welding transformers). The induced error voltage equals $v_s = -A\,dB/dt$, with A the surface area of the conductor loop and B the magnetic induction (Section 7.1.3). The error increases with increasing frequency and loop area.

- For safety reasons, the metal case of a measurement instrument is connected to ground. This measure is sometimes the cause of error injection. Many systems have the instrument ground (circuit ground) connected to the case as well, so to the safety ground. The interconnection of such instruments may cause ground loops (Figure 22.6a).

A varying magnetic field induces a voltage in the ground loop conductor, resulting in a ground loop current. As ground conductors have non-zero resistance (typically in the order of 0.1 Ω), an error voltage is generated in series with the input of the measurement system. The same happens when the systems are connected with both the signal and the return lines (Figure 22.6b). Such loops are sometimes unavoidable, in particular when using coaxial cables for the signals.

- A common ground connection may also introduce error signals (Figure 22.7).

A significant current may flow through the common ground (or return line), due to other instruments that use the same ground as a return line. The error voltage is $I_g R_{ground}$, which is in series with the input signal source.

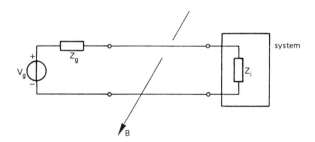

Figure 22.5 Inductive error signal injection due to a varying magnetic field intersecting a loop.

396 Measurement errors

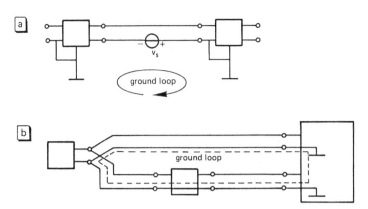

Figure 22.6 Two examples of ground loops: (a) where instruments are interconnected via the safety ground; (b) where instruments are interconnected with signal pairs.

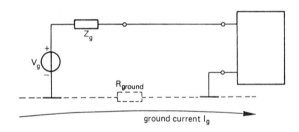

Figure 22.7 Common ground connection may introduce error signals, due to unknown ground currents.

22.2.2 Remedies

In this section, we will give an overview of some measures to reduce or eliminate error signals from interfering with the measurement signal. General measures are:

- removal of the error source;
- isolation of the error source or the measurement system;
- compensation;
- correction;
- separation.

■ Removal of the error source seems trivial, but is very effective. It can only be done when that source can be switched off without a problem. Increasing the distance

between the error source and the measurement system is sometimes very effective.
- Either the measurement system or the error source (or both) can be isolated to prevent interference. Usually, isolation of the measurement system is the most simple solution. We will give some examples of this method.

Example 22.4
The effect of temperature variations on the measurement is eliminated by putting part of the measurement system in a temperature-stabilized compartment. This method is often applied in accurate and stable signal generators and reference sources, where the critical components (the crystal, the Zener diode) are kept at a constant temperature.

Example 22.5
Capacitive error signal injection is drastically reduced by shielding (Figure 22.8).
The shield must be a good conductor and firmly connected to ground; the induced signals flow to ground and cannot reach the input of the measurement system. Likewise, the instrument can be shielded from magnetic error signals as well, using magnetic shielding materials (a metal with a very high permeability and hence a low magnetic resistance).

- Compensation is the most widely used technique in electronics to get rid of unwanted signals and sensitivity. The method is based on two or more simultaneous measurements with, for instance, two sensors arranged so that the sensitivity for unwanted signals is equal, but different for the measurement signal. The error appears as a common-mode signal, one that is eliminated by a differential amplifier with high CMRR (Section 1.2). The compensation method is illustrated with two examples.

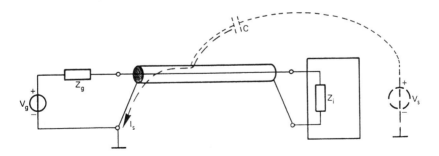

Figure 22.8 A grounded shield reduces capacitive error signal injection from the source into the measurement circuit.

Example 22.6

A Wheatstone bridge (Figure 22.9; see also Example 17.2) contains two strain gauges (Section 7.2.1). Only one of them is sensitive to the strain that has to be measured; however, both are sensitive to temperature changes.

If the temperature coefficients of the strain gauges are identical, full temperature compensation is achieved.

Example 22.7

Interference due to magnetic induction in (ground) loops is reduced drastically by twisting of the conductors (Figure 22.10).

Suppose two adjacent loops have equal surface areas; if further the magnetic induction field is the same, then the two induced voltages are also equal; they have opposite polarity due to the twisting, so the induced voltages cancel.

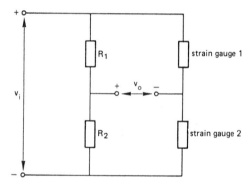

Figure 22.9 A Wheatstone bridge with two strain gauges, for the elimination of temperature sensitivity. One of the gauges is sensitive to strain; both to temperature.

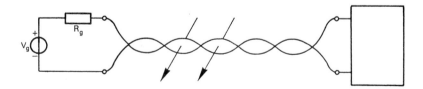

Figure 22.10 Twisting of the conductor pairs results in a reduction of the loop area and hence a reduction of the induced voltage.

- Correction of interfered measurement signals is usually based on the separate measurement of the interfering signal or signals. For instance, the temperature sensitivity of a sensor can be corrected by measuring the temperature separately and correcting the measurement result using this information and the known temperature coefficient of the sensor.
- The separation of measurement signals from the interference signals is performed on the basis of frequency-selective filtering. The method is only applicable when the frequency bands of the measurement signal and the error signal do not overlap. Signals with relatively high frequencies can be cleaned up from drift and other low frequency signals by a simple high-pass filter. When the bands overlap, the signal frequency may be converted to an adequate frequency band using modulation techniques (Chapter 17).

Example 22.8
An optical measurement system is easily interfered with by environmental light (sunlight; 100 Hz lamplight). To eliminate these interference signals, the intensity of the optical source signal is modulated with a relatively high frequency, for instance a few kilohertz. The receiver system converts the optical signals to electrical signals, after which the interference signals are removed with a high-pass filter (Figure 22.11).

Finally, we will discuss two other methods to reduce possible measurement errors. The first is the elimination of the influence of cable impedances in systems that use long interconnection cables between the signal source (a sensor) and the measurement system. In such cases, shielding is necessary to reduce capacitive error signal injection. However, the grounded shield has a relatively large capacitance to ground, which is parallel to the input terminals of the measurement system. The cable impedance has become part of the voltage transfer (Figure 22.12a).

In particular, when the source has a high impedance, the cable impedance has a great effect on the transfer. The method to reduce this effect is active guarding; instead of grounding the shield, it is connected to the output of a buffer amplifier, whose input is the signal voltage. Hence, the voltage across the cable is reduced to zero and no

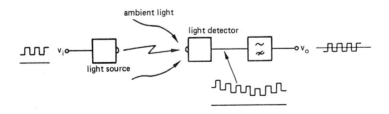

Figure 22.11 Modulation of a high frequency carrier allows the filtering of unwanted low frequency signals; in this way, optical interference signals are eliminated.

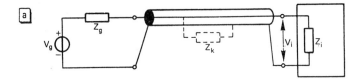

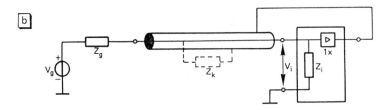

Figure 22.12 (a) The cable impedance Z_k is part of the signal transfer function V_i/V_g; (b) active guarding eliminates the effect of the cable impedance on the signal transfer, because the cable voltage is zero.

current will flow through the cable impedance. The shielding is still effective, because the error signal to the shield flows into the output of the buffer without doing harm to the input signal. Many instruments for the measurement of small currents or with high input impedance have an extra guard connection to realize active guarding.

The second method against interference refers to the elimination of unwanted ground currents (see Figure 22.7). The best remedy is grounding all instruments at a single point, a star configuration (Figure 22.13).

With this way of grounding, no ground currents from other instruments or systems can enter the measurement system.

Unfortunately, in practical situations many error sources can be present at the same time. In particular, the mains network that surrounds the measurement system causes

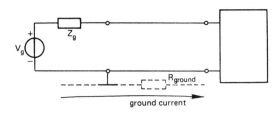

Figure 22.13 The application of a single-point ground connection prevents ground loop currents from entering the measurement system.

a lot of trouble when measuring small signals. Therefore, most instruments are equipped with a metal case and shielded coaxial input cables. Nevertheless, shielding and other measures are seldom perfect. It might even happen that the elimination of one error source introduces a new error, or that the result has become worse because two different error signals were partly compensating for each other. In that case, it is very tempting to renounce the applied measure when this apparent deterioration is observed.

SUMMARY

Types of measurement errors

- The presentation of any measurement result must include an indication of the measurement inaccuracy, for instance through the number of significant digits.
- Three types of errors are mistakes (human errors), systematic errors and random errors. A systematic error is constant for identical experiments: random errors are described in terms of probability.
- The best estimation of the true value of a measured quantity is the mean of all measured data. A measure for the deviation from the true value is the variance.
- The accuracy of a measurement result can be specified in terms of absolute and relative inaccuracy. To find the inaccuracy of a quantity that is the sum (or difference) of various measurement parameters, the absolute errors of those parameters must be added; in the case of a product or the ratio between parameters, the relative errors must be added; in the case of a power function, the relative error of the parameter is raised to the power $|n|$, with n the exponent.

Measurement interference

- Any measurement system is sensitive to interference, undesired signals that enter the system and reduce the accuracy of the measurement.
- The major causes of interference are mechanical vibrations, changes in the ambient temperature, capacitive and inductive error signal injection and errors due to ground currents.
- General remedies against interference are removal, isolation, compensation, correction and separation. Ground errors can be reduced by applying a single ground connection.

Measurement errors

Exercises

Types of measurement errors

22.1 The voltage difference V_{ab} between two points a and b are measured by measuring both voltages at a and b relative to ground: $V_{ab} = V_a - V_b$. The specified relative inaccuracy of the voltmeter is $\pm 1\%$; the measured values are $V_a = 788$ mV and $V_b = 742$ mV.
(a) Calculate the absolute and relative errors in V_{ab}.
(b) Calculate these errors for the case when V_{ab} is measured directly between the two terminals of this voltmeter.

22.2 The measurement of a current is repeated five times. The measured values are 0.62, 6.21, 6.23, 6.18 and 6.24 mA.
(a) Which types of errors are involved: mistakes; systematic errors; random errors?
(b) Find the best estimation of the true value.
(c) Give a method to improve the accuracy of the measurement.
(d) The experiment is repeated with a more precise current meter; the result is 6.25 ± 0.02 mA. What is the systematic error in the previous measurement?

22.3 A sensor voltage v_i from a sensor is amplified using the amplifier depicted in the figure below, with the following specifications: $R_1 = 3.9$ k$\Omega \pm 0.5\%$; $R_2 = 100$ kΩ $\pm 0.5\%$; $|V_{off}| < 1.5$ mV and $|I_{bias}| < 100$ nA. The sensor voltage appears to be $v_i = 0.2$ V $\pm 0.2\%$. Calculate v_o and find the error in this voltage, split into an additive error (in mV) and a multiplicative error (in %). Find the tolerance margins of v_o.

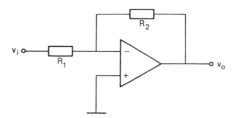

22.4 The resistance value R of a resistor at some distance from the measurement system is measured in three different ways, called the two-wire, three-wire and four-wire method, respectively (see the figure below). In all cases the same measurement system is used, with the following specifications: $I_1 = I_2 = 1$ mA $\pm 0.5\%$; the inaccuracy of the voltmeter is $\pm 0.5\% \pm 0.5$ mV; the resistance r of the wires is between 0 and 3 Ω, the difference between the resistance values of the wires, Δr, is less than 1 Ω. Calculate for each of the three methods the resistance R and the inaccuracy; the measured output voltages v_o are 75 mV, 70 mV and 71 mV, respectively.

Exercises 403

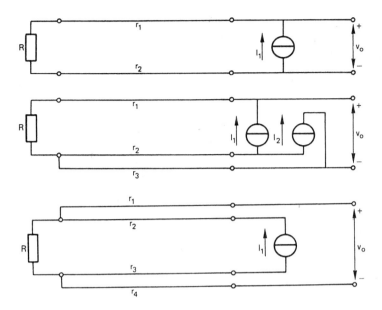

Measurement interference

22.5 A measurement system is subject to capacitive error signal injection from the mains, as is shown in the figure below. Calculate the peak value of the error signal at the input of the measurement system and (if applicable) the signal-to-noise ratio in each of the following cases:
(a) at disconnected signal source (floating input);
(b) at connected signal source;
(c) with a grounded shield which reduces the capacitance C_s to 1% of the original value; at disconnected source;
(d) as (c), with connected source;
(e) what would be the reduction of the interference signal if the shield is not grounded?

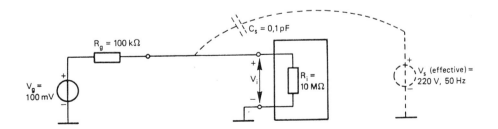

Measurement errors

22.6 A method for the reduction of errors due to capacitive or inductive injection from the mains via the power supply transformer into the measurement system, is the application of a mains filter, as shown in the following figure.
(a) Explain the method.
(b) Explain what happens when the system is connected to the mains without connection to the safety ground terminal.

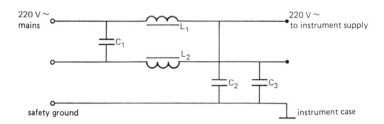

Appendix

A.1 Notation

A.1.1 Symbols

In this book we distinguish between quantities (and parameters) written in capitals and in lower case. In general, quantities and parameters are written in lower-case letters except:

- static quantities;
- complex quantities;
- digital numbers.

Static quantities are DC signals, bias currents (I_{bias}), offset voltages (V_{off}), offset currents (I_{off}) and reference voltages (V_{ref}). Quantities for the biasing of electronic circuits also belong to this category.

The imaginary unit, in mathematical notation i, is written in electrical engineering as j, to avoid confusion with the symbol i for currents, so $j = \sqrt{-1}$.

Digital words are quantities consisting of 1 or more bits, representing a binary number, for example $A = a_3 a_2 a_1 a_0 = 1101$; this can be a binary numerical value ($1101_2 = 13_{10}$), the number of an address or the code for an instruction.

A current is positive in the direction of the arrow; the positive polarity of a voltage is indicated with a + sign. A voltage denoted with two indices indicates the voltage difference between the corresponding points; the first point is positive with respect to the other. For example: V_{AB} is a voltage difference between the points A and B, where A is positive with respect to B (Figure A.1).

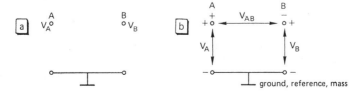

Figure A.1 Notation for voltages. (a) A voltage with one index (V_A, V_B) is the voltage difference as indicated in (b); for both figures $V_{AB} = V_A - V_B = -V_{BA}$.

Consequently, $V_{AB} = -V_{BA}$. A voltage notation with only one index is the voltage difference between the point indicated by the index and a reference point, usually denoted as ground. The indexed point is positive. Figures A.1a and b are different notations for identical voltages and polarities.

It is necessary to mention that these notations refer to the mathematical directions and polarities, and not to the physical values. A current flowing in a particular direction can be denoted as a positive value (the direction of the arrow coincides with the physical direction) as well as a negative value (the arrow is the opposite direction to the physical direction).

A.1.2 *Decimal prefixes*

Table A.1 The prefixes for decimal multiples and submultiples

Factor	Abbreviation	Prefix	Factor	Abbreviation	Prefix
10^{-18}	a	atto	10^1	da	deca
10^{-15}	f	femto	10^2	h	hecto
10^{-12}	p	pico	10^3	k	kilo
10^{-9}	n	nano	10^6	M	mega
10^{-6}	μ	micro	10^9	G	giga
10^{-3}	m	milli	10^{12}	T	tera
10^{-2}	c	centi	10^{15}	P	peta
10^{-1}	d	deci	10^{18}	E	exa

A.1.3 *SI units*

In this book, we use exclusively the notations and units according to the Système International des Unités (SI system). We distinguish base quantities and derived quantities. The base SI quantities are listed in Table A.2.

Table A.2 The base and supplementary SI quantities with corresponding units

Base quantity	SI unit	
length	m	metre
mass	kg	kilogram
time	s	second
electric current	A	ampere
thermodynamic temperature	K	kelvin
luminous intensity	cd	candela
amount of substance	mol	mole
plane angle	rad	radian
solid angle	sr	steradian

The definitions of the base and supplementary units are as follows:

- The metre is the length equal to 1 650 763.73 wavelengths in vacuum of the radiation corresponding to the transition between the levels $2p_{10}$ and $5d_5$ of the krypton 86 atom.
- The kilogram is the unit of mass; it is equal to the mass of the international prototype of the kilogram, at the Bureau International des Poids et Mesures, Sèvres, France.
- The second is the duration of 9 192 631 770 periods of the radiation corresponding to the transition between the two hyperfine levels of the ground state of the caesium 133 atom.
- The ampere is that constant current which, when maintained in two straight parallel conductors of infinite length, of negligible circular cross-section and placed 1 metre apart in a vacuum, would produce between these conductors a force equal to $2 \cdot 10^{-7}$ newtons per metre of length.
- The kelvin unit of thermodynamic temperature is the fraction 1/273.16 of the thermodynamic temperature of the triple point of water.
- The candela is the luminous intensity, in the perpendicular direction, of a surface of 1/600 000 square metre of a black body at the temperature of freezing platinum under a pressure of 101 325 newtons per square metre.
- The mole is the amount of substance of a system which contains as many elementary entities as there are atoms in 0.012 kilogram of carbon 12. When the mole is used, the elementary entities must be specified and may be atoms, molecules, ions, electrons, other particles or specified groups of such particles.
- The radian is the plane angle between two radii which, on the circumference of a circle, cut an arc equal in length to the radius.
- The steradian is the solid angle which has its apex at the centre of a sphere and which describes on the surface of the sphere an area equal to that of a square having as its side the radius of the sphere.

It is not very practical to express all quantities in the base SI units; the unit for electric resistance, for instance, would be expressed as $\text{kg m}^2 \text{s}^{-3} \text{A}^{-2}$. This is the reason for specific SI units for a number of derived quantities (Table A.3).

A.1.4 *Physical constants*

Table A.4 contains the numerical values of the major physical constants utilized in electrical engineering.

A.2 **Examples of manufacturers' specifications**

The two following sections show the complete specifications of two widely used integrated circuits. The first is an example of an analog circuit, an operational amplifier; the second is an example of a digital circuit, a *JK* flipflop. Other integrated

Table A.3 Some derived quantities with the corresponding SI units

Quantity	SI unit		Basic units	
frequency	Hz	hertz		s^{-1}
force	N	newton		$kg\, m\, s^{-2}$
pressure	Pa	pascal	N/m^2	$= kg\, m^{-1}\, s^{-2}$
energy	J	joule	$N\, m$	$= kg\, m^2\, s^{-2}$
power	W	watt	J/s	$= kg\, m^2\, s^{-3}$
charge	C	coulomb		$A\, s$
electrical voltage	V	volt	W/A	$= kg\, m^2\, s^{-3}\, A^{-1}$
electrical resistance	Ω	ohm	V/A	$= kg\, m^2\, s^{-3}\, A^{-2}$
electrical conductance	S	siemens	Ω^{-1}	$= s^3\, A^2\, kg^{-1}\, m^{-2}$
electrical capacitance	F	farad	C/V	$= s^4\, A^2\, kg^{-1}\, m^{-2}$
magnetic inductance	H	henry	Wb/A	$= kg\, m^2\, s^{-2}\, A^{-2}$
magnetic flux	Wb	weber	$V\, s$	$= kg\, m^2\, s^{-2}\, A^{-1}$
magnetic induction	T	tesla	Wb/m^2	$= kg\, s^{-2}\, A^{-1}$
luminous flux	lm	lumen		$cd\, sr$
illuminance	lx	lux		$cd\, sr\, m^{-2}$

Table A.4 Some physical constants

c	speed of light	$(2.997925 \pm 0.000003) \cdot 10^8$	m/s
μ_0	permeability of vacuum	$4\pi \cdot 10^{-7}$	H/m
ε_0	permittivity of vacuum	$(8.85416 \pm 0.00003) \cdot 10^{-12}$	F/m
e, q	electron charge	$(1{,}60207 \pm 0.00007) \cdot 10^{-19}$	C
k	Boltzmann's constant	$(1.3804 \pm 0.0001) \cdot 10^{-23}$	J/K
h	Planck's constant	$(6.6252 \pm 0.0005) \cdot 10^{-34}$	J s
π		3.14159265...	
e		2.71828183...	

circuits are specified in a similar way; they can be found in the data handbooks of the manufacturers.

A.2.1 Specifications of the μA747 (an analog circuit)

LINEAR LSI PRODUCTS

PHILIPS

DUAL OPERATIONAL AMPLIFIER

μA747/747C/SA747C

DESCRIPTION
The 747 is a pair of high performance monolithic operational amplifiers constructed on a single silicon chip. High common mode voltage range and absence of "latch-up" make the 747 ideal for use as a voltage follower. The high gain and wide range of operating voltage provides superior performance in integrator, summing amplifier, and general feedback applications. The 747 is short-circuit protected and requires no external components for frequency compensation. The internal 6dB/octave roll-off insures stability in closed loop applications. For single amplifier performance, see μA741 data sheet.

FEATURES
- No frequency compensation required
- Short-circuit protection
- Offset voltage null capability
- Large common-mode and differential voltage ranges
- Low power consumption
- No latch-up

PIN CONFIGURATIONS

D,F,N PACKAGE

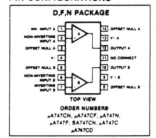

TOP VIEW

ORDER NUMBERS
μA747CN, μA747CF, μA747N,
μA747F, SA747CN, μA747C
μA747CD

ABSOLUTE MAXIMUM RATINGS

PARAMETER	RATING	UNIT
Supply voltage		
μA747	±22	V
μA747C	±18	V
SA747C	±18	V
Internal power dissipation		
H Package	500	mW
N,F Packages	670	mW
Differential input voltage	±30	V
Input voltage	±15	V
Voltage between offset null and V−	±0.5	V
Storage temperature range	−65 to +155	°C
Operating temperature range		
μA747	−55 to +125	°C
μA747C	0 to +70	°C
SA747C	−40 to +85	°C
Lead temperature (soldering, 60 sec)	300	°C
Output short-circuit duration	indefinite	

EQUIVALENT SCHEMATIC

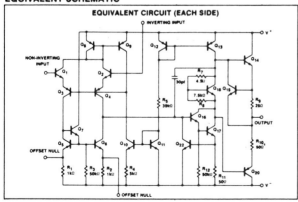

EQUIVALENT CIRCUIT (EACH SIDE)

LINEAR LSI PRODUCTS

DUAL OPERATIONAL AMPLIFIER µA747/747C/SA747C

DC ELECTRICAL CHARACTERISTICS $T_A = 25°C$, $V_S = \pm 15V$ unless otherwise specified.

PARAMETER		TEST CONDITIONS	SA747C Min	SA747C Typ	SA747C Max	UNIT
V_{OS}	Offset voltage	$R_S = 10k\Omega$		2.0	6.0	mV
		$R_S \leq 10k\Omega$, over temperature		3.0	7.5	mV
$\Delta V_{OS}/\Delta T$				10		µV/°C
I_{OS}	Offset current			20	200	nA
		Over temperature			500	nA
$\Delta I_{OS}/\Delta T$				300		pA/°C
I_{BIAS}	Input bias current				500	nA
		Over temperature			1500	nA
$\Delta I_B/\Delta T$				1		nA/°C
V_{OUT}	Output voltage swing	$R_L \geq 2k\Omega$, over temperature	± 10	± 13		V
		$R_L \geq 10k\Omega$, over temperature	± 12	± 14		V
I_{CC}	Supply current			1.7	2.8	mA
		Over temperature		2.0	3.3	mA
	Power consumption			50	85	mW
		Over temperature		60	100	mW
	Input capacitance			1.4		pF
	Offset voltage adjustment range			± 15		V
	Output resistance			75		Ω
	Channel separation			120		dB
PSRR	Supply voltage rejection ratio	$R_S \leq 10k\Omega$, over temperature		30	150	µV/V
A_{VOL}	Large signal voltage gain (DC)	$R_L \geq 2k\Omega$, $V_{OUT} = \pm 10V$	25,000			V/V
CMRR		$R_S \leq 10k\Omega$, $V_{CM} = \pm 12V$ Over temperature	70			dB
I_{SC}			10	25	60	mA

AC ELECTRICAL CHARACTERISTICS $T_A = 25°C$, $V_S = \pm 15V$ unless otherwise specified.

PARAMETER	TEST CONDITIONS	µA747/µA747C/SA747C Min	Typ	Max	UNIT
Transient response	$V_{IN} = 20mV$, $R_1 = 2k\Omega$, $C_1 < 100pf$				
Risetime	Unity gain $CL \leq 100pf$		0.3		µs
Overshoot	Unity gain $CL \leq 100pf$		5.0		%
Slew rate	$RL > 2k\Omega$		0.5		V/µs

LINEAR LSI PRODUCTS

DUAL OPERATIONAL AMPLIFIER   µA747/747C/SA747C

DC ELECTRICAL CHARACTERISTICS $T_A = 25°C$, $V_{CC} = \pm 15V$ unless otherwise specified.[1]

PARAMETER		TEST CONDITIONS	µA747 Min	µA747 Typ	µA747 Max	µA747C Min	µA747C Typ	µA747C Max	UNIT
V_{OS}	Offset voltage	$R_S \le 10k\Omega$		2.0	5.0		2.0	6.0	mV
		$R_S \le 10k\Omega$, over temp.		3.0	6.0		3.0	7.5	mV
$\Delta V_{OS}/\Delta T$				10			10		µV/°C
I_{OS}	Offset current			20	200		20	200	nA
		$T_A = +125°C$		7.0	200				nA
		$T_A = -55°C$		85	500				nA
		Over temperature					7.0	300	nA
$\Delta I_{OS}/\Delta T$				200			200		pA/°C
I_{BIAS}	Input current			80	500		80	500	nA
		$T_A = +125°C$		30	500				nA
		$T_A = -55°C$		300	1500				nA
		Over temperature					30	800	nA
$\Delta I_B/\Delta T$				1			1		nA/°C
V_{OUT}	Output voltage swing	$R_L \ge 2k\Omega$, over temp.	± 10	± 13		± 10	± 13		V
		$R_L \ge 10k\Omega$, over temp.	± 12	± 14		± 12	± 14		V
I_{CC}	Supply current each side			1.7	2.8		1.7	2.8	mA
		$T_A = +125°C$		1.5	2.5				mA
		$T_A = -55°C$		2.0	3.3				mA
		Over temperature					2.0	3.3	mA
	Power consumption			50	85		50	85	mW
		$T_A = +125°C$		45	75				mW
		$T_A = -55°C$		60	100				mW
		Over temperature					60	100	mW
	Input capacitance			1.4			1.4		pF
	Offset voltage adjustment range			± 15			± 15		V
	Output resistance			75			75		Ω
	Channel separation			120			120		dB
PSRR	Supply voltage rejection ratio	$R_S \le 10k\Omega$, over temp.		30	150		30	150	µV/V
A_{VOL}	Large signal voltage gain (DC)	$R_L \ge 2k\Omega$, $V_{OUT} = \pm 10V$	50,000			25,000			V/V
		Over temperature	25,000			15,000			V/V
CMRR		$R_S \le 10k\Omega$, $V_{CM} = \pm 12V$ Over temperature	70			70			dB

LINEAR LSI PRODUCTS

DUAL OPERATIONAL AMPLIFIER   µA747/747C/SA747C

TYPICAL PERFORMANCE CHARACTERISTICS

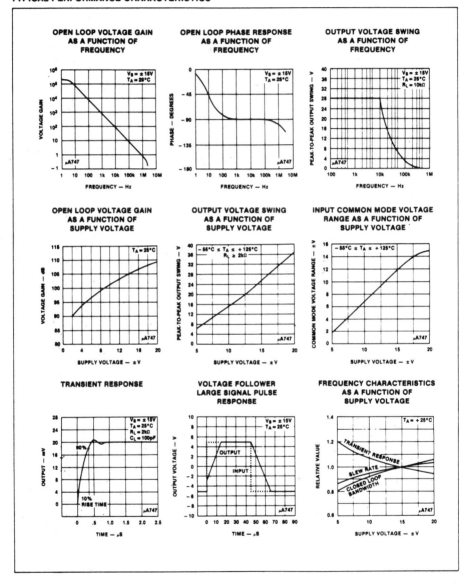

Examples of manufacturers' specifications   **413**

LINEAR LSI PRODUCTS

DUAL OPERATIONAL AMPLIFIER   µA747/747C/SA747C

TYPICAL PERFORMANCE CHARACTERISTICS (Cont'd)

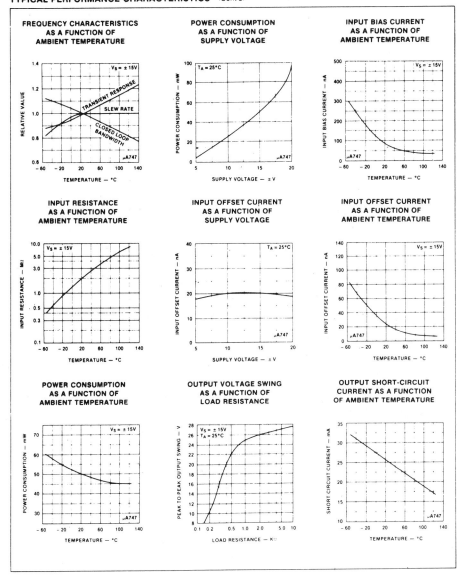

LINEAR LSI PRODUCTS

DUAL OPERATIONAL AMPLIFIER μA747/747C/SA747C

TYPICAL PERFORMANCE CHARACTERISTICS (Cont'd)

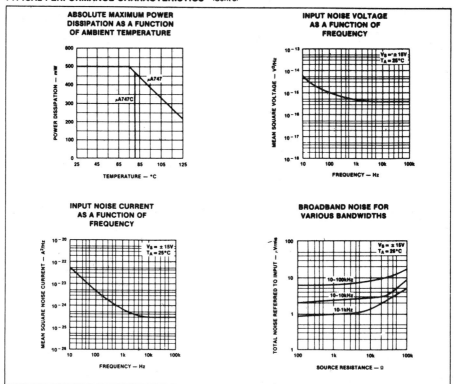

TEST CIRCUITS

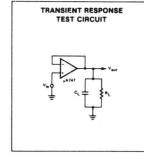

A.2.2 Specifications of the 74HCT73 (a digital circuit)

DUAL JK FLIP-FLOP WITH RESET; NEGATIVE-EDGE TRIGGER

FEATURES

- Output capability: standard
- I_{CC} category: flip-flops

GENERAL DESCRIPTION

The 74HC/HCT73 are high-speed Si-gate CMOS devices and are pin compatible with low power Schottky TTL (LSTTL). They are specified in compliance with JEDEC standard no. 7.

The 74HC/HCT73 are dual negative-edge triggered JK-type flip-flops featuring individual J, K, clock ($n\overline{CP}$) and reset ($n\overline{R}$) inputs; also complementary Q and $\overline{Q}$ outputs.

The J and K inputs must be stable one set-up time prior to the HIGH-to-LOW clock transition for predictable operation.

The reset ($n\overline{R}$) is an asynchronous active LOW input. When LOW, it overrides the clock and data inputs, forcing the Q output LOW and the $\overline{Q}$ output HIGH.

Schmitt-trigger action in the clock input makes the circuit highly tolerant to slower clock rise and fall times.

SYMBOL	PARAMETER	CONDITIONS	TYPICAL HC	TYPICAL HCT	UNIT
$t_{PHL}/$ t_{PLH}	propagation delay $n\overline{CP}$ to nQ $n\overline{CP}$ to $n\overline{Q}$ $n\overline{R}$ to nQ, $n\overline{Q}$	C_L = 15 pF V_{CC} = 5 V	17 15 14	19 18 17	ns ns ns
f_{max}	maximum clock frequency		58	50	MHz
C_I	input capacitance		3.5	3.5	pF
C_{PD}	power dissipation capacitance per flip-flop	notes 1 and 2	30	30	pF

GND = 0 V; T_{amb} = 25 °C; t_r = t_f = 6 ns

Notes

1. C_{PD} is used to determine the dynamic power-dissipation (P_D in μW):

   $P_D = C_{PD} \times V_{CC}^2 \times f_i + \Sigma (C_L \times V_{CC}^2 \times f_o)$ where:

   f_i = input frequency in MHz
   f_o = output frequency in MHz
   $\Sigma (C_L \times V_{CC}^2 \times f_o)$ = sum of outputs
   C_L = output load capacitance in pF
   V_{CC} = supply voltage in V

2. For HC the condition is V_I = GND to V_{CC}
   For HCT the condition is V_I = GND to V_{CC} − 1.5 V

ORDERING INFORMATION/PACKAGE OUTLINES

PC74HC/HCT73P: 14-lead DIL; plastic (SOT-27).
PC74HC/HCT73T: 14-lead mini-pack; plastic (SO-14; SOT-108A).

PIN DESCRIPTION

PIN NO.	SYMBOL	NAME AND FUNCTION
1, 5	$1\overline{CP}$, $2\overline{CP}$	clock input (LOW-to-HIGH, edge-triggered)
2, 6	$1\overline{R}$, $2\overline{R}$	asynchronous reset inputs (active LOW)
4	V_{CC}	positive supply voltage
11	GND	ground (0 V)
12, 9	1Q, 2Q	true flip-flop outputs
13, 8	$1\overline{Q}$, $2\overline{Q}$	complement flip-flop outputs
14, 7, 3, 10	1J, 2J, 1K, 2K	synchronous inputs; flip-flops 1 and 2

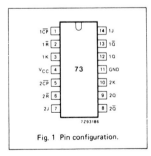

Fig. 1 Pin configuration.

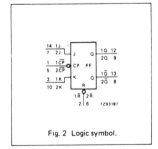

Fig. 2 Logic symbol.

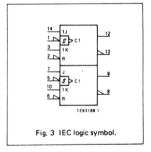

Fig. 3 IEC logic symbol.

Appendix

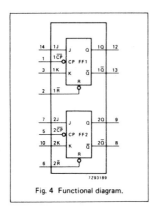

Fig. 4 Functional diagram.

FUNCTION TABLE

OPERATING MODE	INPUTS				OUTPUTS	
	n$\overline{R}$	n$\overline{CP}$	J	K	Q	$\overline{Q}$
asynchronous reset	L	X	X	X	L	H
toggle	H	↓	h	h	$\overline{q}$	q
load "0" (reset)	H	↓	l	h	L	H
load "1" (set)	H	↓	h	l	H	L
hold "no change"	H	↓	l	l	q	$\overline{q}$

H = HIGH voltage level
h = HIGH voltage level one set-up time prior to the LOW-to-HIGH CP transition
L = LOW voltage level
l = LOW voltage level one set-up time prior to the LOW-to-HIGH CP transition
q = lower case letters indicate the state of the referenced output one set-up time prior to the LOW-to-HIGH CP transition
X = don't care
↓ = HIGH-to-LOW CP transition

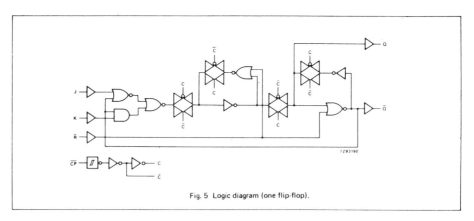

Fig. 5 Logic diagram (one flip-flop).

Examples of manufacturers' specifications

DC CHARACTERISTICS FOR 74HC

For the DC characteristics see chapter "HCMOS family characteristics", section "Family specifications".

Output capability: standard
I_{CC} category: flip-flops

AC CHARACTERISTICS FOR 74HC

GND = 0 V; $t_r = t_f = 6$ ns; $C_L = 50$ pF

SYMBOL	PARAMETER	T_{amb} (°C) 74HC						UNIT	TEST CONDITIONS	
		+25			−40 to +85		−40 to +125			V_{CC} V	WAVEFORMS
		min.	typ.	max.	min.	max.	min.	max.			
$t_{PHL}/$ t_{PLH}	propagation delay $n\overline{CP}$ to nQ		55 20 16	170 34 29		215 43 37		255 51 43	ns	2.0 4.5 6.0	Fig. 6
$t_{PHL}/$ t_{PLH}	propagation delay $n\overline{CP}$ to $n\overline{Q}$		50 18 14	160 32 27		200 40 34		240 48 41	ns	2.0 4.5 6.0	Fig. 6
$t_{PHL}/$ t_{PLH}	propagation delay $n\overline{R}$ to nQ, $n\overline{Q}$		47 17 14	145 29 25		180 36 31		220 44 38	ns	2.0 4.5 6.0	Fig. 7
$t_{THL}/$ t_{TLH}	output transition time		19 7 6	75 15 13		95 19 16		110 22 19	ns	2.0 4.5 6.0	Fig. 6
t_W	clock pulse width HIGH or LOW	80 16 14	25 9 7		100 20 17		120 24 20		ns	2.0 4.5 6.0	Fig. 6
t_W	reset pulse width HIGH or LOW	60 12 10	17 6 5		75 15 13		90 18 15		ns	2.0 4.5 6.0	Fig. 7
t_{rem}	removal time $n\overline{R}$ to $n\overline{CP}$	80 16 14	22 8 6		100 20 17		120 24 20		ns	2.0 4.5 6.0	
t_{su}	set-up time nJ, nK to $n\overline{CP}$	80 16 14	22 8 6		100 20 17		120 24 20		ns	2.0 4.5 6.0	Fig. 6
t_h	hold time nJ, nK to $n\overline{CP}$	4 4 4	−3 −1 −1		4 4 4		4 4 4		ns	2.0 4.5 6.0	Fig. 6
f_{max}	maximum maximum clock pulse frequency	6.0 30 35	18 54 64		4.8 24 28		4.0 20 24		MHz	2.0 4.5 6.0	Fig. 6

DC CHARACTERISTICS FOR 74HCT

For the DC characteristics see chapter "HCMOS family characteristics", section "Family specifications".

Output capability: standard
I_{CC} category: .flip-flops

Note to HCT types

The value of additional quiescent supply current (ΔI_{CC}) for a unit load of 1 is given in the family specifications. To determine ΔI_{CC} per input, multiply this value by the unit load coefficient shown in the table below.

input	unit load coefficient
nJ, nK	0.35
n$\overline{R}$	0.35
n$\overline{CP}$	0.35

AC CHARACTERISTICS FOR 74HCT

GND = 0 V; t_r = t_f = 6 ns; C_L = 50 pF

SYMBOL	PARAMETER	T_{amb} (°C) 74 HCT							UNIT	TEST CONDITIONS	
		+25			−40 to +85		−40 to +125			V_{CC} V	WAVEFORMS
		min.	typ.	max.	min.	max.	min.	max.			
t_{PHL}/t_{PLH}	propagation delay n$\overline{CP}$ to nQ		22	38		48		57	ns	4.5	Fig. 6
t_{PHL}/t_{PLH}	propagation delay n$\overline{CP}$ to n$\overline{Q}$		21	36		45		54	ns	4.5	Fig. 6
t_{PHL}/t_{PLH}	propagation delay n$\overline{R}$ to nQ, n$\overline{Q}$		20	34		43		51	ns	4.5	Fig. 7
t_{THL}/t_{TLH}	output transition time		7	15		19		22	ns	4.5	Fig. 6
t_W	clock pulse width HIGH or LOW	23	12		29		35		ns	4.5	Fig. 6
t_W	reset pulse width HIGH or LOW	18	9		23		27		ns	4.5	Fig. 7
t_{rem}	removal time n$\overline{R}$ to n$\overline{CP}$	12	5		15		18		ns	4.5	
t_{su}	set-up time nJ, nK to n$\overline{CP}$	12	5		15		18		ns	4.5	Fig. 6
t_h	hold time nJ, nK to n$\overline{CP}$	5	0		5		5		ns	4.5	Fig. 6
f_{max}	maximum clock pulse frequency	27	46		22		18		MHz	4.5	Fig. 6

AC WAVEFORMS

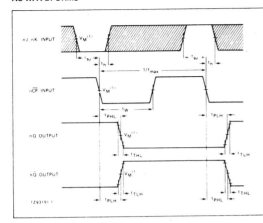

Fig. 6 Waveforms showing the clock (nCP) to output (nQ, nQ̄) propagation delays, the clock pulse width, the J and K to nCP set-up and hold times, the output transition times and the maximum clock pulse frequency.

Note to Fig. 6

The shaded areas indicate when the input is permitted to change for predictable output performance.

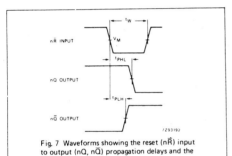

Fig. 7 Waveforms showing the reset (nR̂) input to output (nQ, nQ̄) propagation delays and the reset pulse width.

Note to AC waveforms

(1) HC : $V_M = 50\%$; $V_I =$ GND to V_{CC}.
HCT: $V_M = 1.3$ V; $V_I =$ GND to 3 V.

Answers to exercises

1 Measurement systems

System functions

1.1 See page 3.
1.2 See page 4.
1.3 See page 4.

System specifications

1.4 The factor ½ refers to half the power transfer; as the power is the square of the voltage or current, half the voltage transfer corresponds to $1/\sqrt{2}$ of the power transfer.

1.5 See pages 11, 12.

1.6 $v_c = \frac{1}{2}(v_1 + v_2) = 10.2 \text{ V} \rightarrow v_{oc} = 50 \cdot 10^{-3} \cdot 10.2 = 0.51 \text{ V}$
$v_d = v_1 - v_2 = 0.2 \text{ V} \quad \rightarrow v_{od} = 50 \cdot 0.2 = 10 \text{ V}$
(obviously, the gain of v_c is a factor CMRR smaller than that of v_d; the polarity of v_{oc} is not known).
$v_o = v_{oc} \pm v_{od} = 10 \pm 0.51 \text{ V}$. Hence, $9.49 \le v_o \le 10.51 \text{ V}$.

1.7 $(dV_o/dt)_{max}$ may not exceed the slew-rate.
   (a) $(dV_o/dt)_{max} = \omega \hat{V}_o = 2\pi f \cdot 100 \cdot 0.1 = 20\pi f \le 10 \text{ V}/\mu s = 10^7 \text{ V/s} \rightarrow f \le 10^7/20\pi \cong 160 \text{ kHz}$.
   (b) $(dV_o/dt)_{max} = 2\pi \cdot 10^6 \cdot \hat{V}_o = 2\pi \cdot 10^6 \cdot 100 \cdot \hat{V}_i \le 10^7 \text{ V/s} \rightarrow \hat{V}_i \le 10^7/(2\pi \cdot 10^8) \cong 16 \text{ mV}$.

1.8 $V_{off}(max) = V_{off}(t = 20°) + (\Delta T)_{max} \cdot (t.c.)$:
$V_{off}(max) = 0.5 \text{ mV} + 60 \cdot 5 \text{ } \mu V = 0.8 \text{ mV}$.

1.9 $x_o(\text{linear}) = \alpha x_i \rightarrow \text{non-linearity} = x_o - x_o(\text{linear}) = \beta x_i^2$.
Relative non-linearity $= \beta x_i/\alpha = 0.02 x_i$. The maximum value is $\pm 0.02 \cdot 10 = \pm 20\%$.

2 Signals

Periodic signals

2.1 (a) $x_{pp} = E$; $x_m = |x|_m = \frac{4}{6}E$; $x^2_{rms} = \frac{1}{T}\int_0^T x^2\,dt$

$= \frac{1}{6}\int_1^5 E^2\,dt = \frac{4}{6}E^2 \rightarrow x_{rms} = E\sqrt{\frac{2}{3}}$.

(b) $x_{pp} = \frac{3}{2}E$; $x_m = \frac{1}{6}\left[3\cdot\frac{E}{2} - 2E\right] = -\frac{1}{12}E$;

$|x|_m = \frac{1}{6}\left[3\cdot\frac{E}{2} + 2E\right] = \frac{7}{12}E$;

$x^2_{rms} = \frac{1}{6}\left(\int_1^4 \left(\frac{1}{2}E\right)^2\,dt + \int_4^6 (-E)^2\,dt\right)$

$= \frac{1}{6}\left(3\cdot\frac{1}{4}E^2 + 2E^2\right) = \frac{11}{24}E^2 \rightarrow x_{rms} = E\sqrt{\frac{11}{24}}$.

(c) $x_{pp} = E$; $x_m = |x|_m = \frac{1}{6}\int_0^T x\,dt = \frac{1}{6}\cdot(\text{area } x) = \frac{1}{6}\cdot 2E = \frac{1}{3}E$;

$x = x_1 = -\frac{1}{2}E + \frac{1}{2}Et$ for $1 \leq t \leq 3$, $x = x_2 = \frac{5}{2}E - \frac{1}{2}Et$ for $3 < t \leq 5$;

$x^2_{rms} = \frac{1}{6}\left(\int_1^3 x_1^2\,dt + \int_3^5 x_2^2\,dt\right) \rightarrow x_{rms} = \frac{1}{3}E\sqrt{2}$.

2.2 The crest factor is x_p/x_{rms}; $x_p = A$; $x_{rms} = ((1/T)\int_0^\tau A^2\,dt)^{1/2} = A\sqrt{\tau/T} \rightarrow$ crest factor $= \sqrt{T/\tau}$.

2.3 $\sqrt{T/\tau} \leq 10 \rightarrow \tau \geq T/100$.

2.4 Sine wave signal: $x_p/x_{rms} \approx 1.41$; $x_{rms}/|x|_m \approx 1.11$.
   (a) $V_i = -1.5$ V $\rightarrow V_{ind} = |-1.5|\cdot 1.11 = 1.665$ V.
   (b) The input is a sine wave, so the indication is $V_{ind} = V_{rms}$: $V_{ind} = V_{rms} = \frac{1}{2}\sqrt{2}\hat{V}_i = 1.06$ V.
   (c) $|V_i|_m = 1.5$ V $\rightarrow V_{ind} = 1.11\cdot 1.5 = 1.665$ V.
   (d) $|V_i|_m = 0.5\cdot 1.5$ V $\rightarrow V_{ind} = 1.11\cdot 0.75 = 0.8325$ V.

2.5 $v_s^2 + v_r^2 = v_{sr}^2$, $v_r^2 = (0.75)^2$, $v_{sr}^2 = (6.51)^2$, hence
$v_s = \sqrt{(6.51)^2 - (0.75)^2} = 6.47$.

2.6 (a) a_0 = average value = $\frac{1}{2}E$; $a_n = 0$. This applies to all odd functions.

$$b_n = \frac{2}{T} \int_0^T x(t) \sin n\omega_0 t \, dt, \quad x(t) = Et/T \text{ for } 0 < t < T;$$

$$\int_0^T t \sin n\omega_0 t \, dt = \left[t \cdot \frac{-1}{n\omega_0} \cos n\omega_0 t \right]_0^T$$

$$+ \int_0^T \frac{1}{n\omega_0} \cos n\omega_0 t \, dt = \frac{-T}{n\omega_0} \cos n\omega_0 T + 0 = \frac{-T}{n\omega_0}$$

(because $n\omega_0 T = 2\pi n$).

So: $b_n = \frac{2E}{T^2} \cdot \frac{-T}{n\omega_0} = \frac{-E}{n\pi}$, hence

$$x(t) = \frac{1}{2}E - \frac{E}{\pi}\left(\sin \omega t + \frac{1}{2} \sin 2\omega t + \frac{1}{3} \sin 3\omega t + \cdots \right).$$

(b) T equals half the period of the sine wave: $x(t) = E \sin(\pi t/T)$ for $0 < t < T$.

$\omega_0 = 2\pi/T = 2\omega$.

$$a_0 = \frac{1}{T} \int_0^T x(t) \, dt = \frac{E}{T} \int_0^T \sin \frac{\pi}{T} t \, dt = \frac{E}{T} \left[-\frac{T}{\pi} \cos \frac{\pi}{T} t \right]_0^T = \frac{2E}{\pi}.$$

$$a_n = \frac{2}{T} \int_0^T x(t) \cos n\omega_0 t \, dt = \frac{2E}{T} \int_0^T \sin \frac{\pi t}{T} \cos \frac{2\pi nt}{T} \, dt$$

$$= \frac{2E}{T} \cdot \frac{1}{2} \int_0^T \left(\sin \frac{\pi}{T}(1+2n)t + \sin \frac{\pi}{T}(1-2n)t \right) dt$$

$$= \frac{E}{T} \left[-\frac{\cos(\pi/T)(1+2n)t}{(\pi/T)(1+2n)} - \frac{\cos(\pi/T)(1-2n)t}{(\pi/T)(1-2n)} \right]_0^T$$

$$= -\frac{E}{\pi} \left(\frac{\cos(1+2n)\pi - 1}{1+2n} + \frac{\cos(1-2n)\pi - 1}{1-2n} \right)$$

$$= -\frac{E}{\pi} \left(\frac{-2}{1+2n} + \frac{-2}{1-2n} \right) = \frac{4E}{\pi} \cdot \frac{1}{1-4n^2}, \text{ for } n > 0.$$

$$b_n = \frac{2}{T} \int_0^T x(t) \sin n\omega_0 t \, dt = \frac{2E}{T} \int_0^T \sin \frac{\pi}{T} t \sin \frac{2\pi n}{T} t \, dt$$

from which follows $b_n = 0$ (for all n). This is true for all even functions.

$$x(t) = \frac{2E}{\pi} - \frac{4E}{\pi}\left(\frac{1}{3}\cos 2\omega t + \frac{1}{15}\cos 4\omega t + \frac{1}{35}\cos 6\omega t + \ldots\right), \quad (n\omega_0 = 2n\omega).$$

2.7 The spectral noise power is $P = 4kT = 4 \cdot 1.38 \cdot 10^{-23} \cdot 290 = 1.6 \cdot 10^{-20}$ W/Hz. The bandwidth $B = 10$ kHz, so $P \cdot B = 1.6 \cdot 10^{-16}$ W.
This equals V^2/R, with V the rms value of the noise voltage:
$V = \sqrt{P \cdot B \cdot R} = \sqrt{1.6 \cdot 10^{-16} \cdot 10^4} = 1.26\ \mu V.$

Aperiodic signals

2.8

$$a_0 = \frac{1}{T}\int_0^T x(t)\,dt = \frac{E}{T}\int_0^{T/2} \sin\frac{2\pi}{T}t\,dt = \frac{E}{\pi}.$$

$$a_n = \frac{2}{T}\int_0^{T/2} E\sin\frac{2\pi}{T}t\cos\frac{2\pi n}{T}t\,dt$$

$$= \frac{2E}{T}\cdot\frac{1}{2}\int_0^{T/2}\left(\sin\frac{2\pi}{T}(1+n)t + \sin\frac{2\pi}{T}(1-n)t\right)dt$$

$$= \frac{E}{T}\left[-\frac{\cos(2\pi/T)(1+n)t}{(2\pi/T)(1+n)} - \frac{\cos(2\pi/T)(1-n)t}{(2\pi/T)(1-n)}\right]_0^{T/2}$$

$$= -\frac{E}{2\pi}\left(\frac{\cos\pi(1+n)-1}{1+n} + \frac{\cos\pi(1-n)-1}{1-n}\right).$$

With $\cos\pi(1\pm n) = -\cos n\pi$ this can be written as:

$$a_n = \frac{E}{2\pi}\left(\frac{1}{1+n} + \frac{1}{1-n}\right)(\cos n\pi + 1) = \frac{E}{\pi}\frac{(\cos n\pi + 1)}{1-n^2}, \text{ for } n > 0,\ n \neq 1.$$

$$b_n = \frac{2}{T}\int_0^{T/2} E\sin\frac{2\pi}{T}t\sin\frac{2\pi n}{T}t\,dt$$

$$= -\frac{2}{T}\cdot\frac{E}{2}\int_0^{T/2}\left(\cos\frac{2\pi}{T}(1+n)t - \cos\frac{2\pi}{T}(1-n)t\right)dt$$

$$= -\frac{E}{T}\left[\frac{\sin(2\pi/T)(1+n)t}{(2\pi/T)(1+n)} - \frac{\sin(2\pi/T)(1-n)t}{(2\pi/T)(1-n)}\right]_0^{T/2}.$$

424 Answers to exercises

This is zero for all n, $n \neq 1$.

$$b_1 = \lim_{n \to 1} \frac{E}{T} \left[\frac{\sin(2\pi/T)(1-n)t}{(2\pi/T)(1-n)} - \frac{\sin(2\pi/T)(1+n)t}{(2\pi/T)(1+n)} \right]_0^{T/2}$$

$$= \frac{E}{T} \left[t - \frac{\sin(4\pi/T)t}{(4\pi/T)} \right]_0^{T/2} = \frac{1}{2} E.$$

From a similar calculation it follows that $a_1 = 0$.

$$x(t) = \frac{E}{\pi} + \frac{E}{2} \sin \omega t - \frac{2E}{\pi} \left(\frac{1}{3} \cos 2\omega t + \frac{1}{15} \cos 4\omega t + \frac{1}{35} \cos 6\omega t \ldots \right)$$

2.9 (a) Apply Equation (2.7):

$$\int_{-\infty}^{\infty} |x(t)| \, dt = \int_0^{\infty} e^{-\alpha t} \, dt = \frac{1}{\alpha}, \text{ hence finite.}$$

(b) From Equation (2.6) it follows that:

$$X(\omega) = \int_{-\infty}^{\infty} e^{-\alpha t} e^{-j\omega t} \, dt = \int_0^{\infty} e^{-(\alpha + j\omega)t} \, dt = \frac{1}{\alpha + j\omega}.$$

(c) $|X(\omega)| = \dfrac{1}{\sqrt{\alpha^2 + \omega^2}}$; $\arg X(\omega) = -\tan^{-1} \dfrac{\omega}{\alpha}$.

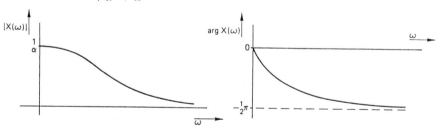

2.10

probability density function   distribution function

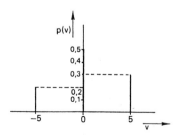 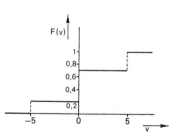

2.11 $$E(y) = \int_{-\infty}^{\infty} yp(y)\,dy = \int_0^{\infty} xp(x)\,dx$$
$$= \int_0^{\infty} x \frac{1}{\sigma\sqrt{2\pi}} e^{-x^2/2\sigma^2}\,dx = \frac{1}{\sigma\sqrt{2\pi}} \cdot \frac{1}{2} \int_0^{\infty} e^{-x^2/2\sigma^2}\,dx^2 = \frac{\sigma}{\sqrt{2\pi}}.$$

3 Networks

Electric networks

3.1 (a) R_2 in series with R_3: 800 Ω; R_1 in parallel with 800 Ω: 5600//800 = 1/(1/5600 + 1/800) = 700 Ω, hence a single resistance of 700 Ω.
(b) Similar to (a): a single self-inductance of 3 mH.
(c) C_2 in series with C_3 is equivalent to a capacitance of 235 μF. C_1 in parallel to 235 μF results in a single capacitance of 236 μF.
(d) A single voltage source of 4.9 V.
(e) A single current source of $1.2 - 125 \cdot 10^{-3} + (-500 \cdot 10^{-6}) = 1.0745$ A.
(f) A single resistance of 200 Ω: first, combine R_1 and R_2, then $(R_1 + R_2)$ parallel to R_3, then $[(R_1 + R_2)//R_3]$ in series with R_4, and so on.

3.2 Due to symmetry, at each node the current splits up into equal parts.

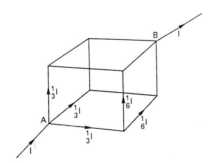

The voltage drop V_{AB} (along an arbitrary trajectory from A to B) is:
$V_{AB} = R_{edge}(I/3 + I/6 + I/3) = (5/6)IR_{edge}$; $R_{cube} = V_{AB}/I = (5/6)R_{edge} = 5/6$ Ω.

3.3 (a) $v_o/v_i = \dfrac{R_2}{R_1 + R_2} = \dfrac{1}{101} \cong 0.01$ (voltage divider).

(b) $i_o/i_i = \dfrac{R_1}{R_1 + R_2} = \dfrac{1}{101} \cong 0.01$ (current divider).

(c) $v_o/v_i = \dfrac{R_4}{R_3 + R_4} \cdot \dfrac{R_2//(R_3 + R_4)}{R_1 + R_2//(R_3 + R_4)} = 0.45 \cdot \dfrac{36.12}{18 + 36.12} = 0.45 \cdot 0.67 = 0.3$.

(Application of the voltage divider rule twice: once to resistances R_3 and R_4 and once to resistances R_1 and $R_2//(R_3+R_4)$.)

3.4 Application of the transformation formula to resistances R_1, R_2 and R_3:

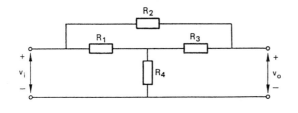
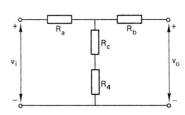

$$R_a = \frac{R_1 R_2}{R_1 + R_2 + R_3} = \frac{2}{3} \text{k}\Omega;$$

$$R_b = R_c = R_a = \frac{2}{3} \text{k}\Omega \text{ (because of symmetry)};$$

$$\frac{v_o}{v_i} = \frac{R_c + R_4}{R_a + R_c + R_4} = 0.8 \text{ (voltage divider)}.$$

3.5 (a) $v_o + RC\dfrac{dv_o}{dt} = v_i.$

(b) $\dfrac{R}{L}v_o + \dfrac{dv_o}{dt} = \dfrac{dv_i}{dt}.$

(c) $v_o + LC\dfrac{d^2 v_o}{dt^2} = LC\dfrac{d^2 v_i}{dt^2}.$

(d) $v_o + RC\dfrac{dv_o}{dt} = RC\dfrac{dv_i}{dt}.$

(e) $v_o + \dfrac{L}{R}\dfrac{dv_o}{dt} = v_i.$

(f) $v_o + LC\dfrac{d^2 v_o}{dt^2} = v_i.$

3.6 $I_c = C\dfrac{dV_c}{dt}$ therefore $\Delta V_c = \dfrac{1}{C}I_c \Delta t + V_c(0) = \dfrac{I_c \Delta t}{C}.$

$$C = \frac{I_c \Delta t}{\Delta V_c} = \frac{10^{-6} \cdot 100}{20} = 5\,\mu\text{F}.$$

Generalized network elements

3.7 Generalized capacitances:
- thermal capacitance

$$q = C_{th} \frac{dT_{ab}}{dt} \quad \text{(see page 45)}$$

- mass

$$F = C_{mech} \frac{dv_{ab}}{dt} \quad \text{(see page 45)}$$

- moment of inertia (ratio between torque and angular acceleration):

$$T_{ab} = J \frac{d\Omega_{ab}}{dt}$$

Generalized self-inductance:
- stiffness

$$v_{ab} = \frac{1}{K} \frac{dF}{dt} \quad \text{(see page 45)}$$

Generalized resistances:
- damping

$$v_{ab} = \frac{1}{b} F \quad \text{(see page 46)}$$

- thermal resistance

$$T_{ab} = R_{th} q \quad \text{(see page 46)}$$

3.8 I-variables: force F; heat flow q and mass flow;
integrated I-variables: electric charge $Q = \int I \, dt$ and heat (energy) $\int q \, dt$;
V-variables: angular velocity Ω and temperature difference T;
integrated V-variable: angular displacement $\varphi = \int \Omega \, dt$.

3.9 The equation of motion of the mechanical system is:

$$F = k \int v \, dt + bv + m \frac{dv}{dt}.$$

The analog equation of the electric system is:

$$i = \frac{1}{L} \int v \, dt + \frac{1}{R} v + C \frac{dv}{dt}.$$

The corresponding electronic network is given below:

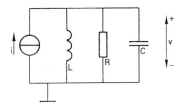

The relation between F and x is:

$$F = kx + b \frac{dx}{dt} + m \frac{d^2 x}{dt^2}.$$

3.10

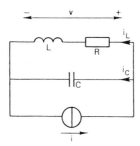

The electric system is described by the following differential equations:

$$v = \frac{1}{C}\int i_c \, dt, \qquad u = i_L R + L\frac{di_L}{dt}, \qquad i_c + i_L = i.$$

Eliminating i_c and i_L results in:

$$LC\frac{d^2v}{dt^2} + RC\frac{dv}{dt} + v = L\frac{di}{dt} + Ri.$$

The self-inductance and the resistance are equivalent to a mechanical spring and damper, connected in series. The capacitance is equivalent to a mass. The current source supplies current to the capacitance which in turn is loaded with (supplies current to) the inductance and the resistance. Analogously, the force is applied to the mass, which is loaded by the spring and the damper.

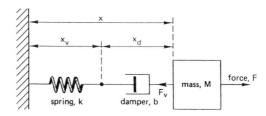

The mechanical system is described by the differential equations:

$$m\frac{d^2x}{dt^2} = F - F_v, \qquad F_v = kx_v = b\frac{dx_d}{dt},$$

$$x_v + x_d = x, \qquad v = \frac{dx}{dt}.$$

Eliminating F_v, x_v and x_d results in:

$$\frac{m}{k}\frac{d^2v}{dt^2} + \frac{m}{b}\frac{dv}{dt} + v = \frac{1}{k}\frac{dF}{dt} + \frac{F}{b}$$

which is equivalent to the equation of the electrical system.

4 Mathematical tools

Complex variables

4.1 Apply the rules for series and parallel impedances:

$Z_s = Z_1 + Z_2 + \ldots$; $Y_p = Y_1 + Y_2 + \ldots$ or $1/Z_p = 1/Z_1 + 1/Z_2 + \ldots$

(a) $R // \dfrac{1}{j\omega C} = \dfrac{R}{1 + j\omega RC}$.

(b) $R_2 // \left(R_1 + \dfrac{1}{j\omega C}\right) = \dfrac{R_2(1 + j\omega R_1 C)}{1 + j\omega(R_1 + R_2)C}$.

(c) $\dfrac{1}{j\omega C_2} // \left(R + \dfrac{1}{j\omega C_1}\right) = \dfrac{1 + j\omega RC_1}{j\omega(C_1 + C_2) - \omega^2 C_1 C_2 R}$.

(d) $R + j\omega L + \dfrac{1}{j\omega C}$.

This is a series resonance circuit, with $Z = 0$ if $R = 0$ and $\omega = 1/\sqrt{LC}$.

(e) $\dfrac{1}{Y}$; $Y = \dfrac{1}{R} + j\omega C + \dfrac{1}{j\omega L} \rightarrow Z = \dfrac{R}{1 + jR(\omega C - 1/\omega L)}$.

This is a parallel resonance circuit, with $Y = 0$ if $1/R = 0$ ($R \rightarrow \infty$) and $\omega = 1/\sqrt{LC}$.

(f) $(R + j\omega L) // \dfrac{1}{j\omega C} = \dfrac{R + j\omega L}{1 + j\omega RC - \omega^2 LC}$.

4.2 Apply for each network the voltage divider rule:

$$\frac{V_o}{V_i} = \frac{Z_2}{Z_1 + Z_2}.$$

(a) $\dfrac{1}{1+j\omega RC}$.

(b) $\dfrac{j\omega L}{R+j\omega L}$.

(c) $\dfrac{j\omega L}{j\omega L + 1/j\omega C} = \dfrac{-\omega^2 LC}{1-\omega^2 LC}$.

(d) $\dfrac{j\omega RC}{1+j\omega RC}$.

(e) $\dfrac{R}{R+j\omega L}$.

(f) $\dfrac{1/j\omega C}{j\omega L + 1/j\omega C} = \dfrac{1}{1-\omega^2 LC}$.

4.3 Simplify the circuit to the next network:

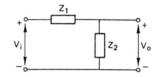

$$Z_1 = \frac{R_1}{1+j\omega R_1 C_1}, \quad Z_2 = \frac{R_2}{1+j\omega R_2 C_2},$$

$$\frac{V_o}{V_i} = \frac{Z_2}{Z_1+Z_2} = \frac{R_2(1+j\omega R_1 C_1)}{R_1(1+j\omega R_2 C_2) + R_2(1+j\omega R_1 C_1)}$$

$$= \frac{R_2}{R_1+R_2} \cdot \frac{1+j\omega R_1 C_1}{1+j\omega R_p(C_1+C_2)}, \quad \text{with } R_p = \frac{R_1 R_2}{R_1+R_2}.$$

The transfer does not depend on the frequency if $R_1 C_1 = R_p(C_1+C_2)$, from which $R_1 C_1 = R_2 C_2$ follows.

4.4 $V_a = \dfrac{RV_i}{R+j\omega L}, \quad V_b = \dfrac{V_i}{1+j\omega RC}$,

$$\frac{V_o}{V_i} = \frac{V_a - V_b}{V_i} = \frac{R}{R+j\omega L} - \frac{1}{1+j\omega RC} = \frac{1}{1+j\omega L/R} - \frac{1}{1+j\omega RC}.$$

$V_o/V_i = 0$, irrespective of V_i, if $L/R = RC$ or $L = R^2 C$.

Laplace variables

4.5 The rules for series and parallel impedances are also applicable in the Laplace domain.

(a) $R // \dfrac{1}{pC} = \dfrac{R}{1 + pRC}$.

(b) $\left(R_1 + \dfrac{1}{pC}\right) // R_2 = \dfrac{R_2(1 + pR_1C)}{1 + p(R_1 + R_2)C}$.

(c) $\left(R_1 + \dfrac{1}{pC_1}\right) // \dfrac{1}{pC_2} = \dfrac{1 + pRC_1}{p(C_1 + C_2) + p^2 C_1 C_2 R^2}$.

(d) $R + pL + \dfrac{1}{pC}$.

(e) $\left(\dfrac{1}{R} + \dfrac{1}{pL} + pC\right)^{-1}$.

(f) $(R + pL) // \dfrac{1}{pC} = \dfrac{R + pL}{1 + pRC + p^2 LC}$.

4.6 Replace $j\omega$ by p and $-\omega^2$ by p^2.

(a) $\dfrac{1}{1 + pRC}$: no zeros, pole $p = -1/RC$.

(b) $\dfrac{pL}{R + pL}$: zero $p = 0$, pole $p = -R/L$.

(c) $\dfrac{p^2 LC}{1 + p^2 LC}$: double zero $p = 0$, two poles $p = \pm j/\sqrt{LC}$.

(d) $\dfrac{pRC}{1 + pRC}$: zero $p = 0$, pole $p = -1/RC$.

(e) $\dfrac{R}{R + pL}$: no zeros, pole $p = -R/L$.

(f) $\dfrac{1}{1 + p^2 LC}$: no zeros, two poles $p = \pm j/\sqrt{LC}$.

4.7 (a) $\dfrac{V_o}{V_i} = \dfrac{pRC}{1 + pRC}$, $V_i(p) = \dfrac{E}{p}$, $V_o(p) = \dfrac{pRC}{1 + pRC} \cdot \dfrac{E}{p} = \dfrac{E}{p + 1/RC}$.

Inverse transformation: $v_o(t) = E e^{-t/RC}$, $t > 0$.

(b) The input signal can be considered as being composed of a positive step voltage E at $t = 0$ and a negative step voltage $-E$ at $t = t_1$.

$$V_i(p) = \frac{E}{p} - e^{-pt_1}\frac{E}{p} \rightarrow V_o(p) = \frac{E}{p + 1/RC} - \frac{e^{-pt_1}E}{p + 1/RC} \rightarrow$$

$v_o(t) = E\,e^{-t/RC}, \; 0 \leq t < t_1,$
$v_o(t) = E\,e^{-t/RC} - E\,e^{-(t-t_1)/RC}, \; t \geq t_1.$

The step at $t = t_1$ equals $-E$ V.

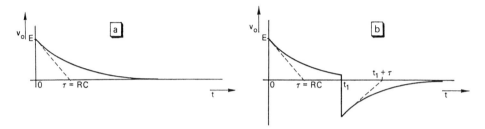

4.8 The differential equation of the system can be found from:

$$v_i = iR + L\frac{di}{dt} + v_o \text{ and } i = C\frac{dv_o}{dt}: \quad v_i = v_o + RC\frac{dv_o}{dt} + LC\frac{d^2v_o}{dt^2}.$$

Using rules (4.5) and (4.8), the Laplace equation is:

$$V_i = V_o + RC(pV_o - v_o(0)) + LC(p^2V_o - pv_o(0) - v'_o(0)).$$

The signal's history (v_i for $t < 0$) is not relevant for the course of v_o for $t \geq 0$, only the boundary conditions are relevant:

$$v_o(0) = 0, \; v'_o(t) = \frac{i(t)}{C} \rightarrow v'_o(0) = \frac{i_o}{C}, \; v_i = 0 \text{ for } t > 0.$$

From this it follows that:

$$V_o(p) = \frac{i_o/C}{p^2 + pR/L + 1/LC}.$$

(a) For $R^2 > 4L/C$ the denominator of the polynomial can be factorized as $(p + a_1)(p + a_2)$, with

$$a_{1,2} = \frac{R}{2L} \pm \sqrt{\frac{R^2}{4L^2} - \frac{1}{LC}}.$$

Hence:

$$V_o = \frac{i_o/C}{(p+a_1)(p+a_2)} = \frac{i_o}{C}\frac{1}{a_1-a_2}\left(\frac{-1}{p+a_1}+\frac{1}{p+a_2}\right).$$

The inverse transformation results in:

$$v_o(t) = \frac{i_o}{C}\frac{1}{a_1-a_2}(-e^{-a_1 t}+e^{a_2 t}),\ t>0.$$

(b) For $R^2 < 4L/C$ this factorization is not possible. V_o can be rewritten as:

$$V_o = \frac{i_o/C}{(p+a)^2+\omega^2},$$

one of the terms in Table 4.1, with $a = R/2L$ and $\omega = 1/LC - R^2/4L^2$. Inverse transformation gives:

$$v_o(t) = \frac{i_o}{\omega C}e^{-at}\sin \omega t,\ t>0.$$

(c) For $R^2 = 4\dfrac{L}{C}$, $\omega^2 = 0$, hence $V_o = \dfrac{i_o/C}{(p+a)^2}$.

Using the transformation $1/p^2 \leftrightarrow t$ and Equation (4.3), this results in:

$$v_o(t) = \frac{i_o}{C}te^{-at},\ t>0.$$

Conclusion:
(a) $R = 400\ \Omega$: $v_o(t)$ is an exponentially decaying voltage.
(b) $R = 120\ \Omega$: $v_o(t)$ is a sinusoidal voltage with exponentially decaying amplitude.
(c) $R = 200\ \Omega$: $v_o(t)$ first increases with time, and then decreases exponentially down to zero.

5 Models

System models

5.1 The source resistance is found by disregarding all sources; the source resistance is the equivalent resistance of the resulting two-pole network.

(a) $V_o = \dfrac{R_2}{R_1+R_2}V = 6\ \text{V};\ R_s = R_3 + R_1//R_2 = 10.025\ \text{k}\Omega$.

(b) $V_o = IR_1 = 90\,\text{mV}$; $R_s = R_1 + R_2 + R_3 = 90\,\Omega$.
(c) Apply the principle of superposition with respect to the two current sources, and the rule for the current divider. The result is:

$$V_o = I_1\left(\frac{R_1 R_2}{R_1 + R_2 + R_3}\right) + I_2(R_2 /\!/ (R_1 + R_3)) \approx 34\,\text{V} + 50.4\,\text{V} = 84.4\,\text{V};$$

$$R_s = R_2 /\!/ (R_1 + R_3) \approx 25.2\,\text{k}\Omega.$$

5.2 $\text{CMRR(DC)} = -90\,\text{dB}$, $20\log\dfrac{V_o}{V_i} = -90$, $\dfrac{V_o}{V_i} = 10^{-4.5} \cong 32\,\mu\text{V/V}$;

$\text{CMRR(1 kHz)} = -80\,\text{dB}$, $20\log\dfrac{V_o}{V_i} = -80$, $\dfrac{V_o}{V_i} = 10^{-4} = 100\,\mu\text{V/V}$;

$\text{SVRR} = -110\,\text{dB}$, $20\log\dfrac{V_o}{V_i} = -110$, $\dfrac{V_o}{V_i} = 10^{-5.5} \cong 3.2\,\mu\text{V/V}$.

5.3 $I_m = \dfrac{R_s}{R_L + R_s} I_s \approx \left(1 - \dfrac{R_L}{R_s}\right) I_s \rightarrow$ relative error

$\approx \dfrac{-R_L}{R_s} < 0.5\% \rightarrow R_L \leqslant 500\,\Omega$.

5.4 (a) $V_s \approx 9.6\,\text{V}$; $\dfrac{R_L}{R_s + R_L} \cdot 9.6 = 8.0\,\text{V} \rightarrow R_s = 2\,\text{k}\Omega$.

(b) $V_o = \dfrac{R_i}{R_s + R_i} \cdot V_s \approx \left(1 - \dfrac{R_s}{R_i}\right) V_s$

$\rightarrow$ the maximum relative error is $-\dfrac{R_s}{R_i} = -\dfrac{2 \cdot 10^3}{10^7} = -0.02\%$.

5.5 See page 70.

5.6 $A_v = \dfrac{v_o}{v_i} = \dfrac{AR_L}{R_o + R_L}$; $i_o = \dfrac{Av_i}{R_o + R_L} = \dfrac{AR_i i_i}{R_o + R_L} \rightarrow A_i = \dfrac{i_o}{i_i}$

$= \dfrac{AR_i}{R_o + R_L}$; $A_p = \dfrac{P_o}{P_i} = \dfrac{v_o i_o}{v_i i_i} = A_v A_i = \dfrac{A^2 R_i R_L}{(R_o + R_L)^2}$.

Signal models

5.7 The next two systems must be equivalent:

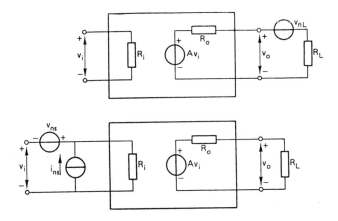

As a criterion for equivalence, take v_o:
At short-circuited inputs:

$$v_o = v_{nL} \frac{R_o}{R_o + R_L} = Av_{ns} \frac{R_L}{R_o + R_L} \rightarrow v_{ns} = \frac{R_o v_{nL}}{R_L A}.$$

At open inputs:

$$v_o = v_{nL} \frac{R_o}{R_o + R_L} = Ai_{ns} R_i \frac{R_L}{R_o + R_L} \rightarrow i_{ns} = \frac{R_o v_{nL}}{R_L R_i A}.$$

5.8 In the preceding exercise, the equivalent noise sources depend on R_L, as part of the output voltage divider. In this exercise, all noise sources originate from the system itself, so the result does not depend on R_L. Therefore, we can put $R_L \rightarrow \infty$ for simplicity.

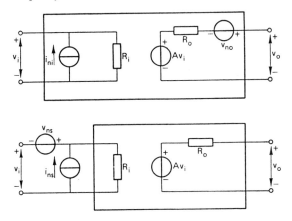

Short-circuited input terminals:

$$v_o = v_{no} = Av_{ns} \to v_{ns} = \frac{v_{no}}{A}.$$

Open input terminals:

$$v_o^2 = (Ai_{ni}R_i)^2 + v_{no}^2 = (Ai_{ns}R_i)^2 \to i_{ns}^2 = i_{ni}^2 + \left(\frac{v_{no}}{AR_i}\right)^2.$$

5.9 The output offset voltage is $A(V_n + I_n R_g)$. At 20°C, $V_n = 1$ mV and $I_n = 10$ nA, hence:
$V_{\text{off, o}} = 10(10^{-3} + 10^{-8} \cdot 10^4) = 11$ mV.
The maximum offset occurs at $T = 50°C$; at that temperature, the offset is:
$V_n = 1$ mV $+ 30 \cdot 10 \, \mu V = 1.3$ mV and $I_n = 10$ nA $\cdot 2^3 = 80$ nA, thus
$V_{\text{off, o}} = 10(1.3 \cdot 10^{-3} + 80 \cdot 10^{-9} \cdot 10^4) = 21$ mV.

5.10 Make a model of the system:

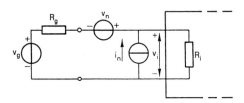

The signal power P_g towards the system is v_i^2/R_i, hence

$$P_g = \left(\frac{R_i}{R_i + R_g} v_g\right)^2 \cdot \frac{1}{R_i}.$$

Both v_n and i_n contribute to the noise power P_n:

$$P_n = \left(\frac{R_i}{R_i + R_g} v_n\right)^2 \cdot \frac{1}{R_i} + \left(\frac{R_g R_i}{R_g + R_i} i_n\right)^2 \cdot \frac{1}{R_i}.$$

$$S/R = \frac{P_g}{P_n} = \frac{v_g^2}{v_n^2 + R_g^2 i_n^2}$$

which is independent of R_i.

6 Frequency diagrams

Bode plots

6.1 (a) $H = \dfrac{j\omega R_2 C}{1+j\omega(R_1+R_2)C}$; $R_2C = 0.5$ ms; $(R_1+R_2)C = 1$ ms.

$\omega \to 0$: $\quad H \to j\omega R_2 C \quad\quad |H| \to \omega R_2 C \quad\quad H$ is positive imaginary, so $\arg H = \pi/2$.

$\omega \to \infty$: $\quad H \to \dfrac{R_2}{R_1+R_2} = \dfrac{1}{2} \quad |H| \to \dfrac{1}{2} \equiv -6$ dB $\quad \arg H \to 0$.

Break point: $\omega(R_1+R_2)C = 1$, or $\omega_k = 10^3$ rad/s.

$\omega = \omega_k$: $\quad H = \dfrac{jR_2/(R_1+R_2)}{1+j} \quad |H| = \dfrac{1}{2\sqrt{2}} \quad \arg H = \dfrac{\pi}{2} - \dfrac{\pi}{4} = \dfrac{\pi}{4}$.

(b) $H = \dfrac{R_2}{R_1+R_2} \cdot \dfrac{1+j\omega\tau_1}{1+j\omega\tau_2}$; $\tau_1 = R_1 C_1 = 2$ ms;

$\tau_2 = \dfrac{R_1 R_2}{R_1+R_2} \cdot (C_1 + C_2) \cong 0.2$ ms; $\quad \dfrac{R_2}{R_1+R_2} \approx \dfrac{1}{20}$; (see Exercise 4.3).

The Bode plot can be composed of the diagrams of $R_2/(R_1+R_2)$ (frequency independent), $1+j\omega\tau_1$ and $1/(1+j\omega\tau_2)$.

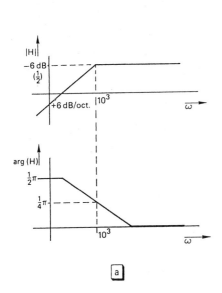

a

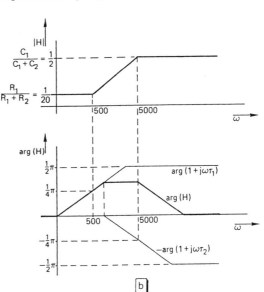

b

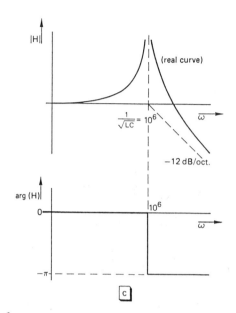

(c) $H = \dfrac{1}{1 - \omega^2 LC}$; see also Exercise 3.5(f).

$\omega \to 0$: $\quad H \to 1 \qquad |H| \to 1 \qquad \arg H \to 0.$

$\omega \to \infty$: $\quad H \to \dfrac{-1}{\omega^2 LC} \qquad |H| \to \dfrac{1}{\omega^2 LC} \qquad \arg H \to -\pi.$

$\omega \to \sqrt{\dfrac{1}{LC}}$: $H \to \infty$, phase not determined.

6.2 $\quad H = \dfrac{1}{1 + j\omega RC - \omega^2 LC} = \dfrac{1}{1 + j2z\omega/\omega_0 - \omega^2/\omega_0^2} \to$

$\omega_0 = \dfrac{1}{\sqrt{LC}} = 10^7$ rad/s.

$\dfrac{2z}{\omega_0} = RC \to z = 0.5.$

$\omega_m = \omega_0 \sqrt{1 - 2z^2} = \tfrac{1}{2}\sqrt{2} \cdot 10^7$ rad/s (see page 85).

$|H(\omega_m)| = \dfrac{1}{2z\sqrt{1-z^2}} = \tfrac{2}{3}\sqrt{3}$ (see page 85).

$|H(\omega_0)| = \dfrac{1}{2z} = 1.$

6.3 $|H| = 1$; $\arg H = -2\tan^{-1}\omega\tau$; $\tau = RC$.
$\omega = 0$: $\arg H = 0$; $\omega = 1/\tau$: $\arg H = -\pi/2$; $\omega \to \infty$: $\arg H = -\pi$ (see the figure below).

6.4 Summing three Bode plots of the function $1/(1+j\omega\tau)$.

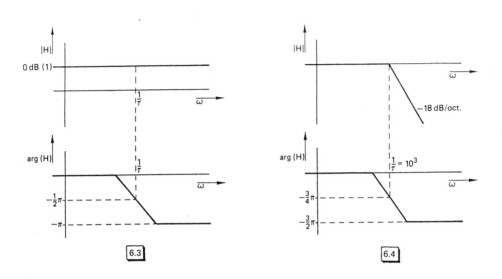

6.5 Summing the Bode plots of the functions $H_1 = 1+j\omega\tau_1$, $H_2 = 1/(1+j\omega\tau_2)$ and $H_3 = 1/(1+j\omega\tau_3)$.

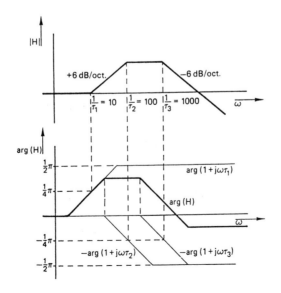

Polar plots

6.6 (a) The polar plot is a circle, with starting point 0 and end point $R_2/(R_1 + R_2)$ (both on the real axis).
The tangent in the starting point makes an angle $\frac{1}{2} \cdot (1 - 0) \cdot \pi$, so the semicircle lies in the upper part of the complex plane.

(b) The polar plot is a circle. The starting point is $R_2/(R_1 + R_2) \approx \frac{1}{20}$ and the end point is $R_2/(R_1 + R_2) \cdot (\tau_1/\tau_2) \approx \frac{1}{2} = C_1/(C_1 + C_2)$.
As the polar plot turns clockwise, it consists of a semicircle in the upper halfplane. If $R_2/(R_1 + R_2) > C_1/(C_1 + C_2)$, the semicircle lies in the lower halfplane.

(c) The function is real for all ω, so the polar plot lies on the real axis. The starting point is 1, the end point is 0, and the tangent in the end point makes an angle $\pi + (0 - 2) \cdot (\pi/2) = 0$.

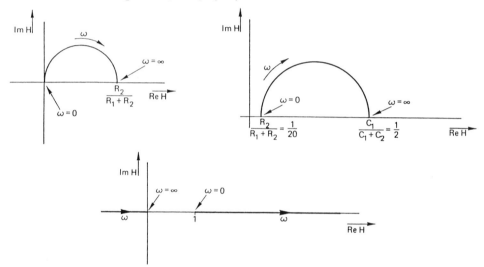

6.7 The network is called an all-pass network because the transfer is the same for all frequencies, and equal to 1; the polar plot lies on the unity circle. The starting point is 1, the end point is -1 and the polar plot turns in a clockwise direction.

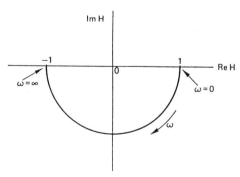

6.8 The polar plot is a semicircle, with starting point 1 and end point τ_1/τ_2 (2, 1 and ½ respectively).

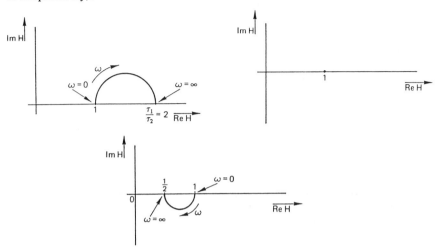

7 Passive electronic components

Passive circuit components

7.1 $i = C\dfrac{dv}{dt}$; $q = Cv$.

7.2 δ is a measure for the dielectric losses in a capacitor; $\tan\delta = I_R/I_C = 1/\omega RC$ (see page 96).

7.3 $v = L\dfrac{di}{dt}$; $\Phi = Li$.

7.4 $B = \mu_0\mu_r H$; $\Phi = \iint \bar{B}\,d\bar{A}$; $[B] = T = V\,s/m^2$; $[H] = A/m$; $[\Phi] = Wb = V\,s$.

7.5

$R_2 = \dfrac{V_2}{I_2}$; $R_1 = \dfrac{V_1}{I_1}$.

$\dfrac{V_2}{V_1} = n$, $\dfrac{I_2}{I_1} = \dfrac{1}{n} \to R_2 = n^2\dfrac{V_1}{I_1} = n^2 R_1$.

Sensor components

7.6 The absolute error is $R_0\beta T^2$; its maximum value (at $T = 100°C$) is $-0.58\,\Omega$.
For $T = 100°C$, in the case of a linear characteristic,
$R = 100(1 + 3.9 \cdot 10^{-3} \cdot 10^2) = 139\,\Omega$.
The non-linearity error is $-0.58/139 \cdot 100\% = -0.42\%$, or $-0.58/0.39 = -1.49°C$ (see Exercise 1.8).

7.7 For a Pt-100, $0.39\,\Omega$ is equivalent to 1 K, so $0.039\,\Omega$ is equivalent to 0.1 K. The resistance of each wire is r, so per sensor the wire resistance is $2r$. $2r < 0.039\,\Omega$, or $r < 20\,m\Omega$.

7.8 The sensitivity of the thermocouple is $40\,\mu V/K$ (see Table 7.3): $0.5\,K \leftrightarrow 20\,\mu V$ and this is just the maximum allowable uncertainty in the offset voltage.

7.9 The common-mode voltage is $\frac{1}{2}v_i$, the differential-mode voltage is $(\Delta R/R)v_i$, hence the smallest voltage that has to be measured is $10^{-6}v_i$. The maximum allowable output voltage due to a common-mode voltage is equal to the output voltage produced by an input differential voltage of $10^{-6}v_i$. So: CMRR $> \frac{1}{2}v_i/10^{-6}v_i = 5 \cdot 10^5$.

7.10

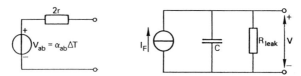

I_F is proportional to the derivative of the force on the sensor. The produced charge is $Q = S_q F$, with S_q the force sensitivity; $I_F = dQ/dt$.

7.11 The slider of the potentiometer divides the resistance R in two parts: αR and $(1-\alpha)R$. The voltage transfer is:

$$\frac{v_o}{v_i} = \frac{R_L // \alpha R}{R_L // \alpha R + (1-\alpha)R} = \frac{\alpha}{1 + (1-\alpha)\alpha R/R_L} \approx \alpha\left(1 - (1-\alpha)\alpha \frac{R}{R_L}\right).$$

The term $(1-\alpha)\alpha R/R_L$ causes the non-linearity (with respect to the transfer $v_o/v_i = \alpha$). This term has a maximum for $\alpha = \frac{1}{2}$; hence the non-linearity is approximately $R/4R_1 = 800/(4 \cdot 10^5) = 2 \cdot 10^{-3}$.

8 Passive filters

First- and second-order RC filters

8.1 $|H| = 1/\sqrt{1 + \omega^2 \tau^2}$, $\arg H = -\tan^{-1}\omega\tau$, $\tau = 10^{-3}$ s.
(a) $\omega^2 \tau^2 = 0.01$ → $|H| \approx 1$, $\arg H = -0.10$ rad;
(b) $\omega^2 \tau^2 = 1$ → $|H| = \frac{1}{2}\sqrt{2}$ (-3 dB frequency or break point), $\arg H = -0.79$ rad;
(c) $\omega^2 \tau^2 = 100$ → $|H| \approx \frac{1}{10}$, $\arg H = -1.47$ rad.

8.2 $|H| = \omega\tau/\sqrt{1+\omega^2\tau^2}$, with $\tau = 0.1$ s.
- (a) For $\omega^2\tau^2 \gg 1$, $|H| \approx 1$: $\omega \gg 1/\tau$, or $\omega \gg 10$ rad/s;
- (b) $|H| \approx 0.1$ for $\omega\tau = 0.1$, or $\omega = 1$ rad/s;
- (c) $|H| \approx 0.01$ for $\omega\tau = 0.01$, or $\omega = 0.1$ rad/s.

8.3 The modulus of the transfer is $|H| = 1/\sqrt{1+\omega^2\tau^2}$.
For the measurement signal, this must be at least 0.97 (namely 3% less than 1). Hence: $1+\omega^2\tau^2 \leq (1/0.97)^2$, or $\omega\tau \leq 0.25$.
The largest attenuation occurs for the highest signal frequency (1 Hz = 2π rad/s), so: $\tau \leq 0.25/2\pi$ s.
The interference signal should be attenuated by at least a factor of 100, so $\omega\tau \geq 100$ (ω is the frequency of the interference signal):

$$\tau \geq \frac{100}{2\pi \cdot 2\text{ kHz}} = \frac{0.05}{2\pi} \text{ s.}$$

The limits for the time constants are:

$$\frac{0.05}{2\pi} \leq \tau \leq \frac{0.25}{2\pi} \text{ s.}$$

8.4 The third-harmonic distortion of a triangular signal is $1/9$. The slope of a low-pass filter of order n is 2^n per octave (a factor of 2 in frequency), or 3^n for a factor of 3 in frequency.
- (a) The attenuation of the third-harmonic relative to the fundamental is 3 (for a first-order filter), so d_3 (the third-harmonic distortion) after filtering equals $1/9 \cdot 1/3 = 1/27 \approx 3.7\%$;
- (b) The attenuation is $(1/3)^2$: $d_3 = 1/9 \cdot (1/3)^2 = 1/81 \approx 1.23\%$;
- (c) The attenuation is $(1/3)^3$: $d_3 = 1/9 \cdot (1/3)^3 = 1/243 \approx 0.41\%$.

8.5 The frequency ratio between the measurement signal and the interference signal is 5. A low-pass filter of order n attenuates the signal with the highest frequency a factor 5^n more than the signal with the lower frequency (see also Exercise 8.4). Requirement: $5^n \geq 100$, so $n \geq 3$.

Filters of higher order

8.6 The order can be found by counting the number of reactive elements (self-inductances, capacitances) of the circuit in its most simple form. The filter type is found by the determination of the transfer for $\omega \to 0$ and $\omega \to \infty$.
- (a) Fourth-order low-pass filter;
- (b) third-order band-pass filter;
- (c) third-order high-pass filter.

8.7 $|H| = \dfrac{1}{\sqrt{1+(1/2)^{2n}}}$.

444 Answers to exercises

(a) $|H| = \dfrac{1}{\sqrt{1 + 1/16}} \approx 0.970 \approx -0.263$ dB;

(b) $|H| = \dfrac{1}{\sqrt{1 + 1/64}} \approx 0.992 \approx -0.070$ dB.

8.8 The transfer function of a network with the structure as shown below is:

$$H = \frac{V_o}{V_i} = \frac{Z_2 Z_4}{Z_1 Z_2 + (Z_1 + Z_2)(Z_3 + Z_4)}.$$

Application of this formula to the given network results in:

$$H = \left[\frac{R_g}{R_L} + 1 + j\omega \left(R_g C_1 + R_g C_2 + \frac{L}{R_L} \right) \right.$$

$$\left. + (j\omega)^2 \left(LC_2 + LC_1 \frac{R_g}{R_L} \right) + (j\omega)^3 R_g C_1 C_2 L \right]^{-1}.$$

Substitution of the numerical values gives:

$$H = \frac{1}{2 + 4j\omega + 4(j\omega)^2 + 2(j\omega)^3}, \text{ so}$$

$$|H| = \frac{1}{[(2 - 4\omega^2)^2 + \omega^2(4 - 2\omega^2)^2]^{1/2}} = \frac{1}{2\sqrt{1 + \omega^6}}.$$

As $|H|$ can be written as $(1 + \omega^{2n})^{-1/2}$, with $n = 3$, this is a Butterworth filter of order 3.

8.9 $H = \dfrac{1}{1 + j\omega L/R - \omega^2 LC}$; $|H|^2 = \dfrac{1}{1 - 2\omega^2 LC + \omega^4 L^2 C^2 + \omega^2 L^2/R^2}.$

This is a Butterworth filter if the term with ω^2 is zero, hence $L = 2R^2 C$. In that case,

$$|H|^2 = \frac{1}{1 + \omega^4 L^2 C^2} = \frac{1}{1 + (\omega/\omega_c)^{2n}}, \text{ where } \omega_c = 1/\sqrt{LC} \text{ and } n = 2;$$

$$C = \frac{1}{\omega_c^2 L} = 10^{-7} \text{ F}; \quad R = \sqrt{\frac{L}{2C}} = 50\sqrt{2}\ \Omega.$$

9 PN diodes

Properties of pn diodes

9.1 $I = I_o(e^{qV/kT} - 1) \approx I_o e^{qV/kT}$

9.2 $r_d = kT/qI$, g (conductance) $= 1/r_d$. Using the rule of thumb: $r_d = 25\,\Omega$ at $I = 1\,\text{mA}$ and $T = 300\,\text{K}$, it follows that:

     at $I = 1\,\text{mA}$: $r_d = 25\,\Omega$, $g = 40\,\text{mA/V}$;
     at $I = \tfrac{1}{2}\,\text{mA}$: $r_d = 50\,\Omega$, $g = 20\,\text{mA/V}$;
     at $I = 1\,\mu\text{A}$: $r_d = 25\,\text{k}\Omega$, $g = 40\,\mu\text{A/V}$.

9.3   The temperature coefficient is about $-2\tfrac{1}{2}\,\text{mV/K}$: a temperature increase of 10 K lowers the diode voltage by 25 mV.

9.4 $V = \dfrac{kT}{q} \ln \dfrac{I}{I_o} = kT \dfrac{\log I/I_o}{\log e} \approx \dfrac{0.025}{0.434} \log \dfrac{I}{I_o} = 0.058 \log \dfrac{I}{I_o}.$

A tenfold increase in I results in a voltage increase of about 58 mV.

9.5   At 0.1 mA, r_s can be neglected with respect to r_d, not, however, at 10 mA. Using the result of Exercise 9.4, the diode voltage V would be $600 + 58 + 58 = 716\,\text{mV}$, if r_s is zero. The measured value is 735 mV, so there is a voltage of 19 mV over r_s. Hence: $r_s = 19\,\text{mV}/10\,\text{mA} = 1.9\,\Omega$. At 0.1 mA, r_d is about 250 Ω, so the assumption that $r_s \ll r_d$ appears to be valid.

Circuits with pn diodes

9.6   The output voltage is limited to either V_k or the Zener voltage V_z. The input voltage is derived from the output voltage.

(a)

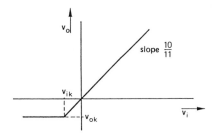

$v_{ok} = -0.5\,\text{V}$;
$v_o = (1000/1100)\,v_i$ as long as the diode is forward biased;
$v_{ik} = (11/10)\,v_{ok} = -0.55\,\text{V}$.

(b) Similar shape; $v_{ok} = 5.5 - 0.5 = 5.0\,\text{V}$;
                             $v_{ik} = (11/10)\,v_{ok} = 5.5\,\text{V}$.
                             $v_{o1} = 5.5\,\text{V}$;
                             $v_{i1} = (11/10)v_{o1} = 6.05\,\text{V}$;
                             $v_{o2} = -0.5\,\text{V}$;
                             $v_{i2} = -0.55\,\text{V}$.

(c)

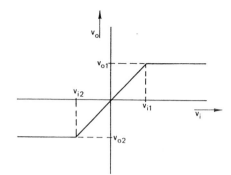

(d) The same shape; $v_{o1} = v_{z1} + v_{k2} = 6.5$ V;
$v_{i1} = 7.15$ V;
$v_{o2} = -v_{z2} - v_{k1} = -10.5$ V;
$v_{i2} = -11.55$ V.

9.7 The circuit is sensitive to the negative peak value. For any input voltage, $v_o - v_i \leq V_k$. The output voltage is $v_o = v_{i,\min} + V_k$, with values -5.4 V, -0.9 V and 0 V, respectively (in the latter case, the diode remains reverse biased).

9.8 The ripple voltage Δv satisfies the equation $\Delta v/(\hat{v}_i - V_k) \approx T/\tau$ (see page 140). For relatively large signal amplitude, V_k can be neglected compared to the amplitude, hence:

$$\frac{\Delta v}{\hat{v}_i} \approx \frac{T}{\tau} = 1\%: \tau = 100T = 100 \cdot 0.2 \text{ ms} = 20 \text{ ms}. \ R = \frac{\tau}{C} = 200 \text{ k}\Omega.$$

9.9 The capacitance values are not relevant here.
  (a) v_o is a sine wave with an amplitude of 5 V. The clamping level is -0.6 V for the negative tops, hence $v_{o,\mathrm{av}} = -0.6 + 5 = 4.4$ V.
  (b) Same as (a); the clamping level is $+5.4 - 0.6 = 4.8$ V, hence $v_{o,\mathrm{av}} = 9.8$ V.
  (c) C_1 and D_1 form a clamping circuit, C_2 and D_2 a peak detector for negative top values. The output voltage of the clamping part is a sine wave with amplitude 5 V, whose positive tops are clamped at $+0.6$ V. Its negative peak is $0.6 - 10 = -9.4$ V, so $V_o = -9.4 + 0.6 = -8.8$ V.

9.10 For proper operation the capacitor must be able to charge, so the diode should be (at least once) forward biased. This happens as soon as $\hat{v}_i > V + V_k$, so for any value of $\hat{v}_i$ the voltage V must satisfy $V < \hat{v}_i - V_k$. The clamping level is $V + V_k$ and there is no condition for the minimum value of V.

9.11 The bridge (without a capacitor) acts as a double-sided rectifier, not as a peak detector.
  (a) The output is a double-sided rectified sine wave.
  $\hat{v}_i = 10\sqrt{2}$ V; $\hat{v}_o = \hat{v}_i - 2V_k \approx 10\sqrt{2} - 1.2 \approx 12.9$ V.
  This is equal to the ripple voltage.

(b) $V_o = |V_i| - 2V_k \approx 8.8$ V.
    This is a DC voltage, so there is no ripple.
(c) Same as (b), the current flows through both other diodes.

9.12 When loaded with a capacitor, the bridge acts as a peak detector.
(a) $V_{o,m} \approx \hat{v}_i - 2V_k = 10\sqrt{2} - 1.2 \approx 12.9$ V.
    $\Delta v = (\hat{v}_i - 2V_k) \cdot T/\tau$, with $T = 10$ ms (double frequency) and
    $\tau = RC = 30$ ms (see page 142). Hence: $\Delta v \approx {}^{10}\!/\!{}_{30} \cdot 12.9 = 4.3$ V.
(b) At a DC input voltage, the capacitor is charged up to $V_i - 2V_k$; the output voltage is 8.8 V (without ripple).
(c) Same as (b).

9.13 The current through the Zener diode should be more than zero, to guarantee a proper Zener voltage. Hence:

$$I_{R1} \geqslant I_{RL} \to \frac{18 - 5.6}{R_1} \geqslant \frac{5.6}{R_L} \to R_L \geqslant \frac{5.6 \cdot 1800}{18 - 5.6} = 813\,\Omega.$$

10 Bipolar transistors

Properties of bipolar transistors

10.1 $I_C = I_0(e^{qv_{BE}/kT} - 1) \approx I_0 e^{qv_{BE}/kT}$.
    The base–emitter junction must be forward biased, the collector–base junction must be reverse biased, for proper operation as a linear amplifying device. For an npn transistor: $V_{BE} \approx 0.6$ V and $V_{BC} \leqslant 0$ V.

10.2 $I_E = I_B + I_C$ and $I_C = \beta I_B$, hence $I_E = (1+\beta)I_B = 800\,\mu A \to$

$$I_B = \frac{800\,\mu A}{1 + 200} \approx 4\,\mu A;\; I_C = I_E - I_B = 796\,\mu A.$$

10.3 The Early effect in a bipolar transistor accounts for the effect of V_C on I_C; ideally, this effect is zero.

10.4 $V_B = \dfrac{10}{10 + 15} \cdot 25 = 10$ V $\to V_E = 10 - 0.6 = 9.4$ V;

$$I_E = \frac{9.4\,V}{4700\,\Omega} = 2\,mA;\; I_C \approx I_E = 2\,mA;$$

$V_o = 25\,V - 2\,mA \cdot 5600\,\Omega = 13.8$ V.

10.5 $V_E(T_1) = -0.6$ V $\to I_{E1} = \dfrac{-0.6\,V - (-10\,V)}{4700\,\Omega} = 2\,mA,\; I_{C1} \approx I_{E1} = 2\,mA;$

$V_E(T_2) = +0.6$ V $\to I_T = \dfrac{10\,V - 0.6\,V}{1880\,\Omega} = 5\,mA;$

Answers to exercises

$I_T = I_{C1} + I_{E2}$ (Kirchhoff's Law) $\to I_{E2} = 5 - 2 = 3$ mA, $I_{C2} \approx I_{E2} = 3$ mA;
$V_o = -10\text{ V} + 3\text{ mA} \cdot 2\text{ k}\Omega = -4$ V.

Circuits with bipolar transistors

10.6 $r_e = 25/I_E\ \Omega$, with I_E the emitter current in mA.

(a) $V_B = V^- + \dfrac{R_2}{R_1+R_2}(V^+ - V^-) = -20 + \dfrac{40}{11} = -16.36$ V,

$V_E = V_B - 0.6 = -16.96$ V;

$I_E = \dfrac{V_E - V^-}{R_E} = \dfrac{-16.96 + 20}{5600} = 0.65\text{ mA} \to r_e = \dfrac{25}{0.65} = 38.5\ \Omega.$

(b) $V_B = V^+ \dfrac{R_2}{R_1+R_2} = 7.5$ V, $V_E = 6.9$ V;

$I_E = \dfrac{V_E}{R_E} = 6.9\text{ mA} \to r_e = 3.6\ \Omega.$

(c) For T_1:

$V_B = \dfrac{R_2}{R_1+R_2}V^+ = \dfrac{47}{147}\cdot 7.5 = 2.4$ V; $V_E = 1.8$ V;

$I_E = \dfrac{V_E}{R_4} = 0.1\text{ mA} \to r_{e1} = 250\ \Omega.$

For T_2: $V_B = V^+ - I_{C1}R_3 = 7.5 - 0.1\cdot 22 = 5.3$ V;
$V_E = V_B + 0.6 = 5.9$ V;

$I_E = \dfrac{V^+ - V_B}{R_5} = \dfrac{1.6}{1500} \approx 1\text{ mA} \to r_{e2} \approx 25\ \Omega.$

10.7 (a) CE stage: $A = -\dfrac{R_C}{R_E} = -\dfrac{33\text{ k}}{5.6\text{ k}} \approx -5.9.$

(b) Emitter follower: $A \approx 1$.

(c) This is a two-stage voltage amplifier with decoupled emitter resistances. As $\beta \to \infty$, the base current through T_2 is zero; the first stage is not loaded by the second stage. In this case: $A = A_1 \cdot A_2$.

$A_1 = -\dfrac{R_3}{r_{e1}} = -88;\quad A_2 = -\dfrac{R_6}{r_{e2}} = -108 \to A = +9504.$

10.8 The transfer of an emitter follower with $\beta \to \infty$ is $\dfrac{R_E}{r_e + R_E} \approx 1 - \dfrac{r_e}{R_E}$.

The relative deviation from 1 is thus $-\dfrac{r_e}{R_E} = -\dfrac{3.6}{1000} = -0.36\%$.

10.9 The transfer of an emitter follower with $\beta \gg 1$ is

$$\dfrac{\beta R_E}{r_b + \beta(r_e + R_E)} \approx 1 - \dfrac{r_e}{R_E} - \dfrac{r_b}{\beta R_E}.$$

The relative deviation from 1 is thus $-\dfrac{r_e}{R_E} - \dfrac{r_b}{\beta R_E} = -0.46\%$.

10.10 The input resistance is $R_1 // R_2 // [r_b + (1+\beta)(r_e + R_E)] \approx R_1 // R_2 // \beta R_E$
= 100 kΩ // 100 kΩ // 100 kΩ = $^{100}\!/_3$ kΩ.

10.11 In Exercise 10.7, the transfer is $A = A_1 \cdot A_2$, with $A_1 = -(R_3/r_{e1})$.
Now, the input resistance r_{i2} of the second stage is parallel to R_3; so replace R_3 in the formulas by $R_3//r_{i2}$. The input resistance is approximately
$r_{i2} = \beta r_{e2} = 2500$ Ω, so $R_3//r_{i2} \approx 2240$ Ω. This is almost a factor of 10 smaller than the value calculated before.

10.12 First, find V_B, then I_E. As β is not infinite, the base current cannot be neglected. A Thévenin equivalent circuit of the input circuit is depicted below (see page 64).

$$V_B = \dfrac{R_2}{R_1 + R_2} V^+ - (R_1//R_2) I_B.$$

Obviously, the base current affects the bias of the transistor. If the second term is small, it can be neglected; not, however, in this example:

$$V_B = \dfrac{R_2}{R_1 + R_2} V^+ - (R_1//R_2) \dfrac{I_E}{\beta} = V_E + 0.6; \quad V_E = I_E R_E;$$

$$I_E \left(R_E + \dfrac{R_1//R_2}{\beta} \right) = \dfrac{R_2}{R_1 + R_2} V^+ - 0.6 \to I_E = \dfrac{7.5 - 0.6}{1000 + 500}$$

= 4.6 mA, r_e = 5.43 Ω.

Usually, the designer makes R_1 and R_2 such that the bias is independent of the base current.

11 Field-effect transistors

Properties of field-effect transistors

11.1 The gate–source voltage at $V_{DS} = 0$, for which the channel is non-conducting (see page 171).

11.2 In this region there are (almost) no free charge carriers (see page 132).

11.3 The gate current of a JFET consists of the reverse currents of the reverse biased junctions (gate–source and gate–drain). In a MOSFET, the gate is completely isolated from the rest of the device by an oxide layer, so I_G is almost zero (smaller than 10^{-12} A).

11.4 $V_G = 0$; $V_{GS} = -4\,\text{V} \rightarrow V_S = 4\,\text{V}$;

$$I_S = \frac{V_S}{R_S} = \frac{4}{1500}; \quad I_D = I_S \text{ (as } I_G = 0\text{)};$$

$$V_D = V^+ - I_D R_D = 20 - \frac{4 \cdot 3300}{1500} = 11.2\,\text{V}.$$

11.5 The maximum resistance of the network (10 kΩ) is obtained for $R_{FET} \rightarrow \infty$, so for $V_{GS} < V_P$: $R = R_1 = 10\,\text{k}\Omega$.
The lowest resistance (1 kΩ) is obtained for $V_{GS} = 0$, $R_{FET} = 800\,\Omega$:
$R = R_1 // (R_2 + R_{FET}) = 1\,\text{k}\Omega \rightarrow R_2 = 311\,\Omega$.

11.6 To avoid any effect of the pn junctions at the source and drain contacts on the signal behaviour of the transistor (see page 176).

11.7 The names refer to the type of charge carriers that are responsible for the conduction in the material. In a bipolar transistor these are both holes and electrons; in an n-channel FET only electrons and in a p-channel FET only holes.

Circuits with field-effect transistors

11.8 $I_D = \frac{5}{3} V_{GS} + 5$ (I_D in mA).

11.9 $V_G = 0$, thus $V_{GS} = -V_S$; $I_D = I_S = V_S/R_S$;
Substitution in the equation of Exercise 11.8 results in: $I_D = \frac{5}{6}$ mA.

11.10 Similar to Exercise 11.9: $I_D = 2.5$ mA;
$V_D = V^+ - I_D R_D = 10 - 2.5 \cdot 1.8 = 5.5\,\text{V}.$

11.11

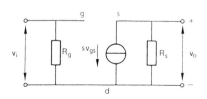

The power supply voltage has no effect on the calculation of the signal properties.
- voltage transfer (at $i_o = 0$):

$$v_o = gv_{gs}R_S = g(v_g - v_s)R_S = g(v_i - v_o)R_S \rightarrow \frac{v_o}{v_i} = \frac{gR_S}{1+gR_S}$$

$$= \frac{4}{1+4} = 0.8;$$

- input resistance (at $i_o = 0$):

$$\frac{v_i}{i_i} = r_i = R_G = 1\,\mathrm{M}\Omega;$$

- output resistance (at $v_i = 0$):

$$v_o = (i_o + gv_{gs})R_S = (i_o - gv_o)R_S \rightarrow \frac{v_o}{i_o} = r_o = \frac{R_S}{1+gR_S} = 400\,\Omega.$$

11.12 (a) T_1 is connected as a source follower. The total source resistance is R in series with the output resistance of T_2 (at the drain). The latter is infinite (current source character) so the voltage transfer of the source follower is 1: the AC voltages at the source of T_1 and the drain of T_2 are both equal to v_i. The DC (average) value of v_o is $V_{GS}(T_1) + I_D R$ lower than that of v_i.
(b) $V_{GS}(T_2) = 0 \rightarrow I_D(T_2) = 2\,\mathrm{mA}$;
$I_D(T_1) = I_D(T_2) = 2\,\mathrm{mA} \rightarrow V_{GS}(T_1) = 0$;
The required voltage drop is 3 V, which is equal to the voltage drop across R: $I_D R = 3$, so $R = 1500\,\Omega$.
(c) The maximum peak voltage depends on the V_{DS}–I_D characteristic of the FET (Figure 11.2c); in any case, all pn junctions must remain reverse biased. From the characteristic it follows that:
for T_1: $V_{DS} > 0$; $V_{S\max} = V_{G\max} = 10\,\mathrm{V} \rightarrow \hat{v}_{i\max} < 3\,\mathrm{V}$;
for T_2: $V_{D\min}(T_2) = V_{S\min}(T_1) - 3 = 0\,\mathrm{V} \rightarrow \hat{v}_{i\max} < 4\,\mathrm{V}$.
$V_{GS}(T_1)$ and $V_{GS}(T_2)$ are 0 V. $V_{DG}(T_1) = 3\,\mathrm{V}$ and $V_{DG}(T_2) = 4\,\mathrm{V}$, hence all pn junctions are reverse biased. The condition for T_2 is always fulfilled if (for T_1) $\hat{v}_{i\max} = 3\,\mathrm{V}$.
From Figure 11.2c it appears that at this large value of the input voltage T_1 operates in the pinch-off region and is not linear anymore. For this reason, the input voltage should be kept lower than the value calculated before.

12 Operational amplifiers

Amplifier circuits with ideal operational amplifiers

12.1 A virtual ground is a point that is at ground potential without being connected to ground. At proper feedback, the voltage between the two inputs of an operational amplifier is zero. At grounded positive terminal the negative terminal is virtually grounded. In a non-inverting configuration, none of the input terminals is grounded.

12.2 (a) Superposition of the separate contributions of V_1 and V_2 to the output voltage:

$$V_o = -\frac{R_3}{R_1}V_1 - \frac{R_3}{R_2}V_2 = -15 \cdot 2 - 12.2 \cdot (-3) = +6.6 \text{ V}.$$

(b) In the ideal case (equal resistances) $V_o = 0$. In the general case,

$$V_o = \frac{R_1 R_4 - R_2 R_3}{R_1(R_3 + R_4)} V_i \text{ (see Figure 12.5);}$$

with $R_{1,2,3,4} = R(1 \pm \varepsilon)$, the maximum output voltage is

$$V_o \approx \frac{4\varepsilon}{2} V_i = 2 \cdot 0.3 \cdot 10^{-2} \cdot 10 = 60 \text{ mV}.$$

(c) $V_o = -\infty$ (in a real situation: limited by the negative power supply voltage).

(d) Superposition of the separate contributions of V_i and I_i to the output:

$$V_o = -\frac{R_2}{R_1}V_i + R_2 I_i = -1 \cdot 0.6 + 15 \cdot 10^3 \cdot 40 \cdot 10^{-6} = 0.$$

(e) The current through the negative input is zero, so the voltage across R_3 is zero as well. The voltage at the negative input is V_o; the circuit behaves as a buffer amplifier. As the output impedance is zero, the load resistor R_4 does not affect the transfer.

$$V_o = \frac{R_2}{R_1 + R_2} V_i = \frac{120}{300} \cdot 5 = 2 \text{ V}.$$

(f) $V_3 = -(R_t/R_1)V_1$ (inverting amplifier); $V_o = V_5$ (buffer amplifier);

$$V_5 = \frac{R_3}{R_2 + R_3} V_3 + \frac{R_2}{R_2 + R_3} V_4 \text{ (superposition).}$$

12 Operational amplifiers

Substitution of the numerical values:

$$V_3 = -\frac{18}{0.6} \cdot 0.3 = -9 \text{ V}; \; V_o = \frac{18}{40} \cdot (-9) + \frac{22}{40} \cdot 1$$
$$= -4.05 + 0.55 = -3.5 \text{ V}.$$

12.3 $R_i = R_1 > 5 \text{ k}\Omega$; take $R_1 = 5.6 \text{ k}\Omega$ (see page 95). $R_2/R_1 = 50$, so $R_2 = 280 \text{ k}\Omega$ (for instance two resistances of 560 kΩ in parallel).

12.4 The circuit must have two inverting inputs and one non-inverting input. Try the next configuration.

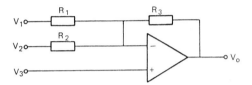

For this circuit: $V_o = -\dfrac{R_3}{R_1} V_1 - \dfrac{R_3}{R_2} V_2 + \left(1 + \dfrac{R_3}{R_1 // R_2}\right) V_3$, with the

conditions: $R_3/R_1 = 10$, $R_3/R_2 = 5$ and $R_3 = R_1 // R_2$.
There is no solution: the gain V_o/V_3 is always larger than the other two gain factors V_o/V_1 and V_o/V_2. The voltage V_3 should first be attenuated, for instance with the voltage divider with resistances R_4 and R_5.

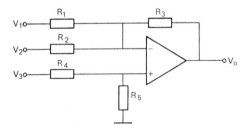

Now: $V_o = -\dfrac{R_3}{R_1} V_1 - \dfrac{R_3}{R_2} V_2 + \left(1 + \dfrac{R_3}{R_1 // R_2}\right)\left(\dfrac{R_5}{R_4 + R_5}\right) V_3$.

We have several options. Take, for instance, $R_3 = 100 \text{ k}\Omega$; then $R_1 = 10 \text{ k}\Omega$ and $R_2 = 20 \text{ k}\Omega$. Further, $R_1 // R_2 = 20/3 \text{ k}\Omega$, so $1 + R_3/(R_1 // R_2) = 16$. The attenuation of the voltage divider must be 8, which can be achieved by (for instance) $R_5 = 10 \text{ k}\Omega$, $R_4 = 70 \text{ k}\Omega$.

12.5 Let the two resistances be R_{b1} and R_{b2}. The common-mode transfer remains 1: the voltage drop across R_a is still zero, hence the voltage drop across R_{b1} and R_{b2} is zero too. If the differential amplifier is perfect, the common-mode transfer is zero.

The differential-mode transfer is found with:

$$V_1' = V_1 + \frac{R_{b1}}{R_a}V_d \text{ and } V_2' = V_2 - \frac{R_{b2}}{R_a}V_d, \text{ from which follows:}$$

$$V_1' - V_2' = V_1 - V_2 + \frac{R_{b1}+R_{b2}}{R_a}V_d = \left(1 + \frac{R_{b1}+R_{b2}}{R_a}\right)V_d.$$

Conclusion: unequal values of R_{b1} and R_{b2} do not affect the CMRR; the resistance values determine the transfer for differential voltages.

Non-ideal operational amplifiers

12.6 Put a voltage source V_{off} in series with the plus input of the operational amplifier. Then:

$$V_o = \left(1 + \frac{R_2}{R_1}\right)V_{off} = \left(1 + \frac{10}{2.2}\right) \cdot 0.4 = 2.22 \text{ mV}.$$

12.7 Without R_3 the output voltage is $V_o = R_2 I_{bias} = 10^4 \cdot 10^{-8} = 0.1$ mV.
$R_3 = R_1 // R_2 = 10 \cdot 2.2/12.2 = 1.8$ kΩ. So $V_o = I_{off}R_2 = 0.01$ mV.

12.8 As the minus terminal is virtually grounded, $R_i = V_i/I_i = R_1 = 2200$ Ω. The amplifier has ideal properties, so $R_o = 0$.

12.9 At unity feedback, the gain is 1; the bandwidth $f_t = 1.5$ MHz. The gain is $A = 10/2.2$, so the bandwidth is equal to $1.5 \cdot 10^6 \cdot 2.2/10 = 330$ kHz.

12.10 The input resistance has the value of R_1. Take $R_1 = 12$ kΩ. The required gain is $-R_2/R_1 = -30$, so $R_2 = 360$ kΩ. The voltage on the slider of R_3 may vary from -15 to $+15$ V; the compensation voltage at the output should vary from $+1.2$ to -1.2 V, so the transfer (for the compensation voltage) must be $-15/1.2 = -12.5$. As the resistance of the potentiometer depends on the position of the slider, so does the transfer. For the slider at the extreme ends of the potentiometer, the transfer is $-R_2/R_4$, so $R_4 = 12.5 \cdot 12 = 150$ kΩ. Take for instance a potentiometer of 100 kΩ; the value is not critical: it determines mainly the current that must be supplied by the power source.

13 Frequency-selective transfer functions with operational amplifiers

Circuits for time-domain operations

13.1 $v_o = \dfrac{1}{C}\displaystyle\int_0^{10}(V_{off}/R + I_{bias})\,dt + V_{off} \approx 10^6(10^{-4}/10^4 + 10^{-8}) \cdot 10 = 0.2$ V.

13 Frequency-selective transfer functions with operational amplifiers

13.2 If both bias currents are equal, the contribution of I_{bias} to the output is eliminated, hence:

$$v_o = \frac{1}{C} \int_0^{10} (V_{off}/R) \, dt = 0.1 \text{ V}.$$

13.3 The output voltage reaches the end value $v_o = (1 + R_o/R)V_{off} = 10.1$ mV.

13.4 (a) The complex transfer function is:

$$H = -\frac{Z_2}{Z_1} = -\frac{R_2}{R_1} \cdot \frac{j\omega R_1 C_1}{1 + j\omega R_1 C_1} \cdot \frac{1}{1 + j\omega R_2 C_2}.$$

The amplitude characteristic is the sum of the next three curves: R_2/R_1 (frequency-independent transfer 100); $j\omega R_1 C_1/(1 + j\omega R_1 C_1)$ (a first-order high-pass characteristic with break point at $f = 1/2\pi R_1 C_1 \approx 10^3$) and $1/(1 + j\omega R_2 C_2)$ (first-order low-pass characteristic with break point at $f = 1/2\pi R_2 C_2 = 10^3$).

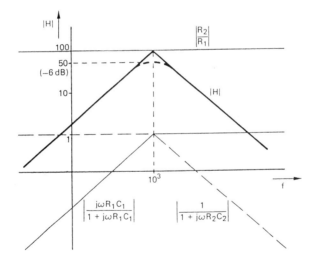

(The deviation from the asymptotic approximation at the break point is -3 dB for the last two curves, so -6 dB (or a factor of 2) for the total transfer, see dashed curve.)

(b) The phase shift follows from the argument of the transfer function (we disregard the minus sign of the inversion):

$$\arg H = \frac{\pi}{2} - \tan^{-1} \omega R_1 C_1 - \tan^{-1} \omega R_2 C_2 = \frac{\pi}{2} - 2\tan^{-1}(10^{-3} f).$$

The deviation from $90°$ is $2\tan^{-1}(10^{-3} f)$; this is more than $10°$ for $f \geq 87$ Hz.

456 Answers to exercises

13.5 The input resistance has the value of R_1, so take $R_1 = 10\,\text{k}\Omega$. The gain in the P-region (the region of high frequencies, where C can be considered as a short-circuit) is $-R_2/R_1$, so $R_2 = 20\,\text{k}\Omega$.
The break point between the I- and P-regions is at $\omega_k = 1/R_2C$. As $f_k > 100$ Hz, $1/2\pi R_2 C > 100$ or $C < 80$ nF. As near the break point the deviation from the integrating character is rather large, we take for instance $C = 10$ nF.

Circuits with high frequency selectivity

13.6 Rewrite the transfer function in the form

$$H = \frac{a_0 + a_1 j\omega + a_2 (j\omega)^2}{1 + j\omega/Q\omega_0 - \omega^2/\omega_0^2}.$$

$$H = \frac{1}{3} \frac{1 + 2j\omega\tau - \omega^2\tau^2}{1 + j\omega\tau/12 - \omega^2\tau^2}, \text{ from which follows:}$$

$\omega_0 = 1/\tau = 10^3$ rad/s,

$\omega\tau/12 = \omega/Q\omega_0 = \omega\tau/Q$, so $Q = 12$,

$|H(\omega \to \infty)| = \dfrac{1}{3}$; $|H(\omega \to 0)| = \dfrac{1}{3}$; $|H(\omega_0)| = \dfrac{1}{3} \cdot \dfrac{2j\omega\tau}{j\omega\tau/12} = 8$.

13.7 $H = \dfrac{Z_2}{Z_1 + Z_2} = \dfrac{j\omega R_2 C_1}{1 + j\omega (R_2 C_1 + R_1 C_1 + R_2 C_2) - \omega^2 R_1 R_2 C_1 C_2}$

$= \dfrac{j\omega\tau}{1 + 3j\omega\tau - \omega^2\tau^2}.$

Comparing with the general expression results in:
$\omega_0 = 1/\tau = 100$ rad/s, or $f_0 = 100/2\pi \approx 15.9$ Hz.
$3\omega\tau = \omega/Q\omega_0 = \omega\tau/Q$, or $Q = \frac{1}{3}$.

13.8 Replace all resistors by capacitors and vice versa. The new transfer function now is:

$$H = \frac{-\omega^2 R_1 R_2 C_1 C_2}{1 + j\omega R_1(C_1 + C_2) - \omega^2 R_1 R_2 C_1 C_2}.$$

$$|H|^2 = \frac{(\omega^2 R_1 R_2 C_1 C_2)^2}{(1 - \omega^2 R_1 R_2 C_1 C_2)^2 + \omega^2 R_1^2 (C_1 + C_2)^2}.$$

The condition for a Butterworth filter is:
$-2R_1 R_2 C_1 C_2 + R_1^2 (C_1 + C_2)^2 = 0$, or $R_2/R_1 = (C_1 + C_2)^2/2C_1 C_2$.
Take $C_1 = C_2 = C$, hence $R_2 = 2R_1$.

13.9 The output voltage of the differential amplifier is denoted as V_x, that of the upper integrator V_y. Now:

$$V_o = -\frac{1}{j\omega R_1 C_1} V_x; \quad V_y = -\frac{1}{j\omega R_2 C_2} V_o;$$

$$V_x = -\frac{R_4}{R_3} V_y + \left(1 + \frac{R_4}{R_3}\right)\frac{R_6}{R_5 + R_6} V_i + \left(1 + \frac{R_4}{R_3}\right)\frac{R_5}{R_5 + R_6} V_o.$$

Elimination of V_x and V_y results in:

$$\frac{V_o}{V_i} = -\frac{(1 + R_3/R_4)R_6}{R_5 + R_6} \cdot j\omega R_2 C_2$$

$$\times \left(1 + j\omega R_2 C_2 \frac{(1 + R_3/R_4)\cdot R_5}{R_5 + R_6} - \omega^2 R_1 R_2 C_1 C_2 R_3/R_4)\right)^{-1}$$

so $\omega_0 = 1/\sqrt{R_1 R_2 C_1 C_2 R_3/R_4}$ and

$$Q = \frac{R_5 + R_6}{(1 + R_3/R_4)R_5} \sqrt{R_1 C_1 R_3/R_2 C_2 R_4}.$$

Q can be adjusted independently of ω_0 by R_5 and/or R_6. To vary ω_0 independently of Q, the condition $R_1 C_1 = R_2 C_2$ must be fulfilled. If further, for simplicity, $R_1 = R_2 = R$, $C_1 = C_2 = C$ and $R_3 = R_4$, then $\omega_0 = 1/RC$ and $Q = (R_5 + R_6)/2R_5$.

14 Non-linear signal processing with operational amplifiers

Non-linear transfer functions

14.1 Error due to the inaccuracy in V_R: $\pm 2\,\text{mV}$;
error due to loading: about $-R_b/(R_i + R_b) = -1/17 \approx -6\%$;
absolute error due to voltage offset: $\pm 2\,\text{mV} \pm (70 \cdot 8\,\mu\text{V}) = \pm 2.56\,\text{mV}$.
absolute error due to input current: $\pm I_b R_b = \pm 5\,\text{mV}$.

14.2 The switching levels are determined by the two possible voltages at the plus terminal of the operational amplifier; these are:

$$\frac{R_2}{R_1 + R_2} V \pm \frac{R_1}{R_1 + R_2} E \approx V \pm \frac{R_1}{R_2} E = -2 \pm 0.01 \cdot 15.$$

E is the modulus of the power supply voltages. The switching levels are -1.85 and $-2.15\,\text{V}$; the hysteresis is $0.3\,\text{V}$.

14.3

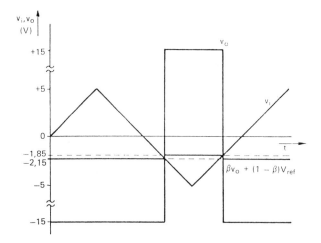

14.4 The break point in the transfer characteristic occurs at the value of v_i for which the current through D_1 (or D_2) is zero. This current equals:

$$\frac{V_{ref1} - V_{ref2}}{R_3} + \frac{v_i - V_{ref2}}{R_1},$$

so the break point occurs for $v_i = V_{ref2} - \frac{R_1}{R_3}(V_{ref1} - V_{ref2})$.

The corresponding output voltage is just equal to V_{ref2}. The output voltage in the linear region (D_2 is forward biased) amounts to

$$-\frac{R_2}{R_3}V_{ref1} - \frac{R_2}{R_1}v_i + \left(1 + \frac{R_2}{R_1 // R_3}\right)V_{ref2}.$$

The characteristic is drawn below.

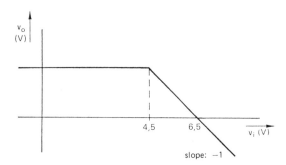

14.5

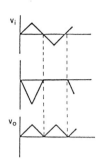 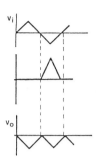

Signals in the circuit of Figure 14.8.   Signals in the case of reversed diodes.

14.6 The break points occur for those values of v_i at which the diodes just become forward biased; the conditions are:

$$\frac{770}{770+2310}V_i = 0.5 \text{ V}; \quad \frac{330}{330+2310}V_i = 0.5 \text{ V and}$$

$$\frac{154}{154+2310}V_i = 0.5 \text{ V, or } 2 \text{ V, } 4 \text{ V and } 8 \text{ V.}$$

The slope for each separate section equals -1, because they amount to $-R_7/R_1$, $-R_7/R_3$ and $-R_7/R_5$, respectively.
The output voltage at the break points is: 0 V; $0 + (-1) \cdot 2 = -2$ V and $-2 + (-2) \cdot 4 = -10$ V. The transfer looks as follows:

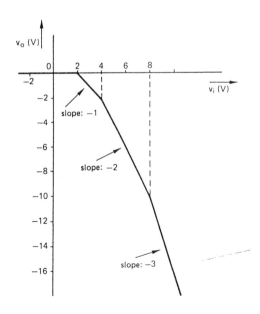

Non-linear arithmetic operations

14.7 $v_{o2} = K_E \exp(v_{i2}/V_E) = K_E \exp(v_{o1}/V_E) = K_E \exp[K_L \ln(v_{i1}/V_L)/V_E]$
$= K_E \exp[\ln(v_{i1}/V_L)^{K_L/V_E}]$; $v_{o2} = v_{i1}$ if $K_L = V_E$ and $K_E = V_R$.

14.8 (a) $v_o = v_x = v_y$; $v_i = K v_x v_y$, hence $v_i = K v_o^2$ or $v_o = \sqrt{v_i/K} = \sqrt{v_i}$.
(b) Both input voltages of the operational amplifier are zero, so: $-v_i/R_1 = (v_o/R_3) + K v_o^2/R_2$.

14.9 $v_o = -10^{-V_q/10} = -10^{-AV_p/10} = -10^{A(\log V_i/10)/10} = -(V_i/10)^{A/10} = -\sqrt{V_i/10}$.

14.10 Weighted addition of $\ln v_1$ and $\ln v_2$, subtraction of $\ln v_3$, then exponential conversion:

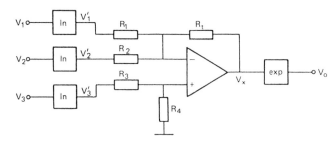

$$V_x = -\frac{R_t}{R_1} v_1' - \frac{R_t}{R_2} v_2' + \left(1 + \frac{R_t}{R_1 // R_2}\right) \frac{R_4}{R_3 + R_4} v_3', \text{ with } v'_{1,2,3} = \ln(v_{1,2,3}).$$

The resistance values must satisfy $R_t/R_1 = 1/3$ and $R_t/R_2 = 2/3$, from which follows: $R_t/(R_1 // R_2) = 1$, hence $R_4/(R_3 + R_4) = 1/2$ or $R_3 = R_4$.

14.11 If the bias currents are zero, then $v_o = \log 1 = 0$. With non-zero bias currents:

$$\log \frac{10^{-5} \pm 10^{-9}}{10^{-4} \pm 10^{-9}} \pm \log 0.1 + \log(1 \pm 10^{-4}) + \log(1 \pm 10^{-5});$$

For $I_1 = 10\,\mu A$ and $I_2 = 100\,\mu A$ the output voltage at zero bias currents is $\log 0.1 = -1$ V. With bias currents this is:

$$\log \frac{10^{-5} \pm 10^{-9}}{10^{-4} \pm 10^{-9}} \approx \log 0.1 + \log(1 \pm 10^{-4}) + \log(1 \pm 10^{-5});$$

the error is thus about $\pm 47.8\,\mu V$.

15 Electronic switching circuits

Electronic switches

15.1 In the on-state, $v_o/v_i = R_L/(R_L + R_g + r_{on}) \approx 1 - R_g/R_L - r_{on}/R_L$.
$R_g/R_L = 0.02\%$, so the error due to r_{on} must be less than 0.08%. Hence, $r_{on}/R_L < 8 \cdot 10^{-4}$, thus $r_{on} < 40\,\Omega$.

In the off-state, $v_o/v_i = R_L/(R_L + R_g + r_{off}) \approx R_L/r_{off}$; this must be less than 10^{-3}, so $r_{off} > 50$ MΩ.

15.2 In the on-state (switch off), $v_o/v_i = R_L/(R_g R_L/r_{off} + R_g + R_L) \approx 1 - R_g/r_{off} - R_g/R_L$; the error due to the switch must be less than 0.08%, so $r_{off} > 12.5$ kΩ.
In the off-state (switch on), $v_o/v_i = r_{on}/(R_g r_{on}/R_L + r_{on} + R_g) \approx r_{on}/R_g$; this must be less than 10^{-3}, hence $r_{on} < 0.01$ Ω.

15.3 In the on-state, $v_o/v_i \approx 1 - R_g/r_{off} - r_{on}/r_{off} - r_{on}/R_L - R_g/R_L$. R_g/R_L contributes 0.02% to the error, so 0.08% remains for the error terms R_g/r_{off}, and r_{on}/R_L. As $r_{off} \gg r_{on}$, the condition is $R_g/r_{off} + r_{on}/R_L < 0.08\%$. Equal partitioning of this error over r_{on} and r_{off} results in $R_g/r_{off} < 4 \cdot 10^{-4}$, and $r_{on}/R_L < 4 \cdot 10^{-4}$, so $r_{off} > 25$ kΩ and $r_{on} < 20$ Ω.
In the off-state, $v_o/v_i \approx r_{on}/r_{off}$, from which follows: $r_{off} > 10^3 r_{on}$; this requirement is fulfilled with the conditions found before.

15.4 Equivalent circuits for the series and shunt switches:

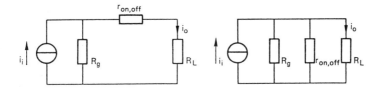

$i_o/i_i = R_g/(R_g + R_L + r_{on/off})$; $i_o/i_i = (R_g // r_{on/off})/(R_g // r_{on/off} + R_L)$;

For the series–shunt configuration:

$$i_o/i_i = \left(1 + \frac{r_{s1}}{R_g} + \frac{R_L}{r_{s2}} + \frac{r_{s1} R_L}{r_{s2} R_g} + \frac{R_L}{R_g}\right)^{-1},$$

with $r_{s1} = r_{on}$, $r_{s2} = r_{off}$ or vice versa.
Assuming $R_g \gg R_L$ and $r_{off} \gg r_{on}$, the next table can be derived:

	r_{on}	r_{off}
series	$\ll R_g$	$\gg R_g$
shunt	$\ll R_L$	$\gg R_L$
series/shunt	$\ll R_g$	$\gg R_L$

15.5 $r_{on} = 2r_d // 2r_d = r_d$. At 1 mA, $r_d = 25$ Ω, so at 5 mA this is 5 Ω.

15.6 In the on-state of the first circuit, r_{on} is in series with the source resistance and V_{off} is in series with the offset voltage of the operational amplifier. The switch increases the offset ($V_{off} + I_{bias} r_{on}$) of the system and increases the load error (relative load error $-r_{on}/R_i$). In the second circuit, the offset V_{off} of the switch is reduced to V_{off}/A at the output of the system, with A the gain of the operational amplifier. The on-resistance of the switch contributes only with r_{on}/A to the output resistance of the circuit. With respect to these parameters, the second circuit is preferred over the first one.

15.7 For $i_s = 0$, $v_{gs} = 0$ (no current through R) hence the FET is conducting, irrespective of the resistance R.
For $i_s = 2$ mA, $v_{gs} = -i_s R$; this must be less than -6 V, so $R > 3$ kΩ.

15.8 To keep the circuit in the on-state, the current through R must be zero irrespective of v_i. This is achieved if the diode is reverse biased, so $v_s > v_{i,max} - V_d$, or $v_s > 3 - 0.6 = 2.4$ V. In the off-state, for each value of v_i, v_{gs} must be less than the lowest pinch-off voltage, hence $v_{gs} < -6$ V. This occurs when the diode is forward biased, so $v_s < v_{i,min} + v_{gs} - V_d$, or $v_s < -3 - 6 - 0.6 = -9.6$ V.

Circuits with electronic switches

15.9 In the on-state, r_{on1} is in series with R_1 and r_{on2} is in series with R_2. Further, $r_{on1} = r_{on} \pm \Delta r_{on1}$, $r_{on2} = r_{on} \pm \Delta r_{on2}$ and $\Delta r \ll r_{on}$.
The differential gain is $A_v \approx R/(R + r_{on1})$.
The common-mode gain is

$$A_c = \frac{v_{oc}}{v_{ic}} = -\frac{R_3}{R_1 + r_{on1}} + \frac{R_4}{R_2 + R_4 + r_{on2}}\left(1 + \frac{R_3}{R_1 + r_{on1}}\right)$$

$$= \frac{R}{R + r_{on1}}\left(\frac{r_{on1} - r_{on2}}{2R + r_{on2}}\right) \approx \frac{R}{R + r_{on1}} \cdot \frac{\pm \Delta r_{on1} \pm \Delta r_{on2}}{2R + r_{on2}}.$$

The CMRR is $\left|\dfrac{A_v}{A_g}\right| = \left|\dfrac{2R + r_{on2}}{\pm 2\Delta r}\right| \pm 5050.$

15.10 In the track mode the output voltage at $v_i = 0$ equals
$v_o = |V_{off}| + |I_{bias}(r_{on} + R_g)| = 250$ μV.
The value of C_H is not relevant here.

15.11 $dv_o/dt = I/C_H = 1$ V/s. $T = 1/f = 10$ ms.
After 1 period (10 ms) $v_o = 8$ V $+ 10$ mV; after 100 periods (1 s)
$v_o = 8 + 1 = 9$ V.

16 Signal generation

Sine wave oscillators

16.1 There are three reasons.
(a) To start the oscillation: first, α must be negative to let the amplitude increase up to the required value; then α should return to zero.

(b) To determine the amplitude: the solution of the differential equation does not depend on the amplitude; it can only be varied by controlling the damping factor α.

(c) To stabilize the amplitude; the factor α is subject to drift, so the amplitude will change even when α is very close to zero.

16.2 Each electronic system generates noise; noise with frequencies close to the oscillation frequency will be amplified for $\alpha < 0$, so the oscillation starts on the system noise.

16.3 (a) The transfer function of two low-pass RC networks in series and loaded with R_3 (the input resistance of the integrator) can be deduced from the next figure, where v_i is the input voltage of the integrator. The transfer is found by applying twice the voltage divider rule (see also Exercise 3.3).

$$H_1 = \frac{Z_1}{R_2 + Z_1} = \frac{1}{2 + j\omega\tau} \quad (\tau = RC)$$

$$H_2 = \frac{Z_2}{R_1 + Z_2} = \frac{2 + j\omega\tau}{3 + 4j\omega\tau - \omega^2\tau^2}$$

$$H = H_1 H_2 = \frac{1}{3 + 4j\omega\tau - \omega^2\tau^2}$$

The transfer of the integrator is $-1/j\omega RC_t$. The product of both transfers must be 1, hence: $1/(3 + 4j\omega\tau - \omega^2\tau^2) \cdot (-1/j\omega RC_t) = 1$. The imaginary part of the left-hand side of this equation is zero, which results in $\omega^2\tau^2 = 3$ or $f = \sqrt{3}/2\pi\tau$. The real part is 1, so $4\omega^2\tau RC_t = 1$. This results, together with the expression of $\omega\tau$ found before, in the condition $C_t = C/12$.

(b) The voltage at the inverting input terminal of the operational amplifier is $\beta_1 v_o$, with $\beta_1 = R_1/(R_1 + R_2)$; the voltage at the non-inverting input is $\beta_2(\omega) v_o$, with $\beta_2(\omega) = j\omega\tau/(1 + 3j\omega\tau - \omega^2\tau^2)$ (see Figure 8.10a). Both voltages are equal, so $\beta_1 = \beta_2(\omega)$. As β_1 is real, the imaginary part of β_2 must be zero, hence $\omega^2\tau^2 = 1$. So $\beta_2 = 1/3$, which is also the value of β_1. Conclusion: $R_2 = 2R_1$; $\omega = 1/\tau = 1/RC$.

(c) The transfer function of both RC sections is found to be $H = (1 - j\omega RC)/(1 + \omega RC)$. The oscillation condition is:

$$\left(\frac{1 - j\omega RC}{1 + j\omega RC}\right)^2 \left(-\frac{R_1}{R_2}\right) = 1,$$

or: $R_1(1 + 2j\omega RC - \omega^2 R^2 C^2) = -R_2(1 - 2j\omega RC - \omega^2 R^2 C^2)$.

From this it follows that $R_1 = R_2$ (constant amplitude); $\omega RC = 1$ (oscillation frequency $f = 1/2\pi RC$).

16.4 From the oscillation condition it follows that $R_4 = 2R_3$ (see page 266), so $R_{\text{th}}(T) = 1 \text{ k}\Omega$. $R_{\text{th}}(T)$ is the resistance value of the NTC when heated up to a temperature T. If $T_0 = 273$ K, the thermistor has a resistance value of $1 \text{ k}\Omega$ at $T = 311$ K. The temperature rise is 11 K, caused by a dissipated power of $\Delta T/100 = 100 \text{ mW} = v^2/R_{\text{th}} \rightarrow \hat{v} = \sqrt{R_{\text{th}} \cdot 110 \cdot 10^{-3}}/\sqrt{2}$ so the output amplitude is 14.8 V.

Voltage generators

16.5 The amplitude of v_2 remains the same, and so does the slope of v_o. The voltage v_o increases (or decreases) up to (or down to) the switching levels of the Schmitt trigger. At increasing R_1, the hysteresis increases as well, so the frequency of the output decreases and the amplitude increases.

16.6 During the upgoing part of v_o, v_2 is negative, hence v_1 is positive ($v_1 = V^+ = 15$ V); during the downgoing part v_1 is negative ($v_1 = V^- = -5$ V). The slope is $-v_2/R_5C = +(R_4/R_3)v_1/R_5C$. The duty cycle is found from the ratio between the charge and discharge time intervals, so the ratio between the slope values. This is 3; the duty cycle of v_1 is $1:4 = 25\%$, that of v_2 (the inverse of v_1) is 75%.

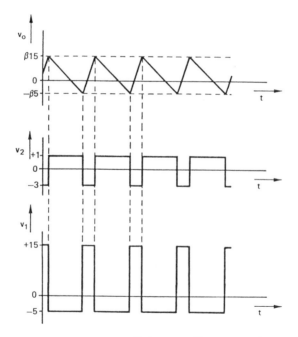

16.7 Amplitude: the positive peak is $\beta V^+ = 7.5$ V; the negative peak is $\beta V^- = -2.5$ V, hence a peak value of 10 V.
Average value: the mean of the peak values, so $v_{o,m} = \frac{1}{2}(\beta V^+ + \beta V^-) = 2.5$ V.

Frequency: the upgoing slope is $3/R_5C = 3 \cdot 10^3$ V/s. The time period of this part is $10(V)/3 \cdot 10^3 (V/s) = 10/3$ ms.
The time interval of the downgoing part is three times as much, so $40/3$ ms.
The frequency is $3/40$ kHz $= 75$ Hz.

16.8 V_{refl} is negative; when the switch is on, v_o increases linearly with time up to a positive value determined by the upper switching level of the Schmitt trigger. This level amounts to

$$V_{ref2} \frac{R_2}{R_1+R_2} + v_s \frac{R_1}{R_1+R_2};$$

for the upper level, $v_s = V^+ = 18$ V, hence the peak value of the ramp voltage is $v_t = 7 \cdot (12/15.9) + 18 \cdot (3.9/15.9) \approx 9.7$ V. The lower switching level is 2.3 V. The slope is V_{refl}/RC; after 1 ms, the peak value should be reached, hence $(V_{refl}/RC) \cdot 10^{-3} = 9.7$, thus $R = 10.3$ kΩ.

16.9 When the switch goes on, v_o is equal to the voltage at the negative input terminal of the operational amplifier, which is zero (virtually grounded). For an average ramp voltage equal to zero, the non-inverting input of the operational amplifier should be connected to a voltage source whose value is equal to the average value of the current situation, hence $-\frac{1}{2}v_t = -4.85$ V.

16.10 Due to the turn-on delay time of the switch, v_o continues rising during the 2 ms after having reached the upper switching level of the Schmitt trigger, with the same slope $V_{refl}/RC = 1000$ V/s. The peak of 9.7 V is reached after 9.7 ms; the switch goes on after 11.7 ms, corresponding to a frequency of 85.5 Hz. The amplitude is $1000 \cdot 11.7 \cdot 10^{-3} = 11.7$ V.

16.11 Take a positive reference voltage V_{refl}, combined with a voltage source in series with the non-inverting input terminal of the operational amplifier that acts as an integrator; the latter should have a value that exceeds the negative switching level of the Schmitt trigger. The switch control circuit should turn on the switch at a voltage $v_s = V^+ = 18$ V, and off for lower values.

16.12 Except for the first period after switching on the circuit, v_c satisfies the equations (see Section 8.1.1):
upgoing part: $\quad v = (V^+ - \beta V^-)(1 - e^{-t/\tau}) + \beta V^-$ \hfill (1)
downgoing part: $\quad v = (\beta V^+ - V^-)e^{-t/\tau} + V^-$ \hfill (2)
Here, the switching occurs at $t = 0$. Equation (1) is valid till $v_c = \beta V^+$; assume this happens at $t = \tau_1$. Then: $\beta V^+ = (V^+ - \beta V^-)(1 - e^{-\tau_1/\tau}) + \beta V^-$, from which follows:

$$-e^{-\tau_1/\tau} = \frac{\beta V^+ - V^+}{V^+ - \beta V^-}, \text{ or: } \tau_1 = \tau \ln \frac{V^+ - \beta V^-}{V^+ - \beta V^+}.$$

Similarly, for the downgoing voltage:

$$\tau_2 = \tau \ln \frac{\beta V^+ - V^-}{\beta V^- - V^-}$$

The duty cycle is $\tau_1/(\tau_1 + \tau_2)$, hence, for a duty cycle of 50%, $\tau_1 = \tau_2$. Thus $V^+ = V^-$, independent of β.

17 Modulation and demodulation

Amplitude modulation and demodulation

17.1 See pages 279 and 284.

17.2 The frequency spectrum of a symmetrical triangular wave contains the fundamental and odd harmonics, so $(2n + 1) \cdot 100$ Hz (n is an integer). Modulation produces sum and difference frequencies of the carrier (5 kHz) and the input signal, hence the modulated signal is composed of components with frequencies 4900 and 5100 Hz, 4700 and 5300 Hz, 4500 and 5500 Hz, and so on.

17.3 The signal bands may not overlap. The bandwidth of one modulated signal is the width over the two sidebands, so $2 \cdot 500 = 1000$ Hz per channel. The system bandwidth should be $12 \cdot 1000 = 12$ kHz.

17.4 Assume $v_i = \hat{v}_i \cos(\omega t + \varphi)$. The bridge output voltage is $v_o = v_i \Delta R/R$, so $\hat{v}_o = \hat{v}_i \Delta R/R$. After amplification by a factor A and multiplication with the synchronous signal $v_s = \hat{v}_s \cos(\omega t + \varphi)$, the result is a signal
$A\hat{v}_o \cos(\omega t + \varphi) \hat{v}_s \cos(\omega t + \varphi) = A(\Delta R/R)\hat{v}_i\hat{v}_s[\frac{1}{2}(1 - \cos 2(\omega t + \varphi))]$.
The DC component is $\frac{1}{2}A\hat{v}_i\hat{v}_s(\Delta R/R) = 2$ V. For $\Delta R/R = -10^{-5}$, this is -2 V.

17.5 The amplified output signal Av_o is multiplied with $s(t)$. Only the term $\cos \omega t$ produces a DC voltage (difference frequency zero). This part of the product equals

$$A\hat{v}_0 \cos \omega t \cdot \frac{4}{\pi} \cos \omega t = A \frac{4}{\pi} \hat{v}_0 \left[\frac{1}{2}(1 - \cos 2\omega t) \right].$$

The DC component is $(2/\pi)A\hat{v}_i(\Delta R/R) = 2/\pi$ V, respectively $-2/\pi$ V.

Systems based on synchronous detection

17.6 All components, multiplied with the reference signal of 15 kHz and resulting in a frequency below 100 Hz, contribute to the output signal. These are components within the frequency band 14 900 to 15 100 Hz.

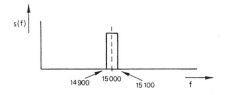

17.7 The quality factor of a band-pass filter is defined as the ratio between the resonance frequency and the bandwidth (Section 13.2.1). According to this definition, the quality factor of the synchronous detector amounts to 15 000/200 = 75.

17.8 $s(t)$ contains the fundamental frequency $1/T = 5\text{ kHz}$ and odd harmonics. All components in v_i, multiplied with $s(t)$ and resulting in a frequency below 200 Hz, contribute to the output signal. These are components within the frequency bands 4800–5200 Hz, 14 800–15 200 Hz, 24 800–25 200 Hz, and so on.

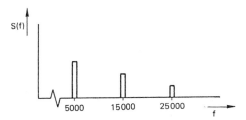

As the amplitude of the components in $\dot{s}(t)$ is inversely proportional to the frequency, the height of the bands is also inversely proportional to the frequency.

17.9 At constant frequency of the VCO, the input signal of H equals $v_i = \hat{v}_i \cos \omega t$ and the output signal is $|H|\hat{v}_i \cos(\omega t + \varphi)$, with $\varphi = \arg H$. Synchronous detection with v_i produces the y signal $v_y = \frac{1}{2}|H|\hat{v}_i^2 \cos \varphi$; synchronous detection with a $\pi/2$ rad shifted input signal ($v_i \sin \omega t$) produces the x signal $v_x = \frac{1}{2}|H|\hat{v}_i^2 \sin \varphi$. As $|H|\cos\varphi = \text{Re}(H)$ and $|H|\sin\varphi = \text{Im}(H)$, the screen shows the value of H at frequency ω in the complex plane. When the frequency of the VCO is controlled by a ramp generator, the spot on the screen describes the polar plot of H (see Section 6.2). A periodic ramp results in a continuous, stable picture on the screen.

17.10 There are no requirements for the reference voltage, because the comparator produces a fixed output amplitude. If the comparator is left out, the output of the synchronous detector depends on the amplitude of V_{ref}. This should have an accurate, stable value.

17.11

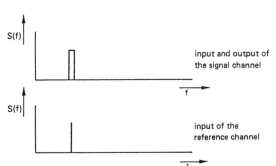

468 Answers to exercises

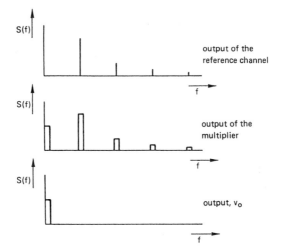

17.12

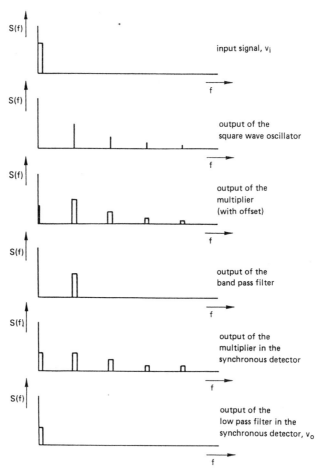

18 Digital-to-analog and analog-to-digital conversion

Parallel converters

18.1
binary	1010111	101111111	100000001	10001111	100010001	1101111	1001001
octal	127	577	401	217	421	157	111
decimal	87	383	257	143	273	111	73
hexadecimal	57	17F	101	8F	111	6F	49

18.2 The MSB is equivalent to 5 V; each next bit is half the preceding bit, so $v_o = 5 + 2.5 + 1.25 + 0.625 + 0.3125 = 9.6875$ V.

18.3 1 LSB $= 1/2^{12} \approx 250$ ppm ($= 250 \cdot 10^{-6}$). Over a temperature range of 80° the inaccuracy amount to $80 \cdot 2 = 160$ ppm, corresponding to $160/250 = 0.64$ LSB.

18.4 See page 306.

18.5 Electronic switches have well-defined and stable on- and off-states; other electronic parameters are much less stable.

18.6 The acquisition time for 1 bit is about $1/f = 5\,\mu$s; a 10 bit word requires $50\,\mu$s.

18.7 $v_o = GV_{\text{ref}}$; V_{ref} is the factor with which G (the input voltage of the DAC) is multiplied.

18.8 The two diodes connected in antiparallel limit the input voltage of the operational amplifier to $+0.6$ and -0.6 V, and protect the input circuit against overvoltage.

Special converters

18.9 One bit requires $1\,\mu$s; neglecting other delay times in the circuit, the conversion time of a 14 bit word is $14\,\mu$s.

18.10
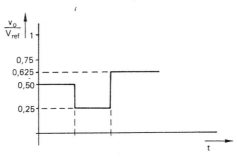

18.11 The output code in the ideal case is 001 (MSB = 0; LSB = 1). The input voltages of the three comparators are $v_i = 0.630$ V, $v_i' = 1.260$ V $v_i'' = 2.520$ V. These voltages are compared with $\tfrac{1}{2}V_{\text{ref}} = 2.500$ V.

(a) The comparators are accurate up to ± 6 mV; in all cases the input voltage difference is more than 6 mV (minimum 20 mV), hence the code is correct.

(b) $v_i' = 2(v_i + V_{\text{off}})$; $v_i'' = 2(v_i' + V_{\text{off}}) = 4v_i + 6V_{\text{off}}$; v_i'' can be between $(2520 + 36)$ mV and $(2520 - 36)$ mV. In the last case the LSB can have the incorrect value 0.

(c) $v'_i = (2 \pm 0.01)v_i$; $v''_i = (2 \pm 0.01)v'_i \approx (4 \pm 0.04)v_i$. v''_i lies between $(2520 + 25.2)$ mV and $(2520 - 25.2)$ mV. In the latter case the LSB can have the incorrect value 0. Apparently, the error depends on the value of v_i.

18.12 During the integration of both v_i and V_{ref} the frequency remains the same; the only requirement is a stable frequency during the conversion time.

18.13 The integration period is not exactly a multiple (here 5) of the interference signal period. Two extreme cases:

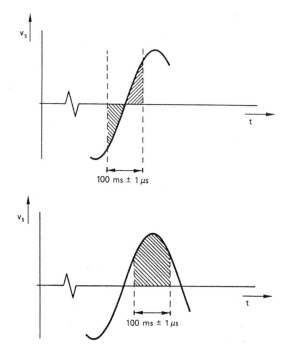

The second situation shows the largest hatched area; the error equals one of these areas (10^{-6} V s) so $\pm 10^{-5}$ (a constant input signal of 1 V that is integrated during 100 ms results in an indication of just 1 V).

19 Digital electronics

Digital components

19.1 (a) $x + xy = x$ (absorption law)
  (b) $\bar{x} + xy = (\bar{x} + x)(\bar{x} + y)$ (distributive law)
  $\qquad = 1 \cdot (\bar{x} + y) = \bar{x} + y$ (negation law and modulus law)
  (c) $\bar{x} + \overline{xy} = x + \overline{(x + \bar{y})}$ (De Morgan's theorem)
  $\qquad = \bar{x} + \bar{y}$ (law of equality)

(d) $x \cdot (x \oplus y) = x \cdot (x\bar{y} + \bar{x}y)$ (can be proved with a truth table)
$= xx\bar{y} + x\bar{x}y$ (distributive law)
$= x\bar{y} + 0 \cdot y$ (law of equality and negation law)
$= x\bar{y}$ (modulus law)

(e) $x \cdot (y + z) + \overline{xyz} = x \cdot (y + z) + x \cdot (\overline{\bar{y} + \bar{z}})$ (De Morgan's theorem)
$= x \cdot (y + z) + x \cdot \overline{(y + z)}$ (negation law)
$= x \cdot (y + z + \overline{y + z})$ (distributive law)
$= x \cdot 1 = x$ (negation law and modulus law).

19.2 (a)

a	b	$a \oplus b$	$ab(a \oplus b)$
0	0	0	0
0	1	1	0
1	0	1	0
1	1	0	0

(b)

a	b	c	$u = a + \bar{b} + \bar{c}$	$v = \bar{a} + \bar{b} + c$	$w = b + c$	uvw
0	0	0	1	1	0	0
0	0	1	1	1	1	1
0	1	0	1	1	1	1
0	1	1	0	1	1	0
1	0	0	1	1	0	0
1	0	1	1	1	1	1
1	1	0	1	0	1	0
1	1	1	1	1	1	1

19.3 $u = 1$ only if $a = 1$ and $b = 0$; $v = 0$ only if $\bar{v} = 1$, thus if $b = 1$ and $a = 0$; $w = 1$ only if $\bar{w} = 0$, thus if $a = b = c = 0$. Further, $y = 0$ only if $v = w = 0$, and $x = 1$ only if $u = y = 1$.

a	b	c	u	v	w	y	x
0	0	0	0	1	1	1	0
0	0	1	0	1	0	1	0
0	1	0	0	0	0	0	0
0	1	1	0	0	0	0	0
1	0	0	1	1	0	1	1
1	0	1	1	1	0	1	1
1	1	0	0	1	0	1	0
1	1	1	0	1	0	1	0

19.4 From the truth table it appears that this flipflop has only a set and reset function. So, the input combination $j = k = 1$ of the JK flipflop must be made impossible. This can be achieved, for instance, by making k the inverse of j, using a NOT gate: $D = J = \bar{K}$ (D is the input).

Logic circuits

19.5 For $e = 1$ at least one of the inputs of each AND gate is 0, so all outputs are 0; $y = 0$ irrespective of the other input values (indicated with – in the table). The AND gates act as switches for d_n. The output is 0 for d_n and 1 if d_n and the other four inputs are 1. This is only true for $e = 1$ and just one combination of s_0, s_1 and s_2, which is different for each gate.

e	$s_2\ s_1\ s_0$	y
1	– – –	0
0	0 0 0	d_0
0	0 0 1	d_1
0	0 1 0	d_2
0	0 1 1	d_3
0	1 0 0	d_4
0	1 0 1	d_5
0	1 1 0	d_6
0	1 1 1	d_7

19.6 The number of gates connected in series increases with increasing numbers of bits, and hence the total delay time increases, in particular for the MSB. Further, the signals are subject to different delay times, so the outputs make several transitions before reaching the steady state.

19.7 FF_0 is in the toggle mode ($J = K = 1$). FF_1 is in the toggle mode if $a_0 = 0$. The modes of FF_2 are listed in the following table.

$a_0\ a_1$	$J\ K\ q$	$a_0\ a_1\ a_2$	decimal value
0 0	0 0 (hold)	1 1 0	3
0 1	0 1 (reset)	0 0 1	4
1 0	0 0 (hold)	1 0 1	5
1 1	1 0 (set)	0 1 1	6
		1 1 0	3
		0 0 1	4
		0 1	5
		. . .	.
		. . .	.
		. . .	.

The flipflops are of the master–slave type, so the values of J and K at the start of the clock pulse determine the output at the end of the clock pulse.

20 Microprocessor systems

Semiconductor memories

20.1 The outputs of tristate buffers can be connected to each other without mutual loading (provided that only one element at the time is selected). This allows the parallel connection of several memory ICs. The chip selection is made by the enable input of the memory IC.

20.2 The diodes avoid the incorrect reading of a 1 (a 'phantom'). The selection of $r_1 k_1$ and $r_1 k_0$ in the PROM of the figure below should result in a 0 and a 1, respectively.

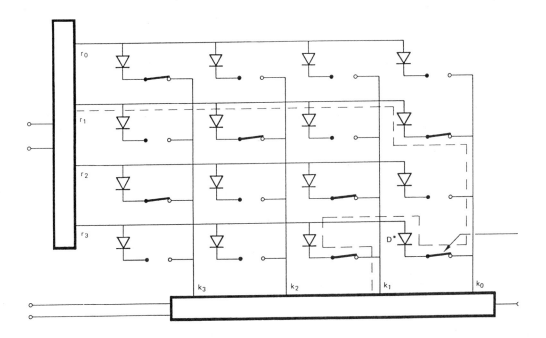

If the diode at location $r_3 k_0$ is left out, selection of r_1 (which means applying a voltage to this row) would result in a voltage on column line k_1, resulting in a false 1.

20.3 It is not possible to make an EPROM with bipolar transistors, using the same principle as with MOSFETs, because of the large base current of a bipolar transistor; the stored charge would disappear very quickly.

20.4 A dynamic memory must be periodically refreshed, which is not necessary for static memories (provided the power supply remains on). CMOS memories have a very low power consumption; the data can be stored for a long time using a small battery.

20.5 (a) Each cell contains 1 bit; 1 kbit is 2^{10} bits. The number of memory cells is thus $64 \cdot 1024 = 65\,536$ or $2^6 \cdot 2^{10} = 2^{16}$.

(b) Each 8 bit word has one address; hence there are $2^{16}/2^3 = 8192$ addresses.

(c) $8192 = 2^{13}$, so 13 bits are required for each address.

20.6 Multiplexing of data and address lines: when reading or writing a word, first the address is connected to the data bus; this address is stored in an internal buffer of the memory chip and the selection can be made. Next the data are connected to the same lines, to read or write them from (or to) the selected locations. In large memories, the addressing can take place in steps, for instance first the 8 most significant address bits, and then the 8 least significant bits, for a full 16 bit address word.

20.7 The address connections are inputs. If these connections are buffered (as is usually the case), they do not load the address bus. The data lines act as input as well as output. If several chips put their data on the data lines, the result cannot be predicted. Further, the chips may be destroyed due to overload.

Structures of microprocessor systems

20.8 See pages 357 and 359.

20.9 (a) For the selection of one out of 2^{16} memory locations 16 bits (or 2 bytes) are required. The loading of the accumulator is usually done by a 1 byte instruction, so in total 3 bytes are needed.
(b) Usually, for this particular operation there is a separate (1 byte) instruction available.
(c) Subtraction from the accumulator: 1 byte; for the number 3A also 1 byte, so 2 bytes in total.
(d) Usually, there is a separate instruction for a conditional jump operation (see Example 20.7). The instruction contains 3 bytes in total.

21 Measurement instruments

Electronic measurement instruments

21.1 The three meters are calibrated for the rms value of sine wave voltages. The correcting factor of meter B is $\pi\sqrt{2}/4 \approx 1.11$ (see also Section 2.1.2). Meter C indicates the peak value of a sine wave, when not corrected for rms values. The correction factor is thus $\frac{1}{2}\sqrt{2}$.
(a) Sine wave: all meters indicate the correct value: $\frac{1}{2}\sqrt{2} \cdot 10 = 7.07$ V.
(b) Square wave: the rms value, the mean of the modulus and the mean of the clamped voltage are all equal to the peak value; hence the indication of the meters is: A: 10 V; B: $1.11 \cdot 10 = 11.1$ V; C: $\frac{1}{2}\sqrt{2} \cdot 10 = 7.07$ V.
(c) Triangle: A triangular voltage whose peak value is $\hat{v}$ has an rms value of $(1.3)\sqrt{3} \cdot \hat{v}$; the mean of its modulus is $\frac{1}{2}\hat{v}$ and the mean of the clamped voltage is equal to the peak value. The indicated values are thus: A: $(1/3)\sqrt{3} \cdot 10 = 5.77$ V; B: $1.11 \cdot 5 = 5.55$ V and C: $\frac{1}{2}\sqrt{2} \cdot 10 = 7.07$ V.

21.2 In the two-wire mode, the resistances of the wires are included in the measured value $R + 2r$, with r the resistance of a single wire. For small values of R, the error would be unacceptably large. In the four-wire mode, a current is applied to the resistance through one pair of wires, whereas the voltage is measured via two other wires. The current through these latter wires is zero, so their resistance is not relevant; neither is the resistance of the first pair of wires, because the voltage is measured across the leads of the resistor.

21.3 See Section 21.1.2. At a large time-base period, the images appear clearly one after another.

21.4 The transfer of the circuit in Figure 21.6a is:

$$\frac{v_i}{v_g} = \frac{R_i}{R_i + R_g} \cdot \frac{1}{1 + j\omega R_p C_s},$$

where $R_p = R_i // R_g = 200\,\text{k}\Omega$ and $C_s = C_i + C_k = 216\,\text{pF}$. At low frequencies, the transfer is $1/1.25 = 0.8$. The break point of the characteristic is $\omega_k = 1/R_p C_s = 2.3 \cdot 10^4$ rad/s, corresponding to 3661 Hz.

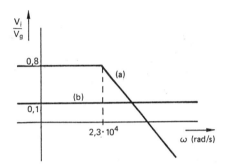

The transfer of the circuit depicted in Figure 21.6b is, under the condition $R_g \ll R$:

$$\frac{v_i}{v_g} = \frac{R_i}{R_i + R} \cdot \frac{1 + j\omega RC}{1 + j\omega R'_p C'_s},$$

with $R'_p = R_i // R = 0.9\,\text{M}\Omega$ and $C'_s = C_k + C_i + C = 240\,\text{pF}$. Apparently, $RC = R'_p C'_s$, so the transfer does not depend on the frequency, and equals $R_i/(R_i + R) = 0.1$.

21.5 Without the attenuator the signal transfer is frequency dependent, corresponding to a low-pass characteristic. This results in the first of the three pictures (see also Figure 8.3). The transfer is compensated by a network with high-pass characteristic (with adjustable capacitance). A too high capacitance results in overcompensation, depicted in the middle of the figure. Only at correct compensation is the transfer independent of the frequency; this corresponds to the rightmost picture.

21.6

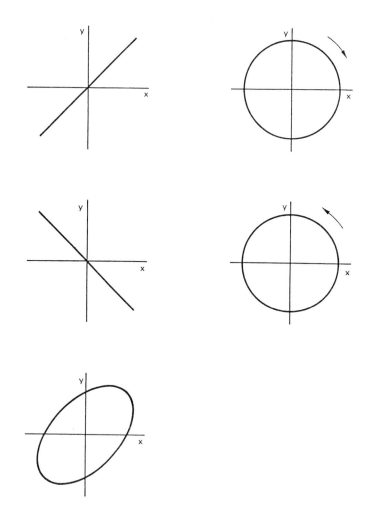

21.7 The transfer of the differential amplifier and the motor are not relevant, due to the feedback. Offset produces only a constant error (shift). All other parameters may have caused the observed non-linearity.

21.8 Logical circuits produce binary signals (0 or 1; high or low). A synchronous noise signal produces a random value (0 or 1) after each clock pulse. There are also noise generators with random transition time.

21 Measurement instruments

21.9

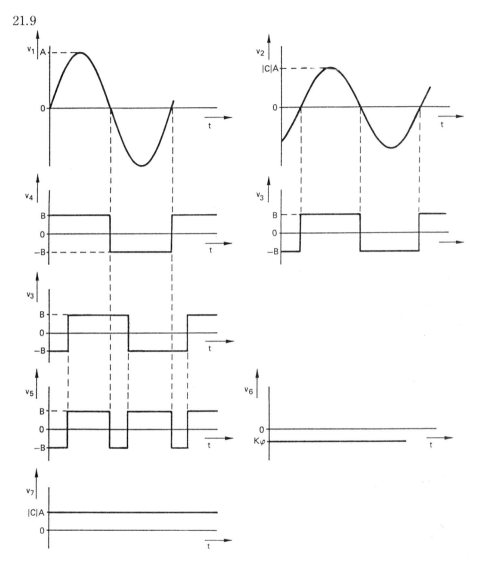

Computer-based measurement instruments

21.10 (a)

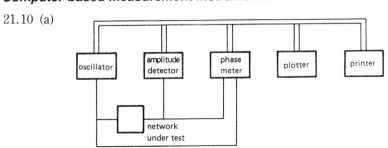

(b) Four decades, spread over 40 cm with a resolution of 0.4 mm, means 1000 steps for a factor of 10^4, or $^{1000}\sqrt{10^4} \approx 1.01$ for each step.
(c) 20 cm with a resolution of 0.4 mm means 500 steps; for 40 dB: 0.08 dB per step.
(d) 500 steps to cover 2π rad: $2\pi/500 = 0.0126$ rad $= 0.72°$ per step.

22 Measurement errors

Types of measurement errors

22.1 (a) The absolute error is $0.01 \cdot (788 + 742) = 15.3$ mV.

The relative error is $\dfrac{15.3}{788-742} = \dfrac{15.3}{46} = 33.3\%$.

(b) The differential voltage is 46 mV; the relative error is 1%; the absolute error of 1% of 46 mV is 0.46 mV.

22.2 (a) The first measurement value shows a deviation of a factor of 10, probably due to a reading mistake. This value will be disregarded. As the measurement is performed repeatedly with the same instrument, systematic errors cannot be recognized. Obviously there are random errors.

(b) The best estimation is the mean (of the four remaining values): $I_m = 24.86/4 = 6.215$ mA. A better notation is 6.2 mA, according to the fluctuations.

(c) The random error can be lowered by a larger number of measurements. Systematic errors cannot be reduced by this method.

(d) The best estimation of the current value is 6.25 mA, the middle of the tolerance band. The difference with the best estimation from the first series of measurements is 0.035 mA. Apparently, the first measurement is subject to a systematic error of about 0.04 mA.

22.3 $v_o = -\dfrac{R_2}{R_1} \cdot v_i = -\dfrac{100}{3.9} \cdot 0.2 = -5.128$ V.

The additive error due to V_off and I_bias is:

$$\Delta v_o = \left(1 + \dfrac{R_2}{R_1}\right) V_\text{off} + I_\text{bias} R_2 = 40 \text{ mV} + 10 \text{ mV} = 50 \text{ mV}.$$

The multiplicative error due to tolerances of v_i, R_1 and R_2 amounts to

$$\dfrac{\Delta v_o}{v_o} = \dfrac{\Delta R_1}{R_1} + \dfrac{\Delta R_2}{R_2} + \dfrac{\Delta v_i}{v_i} = 0.5 + 0.5 + 0.2 = 1.2\%.$$

The maximum error is $50 + 0.012 \cdot 5.13 \cdot 10^3 = 112$ mV. The output voltage has a value somewhere between -5.242 and -5.018 V. The correct indication of the output voltage is -5.13 ± 0.11 V.

22.4 (a) The measured (nominal) resistance is $R = V_o/I_1 = 75\ \Omega$.
The multiplicative error is: $\Delta R/R = \Delta v_o/v_o + \Delta I/I$
$= 0.5 + 0.5 = \pm 1\% \equiv 0.75\ \Omega$, with $\Delta v_o/v_o$ the relative error in v_o.
The total resistance of the two connection wires is between 0 and 6 Ω, denoted as $3 \pm 3\ \Omega$.
The additive error is $\pm(\Delta V_o/I_1)$
$+ 2(r \pm \Delta r) = \pm 0.5 + (3 \pm 3) = 3 \pm 3.5\ \Omega$, where ΔV_o is the absolute error in V_o. In total: $R = 75 - 3 \pm 4.25 = 72 \pm 4.25\ \Omega$.

(b) $V_o = I_1(R + r_1) - I_2 r_3$; assuming $I_1 r_1 = I_2 r_3$, V_o equals $I_1 R$, so the (nominal) resistance is $R = V_o/I_1 = 70\ \Omega$.
The multiplicative error is (as in (a)): $\pm 1\% \equiv \pm 0.7\ \Omega$.
The additive error is:

$$\frac{\Delta V_o}{I_1} + \frac{\Delta(I_1 r_1 - I_2 r_3)}{I_1} = \frac{\Delta V_o}{I_1} + \frac{2(I \Delta r + r \Delta I)}{I}$$

$$\approx \frac{\Delta V_o}{I_1} \pm 2\Delta r = \pm 0.5 \pm 2 = \pm 2.5\ \Omega.$$

Thus: $R = 70\ \Omega \pm 3.2\ \Omega$.

(c) Nominal value: $V_o = I_1 R$, thus $R = V_o/I_1 = 71\ \Omega$.
Multiplicative error (as in (a)): $\pm 1\% \equiv \pm 0.71\ \Omega$.
Additive error: $\pm \Delta V_o/I_1 \equiv \pm 0.5\ \Omega$, so $R = 71\ \Omega \pm 1.2\ \Omega$.
The three measurement methods for the same resistance value have ascending accuracy, as illustrated in the tolerance bands below.

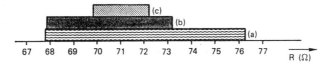

Measurement interference

22.5 The transfer from the error source to the input of the measurement system can be derived using the next circuit model.

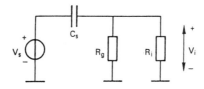

Applying the rules for the complex variables we find:

$$V_i = \frac{R_p}{1/j\omega C_s + R_p} V_s = \frac{j\omega R_p C_s}{1 + j\omega R_p C_s} V_s, \text{ with } R_p = R_g // R_i.$$

In all cases, $\omega R_p C_s \ll 1$ ($\omega = 2\pi f \approx 314$; $R_p < 10^7$ and $C_s < 10^{-13}$) so: $V_i = j\omega R_p C_s V_s$. The rms value of the input voltage due to V_s becomes: $V_i = 2\pi f R_p C_s V_s$. As the peak values are identical to the amplitudes (mean value is zero) the peak value of the input voltage is $\hat{v}_i = 2\pi f R_p C_s \hat{v}_s$. The input voltage due to the signal source V_g is

$$\frac{R_i}{R_i + R_g} V_g = 0.99 V_g.$$

The signal-to-noise ratio is the ratio between the rms values of the input voltages due to V_g and V_s, respectively.

(a) For a disconnected signal source, $R_p = R_i$. So the peak value of V_i is:
$\hat{v}_i = 2\pi \cdot 50 \cdot 10^7 \cdot 10^{-13} \cdot \sqrt{2} \cdot 220 \text{ mV} \approx 97.7 \text{ mV}$.

(b) For a connected signal source, $R_p = R_g // R_i = 99 \text{ k}\Omega$. The rms value and the peak value of V_i (due to V_s) are found to be:
$2\pi \cdot 50 \cdot 99 \cdot 10^3 \cdot 10^{-13} \cdot 220 = 0.68 \text{ mV}$ and 0.97 mV, respectively.

The signal-to-noise ratio becomes: $\dfrac{0.99 \cdot 100 \cdot 10^{-3}}{0.68 \cdot 10^{-3}} = 146$.

(c) The capacitance is reduced to $0.01 C_s = 10^{-15}$ F, resulting in an input voltage with rms value of $2\pi \cdot 50 \cdot 10^7 \cdot 10^{-15} \cdot 220 = 0.69$ mV and a peak value of 0.98 mV.

(d) Similar to (b), the rms value and the peak value of the input voltage (due to V_s) are found to be: $2\pi \cdot 50 \cdot 99 \cdot 10^3 \cdot 10^{-15} \cdot 220 = 6.84 \; \mu\text{V}$ and $9.68 \; \mu\text{V}$, respectively. The corresponding signal-to-noise ratio is

$$\frac{0.99 \cdot 100 \cdot 10^{-3}}{6.84 \cdot 10^{-6}} = 14\,469.$$

(e) The next figure shows a model of the situation: a capacitance between the error voltage source and the signal conductor (here $0.01 C_s$); a capacitance between the error voltage source and the shield (C_a) and a capacitance between the shield and the signal conductor (the cable capacitance C_k). Usually, the dimensions of the shield are much larger than those of the signal conductor, so $C_a \gg C_s$. A typical value for C_k is 100 pF per metre cable length.

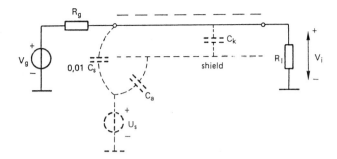

When the shield is grounded, C_a has no effect: error injection only occurs via $0.01C_s$. C_k is parallel to R_i, resulting in a low-pass signal transfer V_i/V_g (see also Figure 21.6), with a break point at 16.1 kHz (for $C_k = 100\,\text{pF}$). If this is not acceptable, a buffer amplifier can be connected close to the signal source, or active guarding can be applied.

When the shield is not grounded, the total capacitance between the error voltage source and the signal conductor is $0.01C_s//(C_a$ in series with $C_k)$ or:

$$0.01C_s + \frac{C_a C_k}{C_a + C_k}.$$

Usually, C_a is much larger than C_s, so the capacitance between the error voltage source and the signal conductor is much larger compared to a non-shielded conductor. Hence, the error voltage at a floating shield is larger than when there is no shield at all.

22.6 (a) The mains filter behaves as a low-pass filter, for signals between the conductors and ground as well as for signals between the conductors. All frequency components above 50 Hz should be filtered out.

(b) Because of the symmetry of the configuration, the voltage between the instrument case and ground is just half the voltage between the two mains connections, hence 110 V.

Index

-3 dB frequency   83
-3 dB point   83
absolute inaccuracy   7, 392
absolute permittivity   96
absolute value   51
AC signals   15
accelerometer (piezoelectric)   111
acceptor   132
access time   350
accumulator   359
accuracy   7
acquisition time   253
across variable   44
active components   38
active guarding   399
active-high   352
active-low   352
actuator   3
AD conversion   2
AD converter   299
ADC   299
adder   332
additive error   73
address bus   357
address decoder   360
address line   349
address selector   360
admittance   53
alias   28
aliasing error   28
alternate mode   369
alternating current   15
ALU   358
AM   279
Ampère's law   98
amplification   3
amplitude control (oscillator)   264
amplitude diagram   82
amplitude modulation   281
amplitude spectrum   22
amplitude stabilization   264
amplitude-time diagram (oscilloscope)   368
amplitude-time diagram (plotter)   373
analogies   44
analog multiplexer   4, 249
analog multiplier   231

analog multiplier (modulator)   284
analog signal   2, 16
analog-to-digital conversion   2
analog-to-digital converter   299
AND   320
AND gate   324
angular displacement sensor   104
antilog converter   230
aperture delay time   253
aperture jitter   253
application-oriented language   363
argument   51, 82
arithmetic and logic unit   358
arithmetic operations   227, 233
assembler language   363
asynchronous counter   336
autopolarity   367
autoscaling   367

band-pass filter   122
band-pass filter (negative feedback)   213
band-pass filter (positive feedback)   211
band-pass filter (second-order)   210
band-reject filter   115
bandwidth   11, 22, 28
bandwidth (of a band-pass filter)   122, 210
base   150
base resistance   154
BCD code   301
Bessel filter   127
bias current   185, 193
binary signal   299
Biot and Savart's law   98
bipolar transistor   150
bipolar transistor (switch)   246
bit   299
bit time   302
blanking   368
Bode plot   82
Bode plot (analyser)   378
Bode relation   128
Boltzman's constant   134
Boolean algebra   319
branche   39
bridged-T filter   123
buffer amplifier   188

bus   356, 380
Butterworth filter   126
Butterworth filter (active)   214
byte   299

calibration chart   384
calibration of rms meters   18
capacitance   39, 96
capacitance (generalized)   44
capacitance (impedance)   53
capacitive error signal injection   394
capacitive sensors   106
capacitor   38, 96
carbon film (resistor)   95
carrier signal   279
cascaded DA converter   313
cathode ray tube   367
CE circuit   157
central processing unit   348
channel (FET)   170
channel resistance   247
characteristic impedance   70
characteristic matching   70
Chebyshev filter   126
chip select   351
chopped mode   369
chopper amplifier   294
chopper-stabilized amplifier   294
clamp circuit   142
clamping circuit   142
clipper   138
clipping   9
clock frequency   307
clock generator   307
CMOS technology   326
CMRR   11, 166, 190
cold junction   108
collector   150
combinatory circuit   324
common emitter circuit   157
common mode rejection ratio   11, 166
common mode signal   11
comparator   220
compensating AD converter   307
compensation (cross-over)   255
compensation (interference)   397
compensation (offset)   195
compensation (temperature)   398
compiler   363
complex Fourier coefficients   24
complex Fourier series   24
complex Fourier transform   24
complex transfer function   53
complex variables   51
computer-based instrument   380
concentration sensor   106
condition flipflop   359
conductance   94
conductivity   94

continuous signals   16
continuous spectrum   26
control line   349
conversion frequency   313
conversion time   302
convolution   26
convolution integral   26
correction (errors)   399
counter   334, 376
counter (decimal)   340
coupling capacitor   157, 178
CPU   348
crest factor   35
cross-over distortion   9
cross-over (interference)   254
crystal oscillator   375
current gain (transistor)   151
current matching   70
current source   39
current source model   65
current-to-voltage converter   186
current-to-voltage transfer   41
current transfer   41
cut-off frequency   116

D flipflop   347
DA conversion   3
DA converter   299
DAC   299
damping factor   263
damping ratio   85
dark current (photodiode)   136
data acquisition   1
data bus   356
data distribution   1
data line   349
data processing   1
DC amplifier (chopper)   294
DC amplifier (stage)   164
DC signal source   374
DC signals   15
DC voltage source   143
De Morgan's theorem   323
dead zone   9
decibel   72, 83
decimal counter   340
decoder   250
delay time (SH circuit)   253
delayed trigger   368
demodulation   3, 288
demultiplexer   251
depletion layer   132
depletion layer (FET)   171
depletion region   132
detection   288
deterministic signals   15
development system   363
dielectric constant   96
dielectric loss   38, 96

differential amplifier   11, 185, 189
differential amplifier (stage)   164
differential equation   40
differential equation (Laplace)   57
differential equation (oscillator)   263
differential mode signal   11
differential multiplexer   249
differential non-linearity   306
differential resistance (diode)   134
differentiating network   120
differentiator   206
digital components   324
digital multiplexer   3, 332
digital signal   2, 16, 299, 319
digital-to-analog conversion   3
digital-to-analog converter   299
diode   131, 133
direct AD converter   312
direct current   15
discrete Fourier transform   20
discrete signals   16
displacement sensor (capacitive)   106
displacement sensor (eddy current)   105
displacement sensor (inductive)   103
distribution function   29
divider   233
donor   131
doping   131
double-sided rectifier   146, 226
double-T filter   123
drain   170
drift   10
droop   253
dual-slope AD converter   314
duty cycle   270
dynamic memory   355

E-12 series   95
EAPROM   353
Early effect   152, 174
eddy current   105
EEPROM   353
electric analog   44
electrolytic capacitor   97
electrometer amplifier   295
electron (free)   131
electronic switches   241
emitter   150
emitter follower   162
enable input   251
envelope   281
EPROM   352
erasable PROM   353
error   389
error propagation   392
error signal sources   74
exclusive OR   320
EXOR   320
EXOR gate   324

expectancy   31
expected value   32
exponential converter   230
external frequency stabilization   198
external trigger   368

failure rate   12
false (F)   319
Faraday's law of induction   98
FET   170
field-effect transistors   170
filtering   3
filters   115
first moment   31
flag   359
flipflop   327
FM   279
FM demodulator   291
force-sensitive sensors   103
force sensor (piezoelectric)   111
four-wire measurement   402
Fourier coefficients   19
Fourier expansion   19
Fourier integral   25
Fourier series   19
Fourier transform   33
frequency band   10
frequency conversion   279
frequency divider   330
frequency meter   376
frequency modulation   279
frequency multiplexing   3, 279
frequency spectrum   19
full-adder   334
function generator   375
fundamental frequency   19
fuse-link PROM   353

gain-bandwidth product   198
gate   171
gauge factor   103
Gaussian distribution function   30, 390
generalized network elements   44
glass fiber transmission   136
Graetz bridge   145, 226
ground   395
ground current   400
ground loop   395
guarding   399
gyrator   39

half-adder   333
handshake   381
harmonic components   19
harmonic oscillator   263, 265
hexadecimal notation   301
high-pass characteristic   210
high-pass filter   115
high-pass filter (first order)   119

Index   **485**

histogram   380
hold capacitor   253
hold range   292
hole   131
humidity sensor   383
hysteresis (Schmitt trigger)   222

I variable   44
IC   38
IEC-625 bus standard   380
IEEE-488 bus standard   380
imaginary part   51, 87
imaginary unit   51
impedance   53
impedance analyser   378
impedance (Laplace domain)   59
impedance mismatch   389
impedance (polar plot)   88
inaccuracy   7, 389
indirect DC amplifier   294
induced voltage   98
induction   97
inductive error signal injection   395
inductive sensors   104
inductor   38, 97
input impedance   65
input interface   348
input offset   10
input/output port   360
input port   41
instruction code   357
instruction decoder   358
instrument ground   395
instrumentation amplifier   191
instrumentation bus   380
integrated circuits   38
integrating AD converter   313
integrating network   118
integrator   202
interfacing   359
interference   394
interference (reduction in ADC)   316
internal frequency compensation   198
interrupt   360
intrinsic silicon   131
inversion layer   176
inversion (logic)   321
inverter (logic)   324
inverting voltage amplifier   187
IO port   360

JFET   170
jitter   222
JK flipflop   327
JK flipflop (specifications)   415
Johnson noise   36
jump instruction   358
junction FET   170
junction FET (switch)   247

junction field-effect transistor   170

kilobit, kilobyte   299
Kirchhoff's rule   39

ladder network   304
Laplace operator   60
Laplace transform   55, 56
Laplace variable   55
LDR   102
leakage current   134
least significant bit   300
least significant digit   6, 389
LED   136
LED display   341
Lenz's law   99
level sensor   106
level shifter   184
light-dependent resistor   102
light-emitting diode   136
light-sensitive diode   135
light-sensitive resistors   102
limiter   138, 224
line spectrum   21
linear displacement sensor   104
linear variable differential transformer   104
Lissajous plot   387
listener   380
load error   69
lock-in amplifier   292
lock-in range   292
log ratio converter   229
logarithmic converter   227
logarithmic transfer   72
logic analyser   379
logic circuits   331
logic equations   322
logic gate   324
logic variables   319
loop   39
loop gain   197
loss angle   97
low-pass characteristic   210
low-pass filter   115
low-pass filter (cascaded)   125
low-pass filter (first order)   116
LSB   300
LSB (unit)   305
lumped element   44
LVDT   104

machine code   363
machine language   362
magnetic field strength   98
magnetic flux   98
magnetic induction   97
mains filter   404
mask ROM   352
master-slave flipflop   327

matching   69
mean absolute value   16
mean time-to-failure   12
mean value   16, 30
measurement error   389
measurement instruments   367
memory   348
memory-mapped IO   360
metal film (resistor)   95
metal wire (resistor)   95
mica capacitor   97
microprocessor   348, 356
mobility   100, 132
modem   362
modulation   3, 279
modulation depth   281, 282
modulus   51, 82
monotony   306
MOS field-effect transistor   174
MOSFET   174
MOSFET (switch)   248
most significant bit   300
MSB   300
MTTF   12
multichannel oscilloscope   369
multimeter   367
multipen plotter   374
multiplexer (digital)   332
multiplexing   3
multiplicative error   73
multiplier   231
multiplying DA converter   318
mutual inductance   99

n-channel FET   170
n-type silicon   131
NAND gate   324
negation   321
network elements   38
node   39
noise   10, 22, 77
noise (comparator)   222
noise generator   375
noise power   77
non-inverting voltage amplifier   188
non-linear transfer   220
non-linearity   8
non-volatile memory   349
normal distribution   390
normal distribution function   30
normally off   176
normally on   176
Norton's theorem   64
NOT   321
notch filter   115, 123, 213
npn transistor   150
NTC   101
NTC-thermistor   101

octal notation   301
off-resistance   243
offset   10, 73
offset voltage   193
Ohm's law   94
on-resistance   243
one-port   41
open loop gain   197
operating range   6
operational amplifier   185
operational amplifier (specifications)   409
OR   320
OR gate   324
oscillation condition   265
oscillator   262
oscilloscope   367
output impedance   65
output interface   348
output port   41
overload protection   138

p-channel FET   170
p-type silicon   132
parallel AD converter   306
parallel converter   299
parallel DA converter   302
parallel enable   341
parallel input-output interface   361
parallel shift register   338
parallel word   301
passive components   38, 94
passive filters   115
PD-characteristic   208
peak detector   140, 288
peak-to-peak detector   143
peak-to-peak value   16, 143
peak value   16
Peltier coefficient   109
Peltier effect   109
Peltier element   109
periodic signals   15, 19
permeability   99
permittivity   96
permittivity of vacuum   96
phase characteristic   84
phase diagram   82
phase-locked loop   291
phase modulation   279
phase-shift oscillator   266
photodiode   135
photoresistive effect   102
photoresistor   102
photoresistor (switch)   245
PI-characteristic   208
PID-characteristic   209
piecewise linear approximation   235
piezoelectric effect   110
piezoelectric sensors   110
piezoelectricity   110

Index

piezoresistive effect   103
pinch-off voltage   171
PIO   361
pipeline effect   313
platinum resistance thermometer   101
PLL   291
plotter   373
pn diode   131
pn diode (switch)   246
pn junction   132
pnp transistor   150
polar plot   86
polar plot (analyser)   378
pole   60
positive feedback   222
potentiometer   103
power control   249
power density spectrum   33
power spectrum   21
power supply voltage   143, 145
power transfer   41, 70
power transistor   152
predetection filter   292
priority (interrupt)   360
probability   29, 390
probability density function   29
probe attenuator   372
program   348
program counter   357
program memory   348, 357
programmable ROM   352
programming   362
PROM programmer   353
Pt-100   101
PTC   102
PTC-thermistor   102
pulse generator   273, 375
pulse height modulation   279
pulse width modulation   279

quality factor   85
quality factor (filter)   210
quantization   2
quantization error   300
quantized signals   16
quantum efficiency (photodiode)   136

RAM   348, 350
ramp generator   270
random-access memory   350
random error   390
read-only memory   349, 350, 352
read-write memory   348, 355
real component   51
real part   87
recombination   132
rectifier   226
rectifier detector   288
reed switch   244

refresh   355
relative bandwidth   85
relative inaccuracy   7, 392
relative permeability   99
reliability   12
resistance   39, 94
resistance (generalized)   44, 45
resistance measurement bridge   286
resistance thermometer   100
resistive sensors   100
resistivity   94
resistor   94
resolution   6
resolution (ADC)   300
resonance   210
resonance (filter)   124
resonance filters   210
resonance frequency   85, 210
reverse current   134
ripple counter   336
ripple (peak detector)   140
rms value   16, 33, 75
rms voltmeter   18
ROM   349, 350, 352
root-mean-square value   16

Sallen and Key filter   214
sample-hold circuit   251
sample/track mode   251
sampled signals   27
sampling   2
sampling frequency   28
sampling oscilloscope   371
sampling theorem   29
SAR   308
saturation   9
saturation line   246
saturation (operational amplifier)   185
saturation region (FET)   173
Schmitt trigger   222
Schmitt trigger (generator)   269
second moment   32
second-order transfer function   85
Seebeck coefficient   108
Seebeck effect   106, 107
Seebeck voltage   107
self-inductance   38, 39, 99
self-inductance (generalized)   44, 45
self-inductance (impedance)   53
semiconductor   94
semiconductor memory   348, 350
sensitivity   7
sensor   2, 100
sequential circuit   327
serial-access memory   350
serial DA converter   310
serial shift register   338
serial word   301
series shunt switch   244, 255

series switch 243
settling time 242, 253
seven-segment code 341
seven-segment decoder 341
seven-segment display 340
Shannon's sampling theorem 29
shielding 397, 399
shift register 336
shunt switch 244
sideband 281
signal conditioning 2
signal generator 262, 374
signal-to-noise ratio 115
silicon 131
sine reshaper 262
sine wave generator 374
sine wave oscillator 262
single sideband modulation 282
single-sided rectifier 36, 146, 226
slew rate limitation 9
small-signal analysis 153
source 170
source follower 179
source impedance 64
specifications 6
spectral power 21
spectral response (LED) 137
spectral response (photodiode) 136
spectrum analyser 377
square power (circuit) 234
square root (circuit) 234
square wave generator 273
SR flipflop 327
standard deviation 32, 33, 390
static memory 355
status register 359
step response (high-pass 1st order) 119
step response (low-pass 1st order) 117
stochastic signals 15, 29
storage oscilloscope 371
strain gauge 103
strain gauge factor 103
subroutine 360
substrate 176
successive approximation AD converter 307
successive approximation register 308
super β transistor 151
superposition 40
supply voltage source 143, 145
suppressed carrier 284
sweep generator 375
sweep oscillator 273
sweep range 274
switch 241
switch (FET) 170
switch modulator 284
synchronous counter 336
synchronous detection 288, 289
system models 64

systematic error 390

T-equivalent circuit 154
tachometer 106
taker 380
telemetry 4
temperature coefficient 10
temperature-sensitive resistors 100
temperature sensor 100
temperature sensor (capacitive) 106
terminal count 341
thermal noise 77
thermistor 101
thermistor (amplitude control) 264
thermocouple 108
thermoelectric sensors 106
thermopile 108
Thévenin's theorem 64
Thomson coefficient 110
Thomson effect 110
through variable 44
thyristor 248
time-base 368
time constant 117
time meter 376
time multiplexer 249
time multiplexing 3
toggle 330
track-hold circuit 251
transconductance 153
transconductance (FET) 174
transducer 2
transduction 2
transfer function (Laplace domain) 59
transformer 39, 97, 99
transient error 254
transient signals 16
transimpedance 41
transistor-transistor logic 326
triac 249
triangle generator 269
triangle voltage generator 269
trigger 368
tristate buffer 310, 351, 356
true (T) 319
true rms value 18
truth table 319
TTL 326
turn-off delay time 242, 253
turn-on delay time 242, 253
twisting 398
two-integrator oscillator 267
two-port network 65
two-terminal network 64

UART 361
undamped angular frequency 85
unity feedback 188
unity gain bandwidth 198

universal synchronous/asynchronous receiver/ transmitter   362
USART   361

V variable   44
varactor   294
varactor amplifier   294
variance   32, 390
VCF   377
VCO   273, 291, 375
Venn diagram   323
VF converter   275
virtual ground   187
voltage amplifier stage   156, 159, 178
voltage comparator   220
voltage-controlled current source   151
voltage-controlled filter   377
voltage-controlled oscillator   273, 291, 375
voltage-controlled resistance   170
voltage divider   43
voltage divider (probe)   372
voltage gain   41
voltage gain (small signals)   160
voltage generator   268
voltage limiter   224

voltage matching   69
voltage source   39, 143
voltage source model   65
voltage stabilization   135
voltage stabilizer   143
voltage-to-current converter   68, 154, 177
voltage-to-current transfer   41
voltage-to-frequency converter   275
voltage transfer   41

Wheatstone bridge   286
white noise   22, 77
Wien bridge   213
Wien filter   213
Wien oscillator   265
write enable   351

x-t plotter   373
x-y plotter   373

Zener breakdown   135
Zener diode   135
Zener voltage   143
zero   60
zero drift   10